Encyclopaedia of Mathematical Sciences
Volume 133

Invariant Theory and Algebraic Transformation Groups IV

Subseries Editors:
R.V. Gamkrelidze V.L. Popov

Preface

During several centuries various reincarnations of projective duality have inspired research in algebraic and differential geometry, classical mechanics, invariant theory, combinatorics, etc. To put it simply, projective duality is the systematic way of recovering a projective variety from the set of its tangent hyperplanes. In this survey I have tried to emphasize that there are many different aspects of projective duality and that it can be studied using a wide range of methods. But at the same time I was pushing hard to minimize the technical details in the hope of writing a text that requires a knowledge of only basic facts from algebraic geometry and the theory of algebraic (or Lie) groups. Most proofs in this book are compilations from various sources.

Projective duality is defined for arbitrary projective varieties and it does not seem natural a priori to consider varieties with symmetries. However, it turns out that many important examples carry the natural action of the algebraic group. This is especially true for projective varieties that have extremal projective properties: self-dual varieties, varieties of positive defect, Severi varieties, Scorza varieties, varieties of small codegree, etc. I have tried to emphasize this phenomenon in this survey. However, one aspect is totally omitted – I decided against including results about dual varieties of toric varieties and A-discriminants. This theory is presented in the beautiful book by Gelfand, Zelevinsky, and Kapranov [GKZ2] (the author always keeps it under his pillow) and I don't feel I have anything to add to it.

Let us say a few words about the contents of this survey.

Chapter 1 is intended as a brief introduction to projective duality. All results here were already well-known in the 19th century. After giving basic definitions in Sect. 1.1 we discuss duality for plane curves in Sect. 1.2. We give parametric equations for dual curves, discuss the connection to the Legendre transform, introduce Plücker formulas, and study curves of degree 2, 3, and 4. In Sect. 1.3 we prove the Reflexivity Theorem. Here we follow the exposition in [GKZ2] and use the interpretation of a conormal variety as a Lagrangian subvariety in the cotangent bundle. We introduce the discriminant as a defining equation for the dual variety and the defect that measures how far the dual variety is from being a hypersurface. We prove that if the defect is positive then the variety is ruled. Finally, in Sect. 1.4 we study the behaviour of a dual variety under projections and introduce the standard notation of algebraic geometry related to divisors and line bundles.

In our survey we are mostly interested in the study of varieties with symmetries, and in Chap. 2 we study projective geometry of homogeneous spaces; more precisely we look at orbit closures for algebraic groups acting on projective spaces with finitely many orbits. In Sect. 2.1 we give the necessary background on algebraic groups and fix notation. In Sect. 2.2 we discuss Pyasetskii pairing, which is an interesting instance of projective duality. We give some standard and a few exotic examples. The next Sect. 2.3 contains the systematic treatment of actions related to gradings of simple Lie algebras. These actions provide a wealth of very important varieties that will be studied throughout this book. Some examples include Severi varieties, smooth self-dual varieties, smooth building blocks for varieties of positive defect, varieties of small codegree, etc. It is quite an interesting phenomenon that varieties with extremal projective properties tend to have maximal symmetries. We finish this chapter with the description of Pyasetskii pairing for actions related to gradings of GL_n called the Zelevinsky involution.

In Chap. 3 we study projectively dual varieties using calculations in local coordinates. In Sect. 3.1 we prove a formula due to Katz that expresses the dimension of a dual variety in terms of the hessian of local equations of a variety. We use it to prove a formula of Weyman and Zelevinsky that expresses the defect of a Segre embedding of a product of two varieties. In Sect. 3.2 we introduce a gadget called the second fundamental form that incorporates these calculations. We prove some results of Griffiths and Harris about the second fundamental form. We finish this section with the description of higher fundamental forms of flag varieties obtained by Landsberg.

In Chap. 4 we study projective constructions related to projective duality but also having a merit of their own. We prove a theorem of Zak and Ran that the Gauss map of a smooth variety is a normalization. We introduce secant and tangent varieties in Sect. 4.2, prove the Terracini Lemma, give a method for the calculation of multisecant varieties of homogeneous spaces, discuss the relationship discovered by Zak between the degree of a dual variety and the order of the variety, and give an overview of old and new results related to the Waring problem for forms. In Sect. 4.3 we discuss theorems of Zak related to the Hartshorne conjecture – theorems on tangencies, on linear normality and on Severi varieties. The main tool is a connectedness theorem of Fulton and Hansen. We finish in Sect. 4.4 by explaining the Cayley trick for Chow forms.

The dual variety is the image of the conormal variety which is the projectivized conormal bundle if the variety is smooth. In Chap. 5 we exploit this relation between duality and vector bundles. In Sect. 5.1 we prove a theorem of Holme and Ein about the defect of a smooth effective very ample divisor. We deduce this result from a theorem of Munoz about the dimension of the linear span of a tangential variety. We discuss the related notion of projective extendability. In Sect. 5.2 we apply Hartshorne's ample vector bundles to prove a theorem of Ein that a dual variety of a smooth complete intersection is a hypersurface. We introduce resultants and prove a classical theorem that they are well-defined. We also explain the Cayley trick for resultants. In the last Sect. 5.3 we describe the "Cayley method" developed by Gelfand, Kapranov, and Zelevinsky. The idea is to show that the dual variety is represented in the derived category by Koszul complexes of jet bundles. The discriminant is then equal to the "Cayley determinant" of a generically exact complex. As an application we deduce some classical formulas for discriminants and their degrees.

In Chap. 6 we discuss about the degree of dual varieties and resultants. We start in Sect. 6.1 by recalling Chern classes and then proving a formula of Katz, Kleiman, and Holme that expresses the degree of a dual variety in terms of Chern classes of the cotangent bundle. We give many examples and generalizations. We prove a theorem of De Concini and Weyman about the formula for the degree with non-negative coefficients. In Sect. 6.2 we discuss formulas for the codegree and ranks related to the Cayley method, such as a formula due to Lascoux.

In Chap. 7 we study varieties with positive defect. In Sect. 7.1 we focus on beautiful theorems of Ein about the normal bundle of a generic contact locus. Since this locus is a projective subspace, it is possible to use the machinery of vector bundles on projective spaces. We prove Ein's theorem that this normal bundle is symmetric and uniform, which explains among other things a parity theorem of Landman. We introduce the Beilinson spectral sequence and use it to calculate the normal bundle to a generic contact locus in small dimensions. Finally, we study dual varieties of scrolls and prove a theorem of Ein that a variety of defect at least 2 is a scroll if and only if the normal bundle to a generic contact locus splits. In Sect. 7.2 we follow [IL] and discuss linear systems of quadrics of constant rank and how they are related to dual varieties via the second fundamental form. In Sect. 7.3 we prove a theorem of Beltrametti, Fania, and Sommese that relates the defect of a projective variety and its Mori-theoretic characteristic called the nef value. We give a brief survey of necessary results from Mori theory. We finish by giving a classification of smooth varieties of positive defect up to dimension 10 obtained by many authors and initiated by Ein. Finally, in Sect. 7.4 we use this connection with Mori theory to classify all flag varieties with positive defect. This approach was developed by Snow in contrast with the original proof of Knop and Menzel that used the Katz dimension formula.

In Chap. 8 we study dual varieties and discriminants of several special homogeneous spaces. We start in Sect. 8.1 by showing how to use standard results of representation theory such as the Borel–Weil–Bott theorem, the BGG homomorphism, identities with Schur functors, and formulas of the Schubert calculus to find the codegree of Grassmannians or full and partial flag varieties. We give a list of formulas for the degree of hyperdeterminants and sketch the proof of a theorem of Zak about varieties of codegree 3. In Sect. 8.2 we generalize the theorem of Matsumura and Monsky about automorphisms of smooth hypersurfaces to automorphisms of smooth very ample divisors on flag varieties. In Sect. 8.3 we study commutative algebras without identities from the "discriminantal" point of view. As a corollary we prove that the algebra of diagonal matrices does not have quasiderivations. In Sect. 8.4 we study anticommutative algebras (nets of skew-symmetric forms). We show that they have beautiful geometric properties related to cubic surfaces, Del Pezzo surfaces, representation theory of S_5, etc. In Sect. 8.5 we show that the discriminant in a simple Lie algebra defined by analogy with the disciminant of a linear operator is equal to the discriminant of the minimal orbit, the so-called adjoint variety. Finally, in Sect. 8.6 we study related questions about schemes of zeros of irreducible homogeneous vector bundles. In particular, we address a question of classifying irreducible homogeneous vector bundles with a trivial line subbundle, find the maximal dimension of an isotropic subspace of a generic symmetric or skew-symmetric form, and study properties of the related Moore–Penrose involution.

In Chap. 9 we study self-dual varieties, i.e. varieties isomorphic to their projectively dual variety. In Sect. 9.1 we consider smooth self-dual varieties. The complete list of these varieties is (conjecturally) surprisingly short. All known varieties are flag varieties so we start by considering this case, where everything follows from the classification of flag varieties of positive defect. After a brief introduction to the Hartshorne conjecture we sketch the proof of the amazing theorem of Ein that gives the complete list of self-dual varieties in the range that is allowed by the Hartshorne conjecture. We also prove a finiteness theorem of Muñoz that uses the distribution of primes to give restrictions on the Beilinson spectral sequence. We finish in Sect. 9.2 by describing results of Popov about self-dual nilpotent orbits.

In the final Chap. 10 we study how the topology of the variety is reflected in singularities of the dual variety. We start in Sect. 10.1 by proving the class formula and its variant due to Landman that relates the degree of the dual variety and the Euler characteristic of the variety and its hyperplane sections. In the singular case this formula was proved by Ernström, but the Euler characteristic has to be substituted by the degree of the Chern–Mather class. In Sect. 10.2 we prove theorems of Dimca, Nemethi, Aluffi and others that multiplicities of the dual variety are given by Milnor numbers (or classes). To give an example we follow Aluffi and Cukierman and calculate multiplicities of the dual variety to a smooth surface. Finally, we give some results of Weyman and Zelevinsky about singularities of hyperdeterminants.

I would like to thank F. Zak for very inspiring discussions of projective geometry and R. Muñoz for his insights about positive defect varieties. I am grateful to S. Keel, V. Popov, A. Kuznetsov, P. Katsylo, D. Saltman, E. Vinberg for many helpful remarks and encouragement. P. Aluffi, F. Cukierman, N.C. Leung, A.J. Sommese and many others sent me lots of comments about the preliminary version of this survey.

Parts of this book were written during my visits to the Erwin Schrödinger Institute in Vienna, Mathematical Institute in Basel, and University of Glasgow. I am grateful to my hosts for the warm hospitality.

Moscow–Edinburgh–Austin,
August 2004

Jenia Tevelev

Contents

1

Introduction to Projective Duality

This chapter is intended as a brief introduction to projective duality. All results here were already well-known in the 19th century. After giving basic definitions in Sect. 1.1 we discuss duality for plane curves in Sect. 1.2. We give parametric equations for dual curves, discuss the connection to the Legendre transform, introduce Plücker formulas, and study curves of degree 2, 3, and 4. In Sect. 1.3 we prove the Reflexivity Theorem. Here we follow the exposition in [GKZ2] and use the interpretation of a conormal variety as a Lagrangian subvariety in the cotangent bundle. We introduce the discriminant as a defining equation for the dual variety and the defect that measures how far the dual variety is from being a hypersurface. We prove that if the defect is positive then the variety is ruled. Finally, in Sect. 1.4 we study the behaviour of a dual variety under projections and introduce the standard notation of algebraic geometry related to divisors and line bundles.

1.1 Projectively Dual Varieties

For any finite-dimensional complex vector space V we denote by $\mathbb{P}(V)$ its projectivization (the set of 1-dimensional subspaces). If $V = \mathbb{C}^{N+1}$ then $\mathbb{P}^N = \mathbb{P}(\mathbb{C}^{N+1})$ is the standard complex projective space.

Let $V^\vee$ be the dual vector space of linear forms on V. Points of the dual projective space $\mathbb{P}(V)^\vee = \mathbb{P}(V^\vee)$ correspond to hyperplanes in $\mathbb{P}(V)$. Conversely, to any point p of $\mathbb{P}(V)$, we can associate a hyperplane in $\mathbb{P}(V)^\vee$, namely the set of all hyperplanes in $\mathbb{P}(V)$ passing through p. Therefore $\mathbb{P}(V)^{\vee\vee}$ is naturally identified with $\mathbb{P}(V)$. More generally, there exists the projective duality between projective subspaces in $\mathbb{P}(V)$ and $\mathbb{P}(V)^\vee$: for any $L \subset \mathbb{P}(V)$, its *projectively dual subspace* $L^* \subset \mathbb{P}(V)^\vee$ parametrizes all hyperplanes that contain L.

Quite remarkably, this projective duality extends to the involutive correspondence between arbitrary subvarieties in $\mathbb{P}^N$ and $(\mathbb{P}^N)^\vee$.

Let $X \subset \mathbb{P}^N$ be an irreducible projective variety with $\dim X = n$. We denote by $T_{x,X}$ the tangent space at any smooth point $x \in X_{sm}$. We can also define the *embedded projective tangent space* $\hat{T}_{x,X} \subset \mathbb{P}^N$ as follows:

$$\hat{T}_{x,X} = \mathbb{P}(T_{v,\mathrm{Cone}(X)}),$$

where $\mathrm{Cone}(X) \subset V$ is the cone over X, v is any non-zero point on the line x, and we consider $T_{v,\mathrm{Cone}(X)}$ as a linear subspace of V (it does not depend on the choice of v).

Definition 1.1 A hyperplane $H \subset \mathbb{P}^N$ is a *tangent hyperplane* of X if $\hat{T}_{x,X} \subset H$ for some $x \in X_{sm}$. The closure of the set of all tangent hyperplanes is called the *projectively dual variety* $X^* \subset (\mathbb{P}^N)^\vee$.

We shall prove later that

Theorem 1.2 (Reflexivity Theorem) $X^{**} = X$.

1.2 Dual Plane Curves

The most classical example of a dual variety is the dual curve C^* of a plane curve $C \subset \mathbb{P}^2$. By definition, generic points of C^* are the tangents to C at smooth points. In this case the Reflexivity Theorem has a simple meaning: the tangent line $\hat{T}_p$ at a smooth point $p \in C$ is the limit of secants $\overline{pq}$ for $q \in C$ as $q \to p$. The point in $\mathbb{P}^2$ that corresponds to the tangent to $C^* \subset (\mathbb{P}^2)^\vee$ at a non-singular point $\hat{T}_p$ is the limit of the intersection points of the tangents $\hat{T}_p$ and $\hat{T}_q$ as $q \to p$. Of course, this point is p.

1.2.1 Parametric Equations

It is easy to write down a parametric representation of the dual curve C^* using a given parametric representation of C. Let x, y, z be homogeneous coordinates on $\mathbb{P}^2$ and p, q, r the dual homogeneous coordinates on $(\mathbb{P}^2)^\vee$. We choose an affine chart $\mathbb{C}^2 = \{z \neq 0\} \subset \mathbb{P}^2$ with affine coordinates x, y (so we set z to be 1). The dual chart $(\mathbb{C}^2)^\vee \subset (\mathbb{P}^2)^\vee$ with coordinates p, q is obtained by setting the coordinate r of $(\mathbb{P}^2)^\vee$ to be -1. Then $(\mathbb{C}^2)^\vee$ consists of all lines in $\mathbb{P}^2$ that do not pass through the origin $(0,0) \in \mathbb{C}^2 \subset \mathbb{P}^2$. Every such line either has an affine equation $px + qy = 1$ or is the line "at infinity" with coordinates $p = q = 0$. Suppose that a local parametric equation of C has the form $x = x(t)$, $y = y(t)$, where t is a local parameter on C, and $x(t)$, $y(t)$ are analytic functions. The dual curve C^* is parametrized by $p = p(t)$, $q = q(t)$, where $p(t)x + q(t)y = 1$ is the tangent line to C at $(x(t), y(t))$. It follows that

$$p(t) = \frac{-y'(t)}{x'(t)y(t) - x(t)y'(t)}, \quad q(t) = \frac{x'(t)}{x'(t)y(t) - x(t)y'(t)}.$$

Applying this formula twice one can prove the Reflexivity Theorem.

1.2.2 Legendre Transformation

Projective duality and the Legendre transformation of classical mechanics are closely related to each other. Let us recall this classical definition in the case of real functions in one variable. Details and generalizations can be found in [Ar].

Definition 1.3 Let $y = f(x)$ be a smooth convex real function, $f''(x) > 0$. The *Legendre transformation* of f is a function $g(p)$ defined as follows. Let $x = x(p)$ be a point at which the graph $y = f(x)$ has slope p. Then

$$g(p) = px(p) - f(x(p)).$$

Equivalently, $x(p)$ is the unique point where the function $F(p, x) = px - f(x)$ has a maximum with respect to x and $g(p) = F(p, x(p))$.

The Legendre transformation is easily seen to be involutive. To link it with projective duality, we need caustic curves. Let us express projective duality entirely in terms of the projective plane $\mathbb{P}^2$. A tangent line to a curve C at some point x is the line which is infinitesimally close to C near x. A point of $(\mathbb{P}^2)^\vee$ is a line $l \subset \mathbb{P}^2$. A curve in $(\mathbb{P}^2)^\vee$ is therefore a 1-parameter family of lines in $\mathbb{P}^2$. For example, a line in $(\mathbb{P}^2)^\vee$ is a pencil x^* of all lines in $\mathbb{P}^2$ passing through a given point $x \in \mathbb{P}^2$. The dual curve C^* is a 1-parameter family of tangent lines to C. Now let us take an arbitrary curve $C' \subset (\mathbb{P}^2)^\vee$ (a 1-parameter family of lines in $\mathbb{P}^2$) and then find a geometric interpretation of the dual curve $C'^* \subset \mathbb{P}^2$. Take some line $l \in C'$. The condition that x^* is tangent to C' at l means that l is a member of a family C' and other lines from C' near l are infinitesimally close to the pencil of lines x^*. This is usually expressed by saying that x is a *caustic point* for the family of lines C'. The set of all caustic points of C' is called the *caustic curve* of C'. This is nothing else but the projectively dual curve C'^*.

By the Reflexivity Theorem, any curve $C \subset \mathbb{P}^2$ is the caustic curve of the family of its tangent lines. Another consequence is that any 1-parameter family of lines in $\mathbb{P}^2$ is a family of tangent lines to some curve C.

The caustic could be found (locally) using the Legendre transformation:

Theorem 1.4 *Consider a family of lines $y = px - g(p)$. Its caustic curve has an equation $y = f(x)$, where f is the Legendre transformation of g.*

Proof. The tangent line to $y = f(x)$ at a point $x(p)$ with slope p is equal to

$$y = p(x - x(p)) + f(x(p)) = px - g(p).$$

It remains to use the involutivity of the Legendre transformation. □

1.2.3 Plücker Formulas

Even if a plane curve $C \subset \mathbb{P}^2$ is smooth, $C^* \subset ((\mathbb{P}^2)^\vee$ is almost always singular. There is a natural map

$$C \to C^*, \quad p \mapsto \hat{T}_p.$$

This map is a resolution of singularities. Moreover, even if C is not smooth, the curves C and C^* are birationally equivalent, in particular they have the same geometric genus g. Indeed, consider the conormal variety

$$I_C \subset \mathbb{P}^2 \times (\mathbb{P}^2)^\vee, \quad I_C = \overline{\{(p,l) \,|\, p \in C_{sm},\ l \in C^*{}_{sm},\ l = \hat{T}_p\}}.$$

Obviously, I_C projects birationally on C. Therefore, by the Reflexivity Theorem, I_C also projects birationally on C^*. So C and C^* are birationally equivalent. If C is smooth then, of course, $I_C = C$.

A line l tangent to C in at least two points is a singular point of C^*. It is known as a *multiple tangent.* If l has exactly two tangency points on C and the intersection multiplicity at each of them is 2, then l is a *bitangent.* A bitangent corresponds to an ordinary double point (node) of C^*.

If $l = \hat{T}_p$ intersects C at p with multiplicity $m \geq 3$, it is again a singular point of C^*. If $m = 3$, and l is not tangent to C at any other point, then p is called an *inflection point* (or flex) of C. l is a cuspidal point (or cusp) of C^*.

Now we can introduce a class of "generic" curves with singularities, which is preserved by the projective duality. Namely, we say that C is generic if both C and C^* have only double points and cusps as their singularities. Suppose that C is generic in this sense. Let d, g, κ, δ, b, f be the degree, the geometric genus, the number of cusps, the number of double points, the number of bitangents, and the number of flexes of C. Let d^*, g^*, κ^*, δ^*, b^*, f^* be the corresponding numbers for C^* (d^* is also called the *class* of C). Then by the Reflexivity Theorem we have the following

Proposition 1.5 $g = g^*$, $\kappa = f^*$, $\delta = b^*$, $b = \delta^*$, $f = \kappa^*$.

It turns out that there is another remarkable set of equations linking these numbers. It was discovered by Plücker and Clebsch. The proof can be found in [GH].

Theorem 1.6

$$g = \frac{1}{2}(d^* - 1)(d^* - 2) - b - f,$$

$$g = \frac{1}{2}(d - 1)(d - 2) - \delta - \kappa,$$

$$d = d^*(d^* - 1) - 2b - 3f,$$

$$d^* = d(d - 1) - 2\delta - 3\kappa.$$

1.2.4 Curves of Small Degree

Conics

Let $C \subset \mathbb{P}^2$ be a smooth conic given by the equation

$$(Ax, x) = \sum_{i,j=1}^{3} a_{ij} x_i x_j = 0,$$

where $A = (a_{ij})$ is a non-degenerate symmetric 3×3 matrix. The tangent line to C at $x_0 \in C$ is given by $(Ax_0, x) = 0$. Hence the point $\xi \in (\mathbb{P}^2)^\vee$ corresponding to this tangent line has homogeneous coordinates Ax_0, which implies that $(A^{-1}\xi, \xi) = 0$. Therefore $C^* \subset (\mathbb{P}^2)^\vee$ is also a smooth conic defined by A^{-1}. Conics are the only smooth plane curves having smooth duals.

Cubics

Let $C \subset \mathbb{P}^2$ be a smooth cubic curve. By the Bézout Theorem, C^* has only flexes as singularities and the Plücker formulas are always applicable. Hence C has no bitangents and exactly 9 flexes. The dual curve C^* is a very special curve of degree 6 with 9 cusps and no double points. A beautiful determinantal formula for it was found by Schläfli. Let x_1, x_2, x_3 be homogeneous coordinates in $\mathbb{P}^2$ and p_1, p_2, p_3 the dual coordinates in $(\mathbb{P}^2)^\vee$. Let $f(x_1, x_2, x_3) = 0$ be the homogeneous equation of C and $F(p_1, p_2, p_3) = 0$ the homogeneous equation of C^*. Consider the polynomial

$$V(p, x) = \begin{vmatrix} 0 & p_1 & p_2 & p_3 \\ p_1 & \frac{\partial^2 f}{\partial x_1 \partial x_1} & \frac{\partial^2 f}{\partial x_1 \partial x_2} & \frac{\partial^2 f}{\partial x_1 \partial x_3} \\ p_2 & \frac{\partial^2 f}{\partial x_2 \partial x_1} & \frac{\partial^2 f}{\partial x_2 \partial x_2} & \frac{\partial^2 f}{\partial x_2 \partial x_3} \\ p_3 & \frac{\partial^2 f}{\partial x_3 \partial x_1} & \frac{\partial^2 f}{\partial x_3 \partial x_2} & \frac{\partial^2 f}{\partial x_3 \partial x_3} \end{vmatrix}.$$

Clearly $V(p, x)$ has degree 2 in x. Then Schläfli's formula is as follows:

$$F(p_1, p_2, p_3) = \begin{vmatrix} 0 & p_1 & p_2 & p_3 \\ p_1 & \frac{\partial^2 V}{\partial x_1 \partial x_1} & \frac{\partial^2 V}{\partial x_1 \partial x_2} & \frac{\partial^2 V}{\partial x_1 \partial x_3} \\ p_2 & \frac{\partial^2 V}{\partial x_2 \partial x_1} & \frac{\partial^2 V}{\partial x_2 \partial x_2} & \frac{\partial^2 V}{\partial x_2 \partial x_3} \\ p_3 & \frac{\partial^2 V}{\partial x_3 \partial x_1} & \frac{\partial^2 V}{\partial x_3 \partial x_2} & \frac{\partial^2 V}{\partial x_3 \partial x_3} \end{vmatrix}.$$

The proof can be found in [GKZ2].

Quartics

Let $C \subset \mathbb{P}^2$ be a generic smooth quartic curve. By the Plücker formulas, C has genus 3, 24 flexes, and 28 bitangents. There are two visual descriptions of these bitangents. The first classical approach is to realize the quartic curve as a 'shade' of a generic cubic surface $S \subset \mathbb{P}^3$. Namely, suppose that $x \in S$ is a generic point. Consider the projective plane $\mathbb{P}^2$ of lines in $\mathbb{P}^3$ passing through x. Then lines $l \in \mathbb{P}^2$ that are tangent to S form a quartic curve C. The 28 bitangents are given by projections of 27 lines on S plus one extra line, the projection of $\hat{T}_{x,S}$.

Another classical description is less well-known: Consider a generic 3-dimensional linear system L of quadrics in $\mathbb{P}^3$. Singular members of this linear system give rise to a quartic curve in $\mathbb{P}(L) = \mathbb{P}^2$. It can be shown that any generic quartic curve can be obtained this way. By the Bézout Theorem, L has $2^3 = 8$ base points in $\mathbb{P}^3$. For any two base points p, q let $l_{pq} \subset \mathbb{P}(L)$ be the line formed by all quadrics that contain not only points p and q, but also the line $[pq]$ connecting them. Thus we have 28 lines $l_{pq} \subset \mathbb{P}(L)$. Let us show that these lines are bitangents to C.

Fix a basis $\{e_1, e_2, e_3, e_4\}$ in $\mathbb{C}^4$ such that e_1 belongs to p and e_2 belongs to q. Then quadrics from l_{pq} have the form

$$G = \begin{pmatrix} 0 & A^T \\ A & B \end{pmatrix},$$

and therefore

$$\det G = (\det A)^2.$$

This means exactly that l_{pq} intersects C in two double points, i.e. l_{pq} is a bitangent to C.

1.3 Reflexivity Theorem

Now let us formulate the general version of the Reflexivity Theorem.

Theorem 1.7

(i) *For any irreducible projective variety* $X \subset \mathbb{P}^N$, *we have* $X^{**} = X$.
(ii) *More precisely, if* $z \in X_{sm}$ *and* $H \in X^*{}_{sm}$, *then* H *is tangent to* X *at* z *iff* z, *regarded as a hyperplane in* $(\mathbb{P}^N)^\vee$, *is tangent to* X^* *at* H.

In the proof of the Reflexivity Theorem we follow [GKZ2] and deduce this result from the classical theorem of symplectic geometry. Other proofs, including the subtleties of a prime characteristic case, can be found in [Se], [M], [Wa].

1.3.1 Proof of the Relexivity Theorem

We start with some standard definitions.

Definition 1.8 If X is a smooth algebraic variety, then T_X denotes the tangent bundle of X. If $Y \subset X$ is a smooth algebraic subvariety, then T_Y is a subbundle in $T_X|_Y$. The quotient is called the normal bundle of Y in X, denoted by $N_{Y,X}$ (or simply N_Y if X is clear). By taking dual bundles we obtain the cotangent bundle $T_X^\vee$ and the conormal bundle $N_{Y,X}^\vee$. The conormal bundle can be naturally regarded as a subvariety of $T_X^\vee$.

Suppose that $X \subset \mathbb{P}^N$ is an irreducible subvariety.

Definition 1.9 Consider the set $I_X^0 \subset \mathbb{P}^N \times (\mathbb{P}^N)^\vee$ of pairs (x, H) such that $x \in X_{sm}$ and H is the hyperplane tangent to X at x. Let I_X be the Zariski closure of I_X^0. Then I_X is called the *conormal variety* of X.

The first projection $\mathrm{pr}_1 : I_X^0 \to X_{sm}$ makes I_X^0 a projective bundle over X_{sm} whose fibers are projective subspaces of dimension $N - n - 1$. Therefore I_X is an irreducible variety of dimension $N - 1$. The dual variety X^* is the image of the second projection $\mathrm{pr}_2 : I_X \to (\mathbb{P}^N)^\vee$. Hence X^* is an irreducible variety. Let us show that $I_X^0 = \mathbb{P}(N^\vee_{X_{sm},\mathbb{P}^N})$. Indeed, the choice of a hyperplane $H \subset \mathbb{P}^N$ tangent to X_{sm} at x is equivalent to the choice of a hyperplane $T_{x,H}$ in the tangent space $T_{x,\mathbb{P}^N}$, which contains $T_{x,X}$. The equation of this hyperplane is an element of $N^\vee_{X_{sm},\mathbb{P}^N}$ at x.

The Reflexivity Theorem can be reformulated as follows:

$$I_X = I_{X^*}. \tag{1.1}$$

Let $\mathbb{P}^N = \mathbb{P}(V)$ and $(\mathbb{P}^N)^\vee = \mathbb{P}(V^\vee)$. We take

$$Y = \mathrm{Cone}(X) \subset V, \quad Y^* = \mathrm{Cone}(X^*) \subset V^\vee.$$

Denote by $\mathrm{Lag}(Y)$ the closure of $N^\vee_{Y_{sm},V}$ in $T_V^\vee$ (we are about to show that $\mathrm{Lag}(Y)$ is a Lagrangian subvariety).

$T_V^\vee$ is canonically identified with $V \times V^\vee$. Denote by $\mathrm{pr}_1, \mathrm{pr}_2$ the projections of this product to its factors. Then Y^* coincides with $\mathrm{pr}_2(\mathrm{Lag}(Y))$. Therefore, (1.1) can be reformulated as follows:

$$\mathrm{Lag}(Y) = \mathrm{Lag}(Y^*), \tag{1.2}$$

where we identify $T_V^\vee$ and $T_{V^\vee}^\vee$ with $V \times V^\vee$.

Recall that a smooth algebraic variety M is called symplectic if it has a symplectic structure, i.e. a differential 2-form ω such that

- ω is closed, $d\omega = 0$.
- For any $p \in M$, the restriction of ω on $T_{p,M}$ is non-degenerate.

In this case $\dim M$ is necessarily even. An irreducible closed subvariety $\Lambda \subset M$ is called *Lagrangian* if $\dim \Lambda = \dim M/2$ and the restriction of ω to Λ_{sm} vanishes as a 2-form (is totally isotropic).

The cotangent bundle $T^\vee_X$ of a smooth algebraic variety X carries a canonical symplectic structure defined as follows. Let $(x_1, \dots, x_n)$ be a local coordinate system in X. Let ξ_i be the fiberwise linear function on $T^\vee_X$ given by the pairing of a 1-form with the vector field $\partial/\partial x_i$. Then $(x_1, \dots, x_n, \xi_1, \dots, \xi_n)$ forms a local coordinate system in $T^\vee_X$. The form ω is defined by

$$\omega = \sum_{i=1}^{n} d\xi_i \wedge dx_i. \tag{1.3}$$

It easy to give an equivalent definition of ω without any coordinate systems. Let $\pi : T^\vee_X \to X$ be the canonical projection. Let $p \in T^\vee_X$ and $v \in T_{p,T^\vee_X}$ be a vector tangent to $T^\vee_X$ at p. Then $\nu(v) = p(\pi_* v)$ is a canonical 1-form on $T^\vee_X$ and $\omega = d\nu$.

An example of a Lagrangian subvariety in $T^\vee_X$ can be obtained as follows. Let $Y \subset X$ be an irreducible subvariety and let

$$\mathrm{Lag}(Y) = \overline{N^\vee_{Y_{sm},X}}$$

be the closure in $T^\vee X$. Clearly, $\mathrm{Lag}(Y)$ is a conical subvariety (invariant under dilations of fibers of $T^\vee_X$).

Theorem 1.10

(i) $\mathrm{Lag}(Y)$ *is a Lagrangian subvariety.*
(ii) *Any conical Lagrangian subvariety has the form* $\mathrm{Lag}(Y)$ *for some irreducible subvariety* $Y \subset X$.

Proof. Let us show that $\mathrm{Lag}(Y)$ is Lagrangian. Clearly,

$$\dim \mathrm{Lag}(Y) = \dim Y + (\dim X - \dim Y) = \dim T^\vee_X/2.$$

So we need only verify that $\omega|_{\mathrm{Lag}(Y)} = 0$. It suffices to restrict to points over Y_{sm}. Let $x_1, \dots, x_n$ be a local coordinate system on X such that Y is locally defined by equations $x_1 = \dots = x_r = 0$. Then the fibers of the conormal bundle over points of Y are generated by 1-forms $dx_1, \dots, dx_r$. Hence $\xi_{r+1} = \dots = \xi_n = 0$ on $N^\vee_{Y_{sm},X}$ and by (1.3) we see that $\omega = 0$ on $N^\vee_{Y_{sm},X}$. Therefore $\mathrm{Lag}(Y)$ is indeed Lagrangian.

Suppose now that $\Lambda \subset T^\vee_X$ is a conical Lagrangian subvariety. We take $Y = \pi(\Lambda)$, where $\pi : T^\vee_X \to X$ is the projection, and claim that $\Lambda = \mathrm{Lag}(Y)$. It suffices to show that $\Lambda \subset \mathrm{Lag}(Y)$, because Λ and $\mathrm{Lag}(Y)$ are irreducible varieties of the same dimension. In turn, to prove that $\Lambda \subset \mathrm{Lag}(Y)$, it suffices to check that, for any $y \in Y_{sm}$, the fiber $\pi^{-1}(y) \cap \Lambda$ is contained in the conormal space $(N^\vee_{Y,X})_y$. Let ξ be any covector from $\pi^{-1}(y) \cap \Lambda$. Since $T^\vee_{y,X}$ is a vector space, we can regard ξ as a "vertical" tangent vector to $T^\vee_X$ at $y \in X \subset T^\vee_X$, where we identify X with the zero section of π. Since Λ is conical, $\xi \in T_{\xi,\Lambda}$. Since Λ is Lagrangian, ξ is orthogonal with respect to ω to any tangent vector from $T_{\xi,\Lambda}$ and, hence, to any tangent vector $v \in T_{y,Y}$. But by (1.3) it is easy to see that this is equivalent to $\xi \in (N^\vee_{Y,X})_y$. □

Now we can prove the Reflexivity Theorem:

Proof of the Reflexivity Theorem

We are going to verify (1.2). The identification $T_V^\vee = V \times V^\vee = T_{V^\vee}^\vee$ preserves the canonical symplectic structure up to a sign. Therefore $\mathrm{Lag}(Y)$ is still Lagrangian as a subvariety of $T_{V^\vee}^\vee$. Moreover, since $\mathrm{Lag}(Y) \subset V \times V^\vee$ is invariant under dilations of V and $V^\vee$, $\mathrm{Lag}(Y)$ is a conical Lagrangian variety of $T_{V^\vee}^\vee$. Therefore, by Theorem 1.10, $\mathrm{Lag}(Y) = \mathrm{Lag}(Z)$, where Z is the projection of $\mathrm{Lag}(Y)$ on $V^\vee$. But this projection coincides with Y^*. Hence $\mathrm{Lag}(Y) = \mathrm{Lag}(Y^*)$. □

1.3.2 Defect and Discriminant

We should expect that in 'typical' cases X^* is a hypersurface. Having this in mind, we give the following definition:

Definition 1.11 $\operatorname{def} X := \operatorname{codim}_{(\mathbb{P}^N)^\vee} X^* - 1$ is called the *defect* of X.

If $\operatorname{def} X = 0$ then X^* is defined by an irreducible polynomial Δ_X.

Definition 1.12 Δ_X is called the *discriminant* of X.

Δ_X is defined up to a scalar multiple. If $\operatorname{def} X > 0$ then we set $\Delta_X = 1$.

Let us give an equivalent definition of Δ_X. Suppose that $x_1, \ldots, x_{n+1}$ are some local coordinates on $\mathrm{Cone}(X) \subset V$. Any $f \in V^*$, a linear form on V, being restricted to $\mathrm{Cone}(X)$ becomes an algebraic function in $x_1, \ldots, x_{n+1}$. Then Δ_X is just an irreducible polynomial, which vanishes at $f \in V^*$ whenever the function $f(x_1, \ldots, x_{n+1})$ has a multiple root, that is, vanishes at some $v \in \mathrm{Cone}(X)$, $v \neq 0$, together with all first derivatives $\partial f / \partial x_i$.

Example 1.13 Consider the projective space $\mathbb{P}^N = \mathbb{P}(V)$ with homogeneous coordinates $z_0, \ldots, z_N$, and let $X \subset \mathbb{P}^N$ be the rational normal curve

$$(x^N : x^{N-1}y : x^{N-2}y^2 : \ldots : xy^{N-1} : y^N),\ x, y \in \mathbb{C},\ (x, y) \neq (0, 0)$$

(the image of the Veronese embedding $\mathbb{P}^1 \subset \mathbb{P}^N$). Any linear form $f(z) = \sum a_i z_i$ is uniquely determined by its restriction to $\mathrm{Cone}(X)$, which is a binary form $f(x, y) = \sum a_i x^{N-i} y^i$. Therefore $f \in \mathrm{Cone}(X^*)$ if and only if $f(x, y)$ vanishes at some point $(x_0, y_0) \neq (0, 0)$ (so $(x_0 : y_0)$ is a root of $f(x, y)$) with its first derivatives (so $(x_0 : y_0)$ is a multiple root of $f(x, y)$). It follows that Δ_X is the classical discriminant of a binary form.

The following Theorem is an easy corollary of the Reflexivity Theorem. It allows us to find singular points of hyperplane sections of smooth projective varieties.

Theorem 1.14 *Suppose that $X \subset \mathbb{P}^N$ is smooth and $X^* \subset (\mathbb{P}^N)^\vee$ is a hypersurface. Let $z_0, \ldots, z_N$ be homogeneous coordinates on $\mathbb{P}^N$ and $a_0, \ldots, a_N$ the dual homogeneous coordinates on $(\mathbb{P}^N)^\vee$. Suppose that $f = (a_0 : \ldots : a_N)$ is a smooth point of X^*. Then the hyperplane section $\{f = 0\}$ of X has a unique singular point with coordinates given by*

$$\left(\frac{\partial \Delta_X}{\partial a_0}(f) : \ldots : \frac{\partial \Delta_X}{\partial a_N}(f)\right).$$

Proof. Let $H \subset \mathbb{P}^N$ be the hyperplane corresponding to f. By the Reflexivity Theorem, H is tangent to X at z if and only if the hyperplane in $(\mathbb{P}^N)^\vee$ corresponding to z is tangent to X^* at f. Since X^* is smooth at f, such a point z is unique and is given by $z_i = \frac{\partial \Delta_X}{\partial a_i}(f)$. □

Consider the diagram of projections

$$X \overset{\mathrm{pr}_1}{\longleftarrow} I_X \overset{\mathrm{pr}_2}{\longrightarrow} X^*.$$

Theorem 1.15

(i) *If X is smooth then I_X is smooth.*
(ii) *If X^* is a hypersurface then pr_2 is birational.*
(iii) *If X is smooth and X^* is a hypersurface then pr_2 is a resolution of singularities.*

Proof. The map pr_1 is a projective bundle over X_{sm}. Therefore, if X is smooth then I_X is smooth. If X^* is a hypersurface then $\dim X^* = \dim I_X = N - 1$. Since pr_2 is generically a projective bundle, it is birational. Finally, (iii) follows from (i) and (ii). □

Typically the dual variety $X^* \subset (\mathbb{P}^N)^\vee$ is a hypersurface. Namely, we shall see that if $\operatorname{def} X > 0$ then X is a ruled variety.

Definition 1.16 A projective variety X is called *ruled* in projective subspaces of dimension r if for any $x \in X$ there exists a projective subspace L such that $x \in L \subset X$ and $\dim L = r$.

Definition 1.17 For any projective variety $X \subset \mathbb{P}^N$ and a hyperplane H, the singular locus of $X \cap H$ is called the *contact locus* of X and H.

Theorem 1.18 *Suppose that* $\operatorname{def} X = r \geq 1$. *Then*

(i) *X is ruled in projective subspaces of dimension r.*
(ii) *If X is smooth then the contact locus $(X \cap H)_{sing}$ is a projective subspace of dimension r for any $H \in (X^*)_{sm}$. The union of these projective subspaces is dense in X.*

Proof. By the Reflexivity Theorem 1.7, (i) is equivalent to the following: if $\operatorname{codim} X = r + 1$ then X^* is ruled in projective subspaces of dimension r. By a standard closedness argument it is sufficient to check this property only for some Zariski open dense subset of X^*. The condition for a hyperplane H to be tangent to X at $x \in X_{sm}$ is that H contains $\hat{T}_{x,X}$. For a given x, all H with this property form a projective subspace of dimension r. But the set of hyperplanes of X^* tangent to X at some smooth point obviously contains a Zariski open subset of X^*.

(ii) is proved by the same argument. □

Example 1.19 Suppose that $X \subset \mathbb{P}^N$ is a non-linear curve. Then $\operatorname{def} X = 0$. Indeed, X obviously could not contain a projective subspace $\mathbb{P}^k$ for $k > 0$.

Example 1.20 Let us give the minimal possible example of a smooth variety with positive defect. Let $V = \mathrm{Mat}_{2,3}$ be the space of 2×3 matrices. Then $V^\vee = \mathrm{Mat}_{3,2}$ and the pairing is given by the trace of the product. Let $X \subset \mathbb{P}(V)$ and $X^* \subset \mathbb{P}(V)^\vee$ be the projectivizations of the varieties of matrices of rank less than 2. Then X and X^* are smooth, projectively dual to each other, and both isomorphic to $\mathbb{P}^1 \times \mathbb{P}^2$. Therefore $\operatorname{def} X = \operatorname{codim} X^* - 1 = 1$. For any $H \in X^*$, the intersection $X \cap H$ is the union of a smooth quadric surface and a plane. Their intersection, i.e. $(X \cap H)_{sing}$, is a line. See also 2.11.

1.4 Projections and Linear Normality

1.4.1 Projections

Let $P = \mathbb{P}(V)$ be an N-dimensional projective space and $L \subset P$ a projective subspace of dimension k, $L = \mathbb{P}(U)$, $U \subset V$. The quotient projective space $P/L = \mathbb{P}(V/U)$ has, as points, $(k+1)$-dimensional projective subspaces in P containing L. The *projection with center* L is the map

$$\pi_L : P \setminus L \to P/L$$

which takes any point $x \in P \setminus L$ to the $(k+1)$-dimensional projective subspace spanned by x and L. It corresponds to the projection $V \to V/U$. P/L can be identified with any $(N-k-1)$-dimensional projective subspace $K \subset P$ not intersecting L. Then π_L sends $x \in P \setminus L$ to the unique intersection point of K and the $(k+1)$-dimensional projective subspace spanned by x and L. Clearly, the dual projective space $(P/L)^\vee$ is canonically embedded in $P^\vee$ as the set of hyperplanes containing L, and so it coincides with L^*, the dual variety of L.

Theorem 1.21

(i) *Let* $X \subset P = \mathbb{P}(V)$ *be an irreducible subvariety not intersecting a subspace* $L = \mathbb{P}(U)$ *and such that* $\dim X < \dim P/L$. *Then*

$$(\pi_L(X)^* \subset L^* \cap X^*.$$

The discriminant $\Delta_{\pi_L(X)}$ *(as a polynomial on* $(V/U)^\vee \subset V^\vee$*) is a factor of the restriction of* Δ_X *to* $(V/U)^\vee$.

(ii) *Suppose further that* $\pi_L : X \to \pi_L(X)$ *is an isomorphism. Then*

$$(\pi_L(X)^* = L^* \cap X^*, \text{ and } \Delta_{\pi_L(X)} = \Delta_X|_{(V/U)^\vee}.$$

Proof. A hyperplane in P/L is just a hyperplane in P containing L. It is clear that, if a hyperplane $H \subset P/L$ is tangent to $\pi_L(X)$ at some smooth point $y = \pi_L(x)$, where $x \in X$ is also smooth, then H as a hyperplane in P is tangent to X at x. This proves (i).

Suppose now that π_L is an isomorphism. The same argument as above shows that, if $X_0^* \subset X^*$ (resp. $\pi_L(X)_0^* \subset \pi_L(X)^*$) is a dense subset of hyperplanes tangent to X (resp. $\pi_L(X)$) at some smooth point, then $\pi_L(X)_0^* = L^* \cap X_0^*$, since π_L induces an isomorphism $X_{sm} \to \pi_L(X)_{sm}$. It suffices to show that $L^* \cap X^* = \overline{L^* \cap X_0^*}$. Here the assumption that π_L is an isomorphism is crucial, because otherwise critical points of π_L may produce extra components of $L^* \cap X^*$.

Consider a germ of a curve $(x(t), H(t))$ in the conormal variety I_X such that $x(t) \in X_{sm}$ for $t \neq 0$ and $H(0)$ contains L. We claim that there is a germ $(x(t), H'(t))$ such that $H'(0) = H(0)$ and and $H'(t)$ contains L for any t. Since π_L is an isomorphism, it induces an isomorphism of Zariski tangent spaces. Therefore $\hat{T}_{x(0),X}$ does not intersect L. Let $T_0 \subset \hat{T}_{x(0),X}$ be a limit position of embedded tangent Zariski spaces $T_t = \hat{T}_{x(t),X}$ as $t \to 0$. Then T_t does not intersect L for any t. Therefore we may consider a family T'_t of $(\dim L + \dim X + 1)$-dimensional projective subspaces such that for any t we have $L \subset T'_t$ and for $t \neq 0$ the subspace T'_t is tangent to X at $x(t)$. Namely, T'_t is a projective subspace spanned by L and T_t. Since $\dim X < \dim P/L$, we can embed T'_t into a family of hyperplanes H'_t with same properties. □

1.4.2 Degenerate Varieties

Definition 1.22 An irreducible subvariety $X \subset \mathbb{P}^N$ is called *non-degenerate* if X is not contained in any hyperplane H.

The next theorem shows that we may restrict ourselves to non-degenerate varieties while studying dual varieties and discriminants.

Theorem 1.23 *Let* $X \subset \mathbb{P}^N$ *be an irreducible subvariety.*

(i) *Assume that* X *is contained in a hyperplane* $H = \mathbb{P}^{N-1}$. *If* $X^{*\prime}$ *is the dual variety of* X, *when we consider* X *as a subvariety of* $\mathbb{P}^{N-1}$, *then* X^* *is the cone over* $X^{*\prime}$ *with vertex* p *corresponding to* H.
(ii) *Conversely, if* X^* *is a cone with vertex* p, *then* X *is contained in the corresponding hyperplane* H.

Proof. If $H' \neq H$ is a tangent hyperplane of X then $H \cap H'$ is a tangent hyperplane of X in $\mathbb{P}^{N-1}$. Conversely, if T is a tangent hyperplane of X in $\mathbb{P}^{N-1}$ then each hyperplane H' in $\mathbb{P}^N$ containing T is tangent to X. Therefore X^* is the cone over $X^{*\prime}$. This proves (i).

By the Reflexivity Theorem, we also have (ii). Namely, each hyperplane which is tangent to X^* at a smooth point necessarily contains p. Therefore $X = X^{**}$ is contained in the hyperplane corresponding to p. □

In terms of discriminants, Theorem 1.23 can be reformulated as follows. Consider a surjection $\pi : V \to U$. Then we have an embedding $i : \mathbb{P}(U^\vee) \hookrightarrow \mathbb{P}(V^\vee)$. Let $X \subset \mathbb{P}(U^\vee)$. Then Δ_X is a polynomial function on U. If we consider X as a subvariety in $\mathbb{P}(V^\vee)$ then $\Delta_{i(X)}$ is a function on V. By Theorem 1.23 these polynomial functions are related as follows:

$$\Delta_{i(X)}(f) = \Delta_X(\pi(f)).$$

In other words, $\Delta_{i(X)}$ does not depend on some of the arguments and forgetting these arguments gives Δ_X.

1.4.3 Linear Normality

Theorems 1.21 and 1.23 show that in order to study dual varieties and discriminants it suffices to consider only projective varieties X that are non-degenerate and not equal to a non-trivial projection. These projective varieties are called linearly normal. To give a more intrinsic definition of linearly normal varieties we need to recall the correspondence between invertible sheaves, linear systems, and projective embeddings.

An invertible sheaf on an algebraic variety X is simply the sheaf of sections of some algebraic line bundle. For example, the structure sheaf of regular functions $\mathcal{O}_X$ corresponds to the trivial line bundle. Usually we shall not distinguish notationally between invertible sheaves and line bundles. Invertible sheaves form the group $\mathrm{Pic}(X)$ with respect to the tensor product. A Cartier divisor on X is a family (U_i, g_i), $i \in I$, where U_i are open subsets of X covering X, and g_i are rational functions on U_i such that g_i/g_j is regular on each intersection $U_i \cap U_j$. The functions g_i are called local equations of the divisor. More precisely, a Cartier divisor is an equivalence class of such data. Two collections (U_i, g_i) and (U_i', g_i') are equivalent if their union is still a divisor. Cartier divisors can be added by multiplying their local equations. Thus they form a group, denoted by $\mathrm{Div}(X)$.

If each local equation g_i is regular on U_i, then we say that the divisor D is effective, and we write $D \geq 0$. The subschemes $\{g_i = 0\}$ of the U_i can then be glued together into a subscheme of X, also denoted by D. Therefore, effective Cartier divisors can be identified with locally principal subschemes of X (locally given by one equation). Any non-zero rational function $f \in \mathbb{C}(X)$ determines a principal Cartier divisor (f). Principal divisors form a subgroup of $\mathrm{Div}(X)$.

Let $\mathcal{K}_X$ denote the sheaf of rational functions on X, $\mathcal{K}_X(U) = \mathbb{C}(U)$. To every Cartier divisor $D = (U_i, g_i)_{i \in I}$ we can attach a subsheaf $\mathcal{O}_X(D) \subset \mathcal{K}_X$. Namely, on U_i it is defined as $g_i^{-1}\mathcal{O}_{U_i}$. On the intersection $U_i \cap U_j$ the sheaves $g_i^{-1}\mathcal{O}_{U_i}$ and $g_j^{-1}\mathcal{O}_{U_j}$ coincide, since g_i/g_j is invertible. Therefore these sheaves can be pasted together into a sheaf $\mathcal{O}_X(D) \subset \mathcal{K}_X$. For instance, $\mathcal{O}_X(0) = \mathcal{O}_X$ and $\mathcal{O}_X(D_1 + D_2) = \mathcal{O}_X(D_1) \otimes \mathcal{O}_X(D_2)$. The sheaves $\mathcal{O}_X(D)$ are invertible. In fact, multiplication by g_i defines an isomorphism ("trivialization") $\mathcal{O}_X(D)|_{U_i} \simeq \mathcal{O}_{U_i}$. This gives a surjective homomorphism $\mathrm{Div}(X) \to \mathrm{Pic}(X)$. Its kernel consists of principal divisors.

A non-zero section of $\mathcal{O}_X(D)$ is a rational function f on X such that fg_i is regular on the U_i for any i, in other words, such that the divisor $(f) + D$ is effective. If D itself is effective, the sheaf $\mathcal{O}_X(D)$ has a canonical section s_D, which corresponds to the constant function 1. This can also be reformulated as follows. Let $s \in H^0(X, \mathcal{L})$ be a non-trivial global section of an invertible sheaf $\mathcal{L}$. After choosing some trivializations $\phi_i : \mathcal{L}_{U_i} \simeq \mathcal{O}_{U_i}$ on a covering (U_i), we obtain an effective divisor $(U_i, \phi_i(s_i))$, the scheme of zeros of s, which we denote by $Z(s)$. For instance, if D is effective then $Z(s_D) = D$. So basically any effective divisor is defined by one equation $s = 0$. However, s is not a function, but a section of an invertible sheaf. If s' is another non-zero global section of $\mathcal{L}$ then the divisors $Z(s')$ and $Z(s)$ differ by the divisor of a rational function s/s'. One also says that they are linearly equivalent.

The sheaf $\mathcal{O}_X(-D)$, for D effective, is an ideal sheaf in $\mathcal{O}_X$. It defines D as a subscheme.

Example 1.24 Let us recall the construction of invertible sheaves on projective spaces $\mathbb{P}(V)$. All these sheaves have the form $\mathcal{O}(d)$, where $\mathcal{O}(d)$ is the sheaf of homogeneous functions of degree d on $\mathbb{P}(V)$. More precisely, let $\pi : V \setminus \{0\} \to \mathbb{P}(V)$ be the canonical projection. If $U \subset \mathbb{P}(V)$ is a Zariski open set, then the sections of $\mathcal{O}(d)$ over U are, by definition, regular functions f on $\pi^{-1}(U) \subset V$, which are homogeneous of degree d: $f(\lambda v) = \lambda^d f(v)$. It is well-known that these sheaves are invertible and any invertible sheaf has the form $\mathcal{O}(d)$ for some d. For example, $\mathcal{O}(-1)$, as a line bundle, is the tautological line bundle. Its fiber over a point of $\mathbb{P}(V)$ represented by a 1-dimensional subspace l is l itself. If $d < 0$ then $H^0(\mathbb{P}, \mathcal{O}(d)) = 0$. If $d \geq 0$ then $H^0(\mathbb{P}, \mathcal{O}(d)) = \mathrm{Sym}^d\, V^\vee$, homogeneous polynomials of degree d. For any non-zero $s \in H^0(\mathbb{P}, \mathcal{O}(d))$, the corresponding effective divisor $Z(s)$ is just the hypersurface defined by the polynomial s. In particular, hyperplanes in $\mathbb{P}(V)$ correspond to global sections of $\mathcal{O}(1)$.

Suppose now that X is an irreducible subvariety in $\mathbb{P}^N$. Then $\mathcal{O}(d)$ can be restricted on X, giving a sheaf $\mathcal{O}_X(d)$. In particular, we have a restriction homomorphism of global sections:

$$V^\vee = H^0(\mathbb{P}^N, \mathcal{O}(1)) \xrightarrow{\mathrm{res}} H^0(X, \mathcal{O}_X(1)).$$

Clearly this map is not injective if and only if X is degenerate. So suppose now that X is non-degenerate. In general, res is not surjective either. Its image is a vector subspace $W \subset H^0(X, \mathcal{O}_X(1))$ with the following obvious property: for any $x \in X$ there exists a section $s \in W$ such that $s(x) \neq 0$. Divisors of the form $Z(s)$, $s \in W$, are just hyperplane sections $X \cap H$ for various hyperplanes H.

In general, if X is a projective variety with an invertible sheaf $\mathcal{L}$, then a family of divisors

$$|W| = \{Z(s) \mid s \in W \subset H^0(X, \mathcal{L})$$

is called a *linear system* of divisors. If $W = H^0(X, \mathcal{L})$, the linear system is called *complete.* It is denoted by $|\mathcal{L}|$, or by $|D|$ if $\mathcal{L} = \mathcal{O}(D)$. Every effective divisor linearly equivalent to D appears in $|D|$ exactly once.

$|W|$ is called *base-point free* if the intersection of all its divisors is empty, or, equivalently, if for any $x \in X$ there exists $s \in W$ such that $s(x) \neq 0$. As we have seen, if $X \subset \mathbb{P}(V)$ is a projective variety then the image of $\mathrm{res} : V^\vee \to H^0(X, \mathcal{O}_X(1))$ defines a base-point free linear system. Conversely, any base-point free linear system $|W|$ defines a morphism into a projective space. Namely, for any $x \in X$ we have a hyperplane $\Gamma(x) \subset W$ consisting of all sections vanishing at x. This gives a regular map $\Gamma : X \to \mathbb{P}(W^*)$. If Γ is an embedding then W is identified with the image of a restriction map $W \to H^0(X, \mathcal{O}_X(1))$ and $\mathcal{O}_X(1)$ is identified with $\mathcal{L}$. We therefore need a final definition.

Definition 1.25 A linear system $|W|$ is called *very ample* if it defines an embedding in a projective space. An invertible sheaf $\mathcal{L}$ is called very ample if the complete linear system $|\mathcal{L}|$ is very ample. $\mathcal{L}$ is called *ample* if some tensor power $\mathcal{L}^{\otimes m}$, $m > 1$, is very ample.

In particular, we see that non-degenerate projective embeddings of X are in a $1-1$ correspondence with very ample linear systems of divisors.

Now we can describe linearly normal varieties.

Theorem 1.26 *The follolowing statements are equivalent:*
(i) *A projective variety $X \subset \mathbb{P}(V)$ is linearly normal.*
(ii) $\mathrm{res} : V^\vee \to H^0(X, \mathcal{O}_X(1))$ *is an isomorphism*
(iii) *The embedding $X \subset \mathbb{P}(V)$ is given by a complete linear system corresponding to some very ample invertible sheaf $\mathcal{L}$ on X.*

Proof. Clearly res is injective if and only if X is non-degenerate. Suppose that res is not surjective. Then the linear system $|\mathcal{O}_X(1)|$ gives a non-degenerate embedding of X into the projectivization of a higher-dimensional vector space $H^0(X, \mathcal{O}_X(1))^\vee$. The initial embedding is obtained from this one by the projection with center $\mathbb{P}(\mathrm{Im}(\mathrm{res})^*)$. Conversely, if X is a non-trivial isomorphic projection of a non-degenerate variety $\tilde{X}$ in a larger projective space $\mathbb{P}(U)$ then we have a proper embedding $V^\vee \subset U^\vee$. The map $U^\vee \to H^0(X, \mathcal{O}_{\tilde{X}}(1))$, given by restricting linear functions from $\mathbb{P}(U)$ to $\tilde{X}$, is an injection, since $\tilde{X}$ is non-degenerate. Thus the map res is not surjective, being the composition $V^\vee \hookrightarrow U^\vee \to H^0(X, \mathcal{O}_X(1))$. □

We see that in order to study dual varieties and discriminants we can restrict ourselves to linearly normal varieties. We adopt the following definition:

Definition 1.27 A pair $(X, \mathcal{L})$ of a projective variety X and a very ample invertible sheaf $\mathcal{L}$ on it is called a *polarized variety.*

Any polarized variety admits a canonical embedding in a projective space with a linearly normal image, namely, the embedding given by the complete linear system $|\mathcal{L}|$. Therefore we may speak about dual varieties, defect, discriminants, etc. of polarized varieties.

2

Actions with Finitely Many Orbits

In our survey we are mostly interested in the study of varieties with symmetries, and so in this chapter we study projective geometry of homogeneous spaces; more precisely we look at orbit closures for algebraic groups acting on projective spaces with finitely many orbits. In Sect. 2.1 we give the necessary background on algebraic groups and fix notation. In Sect. 2.2 we discuss Pyasetskii pairing, which is an interesting instance of projective duality. We give some standard and a few exotic examples. The next Sect. 2.3 contains the systematic treatment of actions related to gradings of simple Lie algebras. These actions provide a wealth of very important varieties that will be studied throughout this book. Some examples include Severi varieties, smooth self-dual varieties, smooth building blocks for varieties of positive defect, varieties of small codegree, etc. It is quite an interesting phenomenon that varieties with extremal projective properties tend to have maximal symmetries. We finish this chapter with the description of Pyasetskii pairing for actions related to gradings of GL_n called the Zelevinsky involution.

2.1 Algebraic Groups

In this section we recall basic facts about algebraic groups and homogeneous spaces. Details can be found in [Hu, OV1].

Linear Algebraic Groups

An algebraic variety G with a group structure is called an *algebraic group* if multiplication and inverse maps

$$G \times G \to G,\ (x,y) \mapsto x \cdot y, \quad G \to G,\ x \mapsto x^{-1}$$

are morphisms of algebraic varieties. A subgroup $H \subset G$ is *algebraic* if H is an algebraic subvariety. In the same way we define algebraic homomorphisms.

Images and kernels of algebraic homomorphisms are algebraic subgroups. If $H \subset G$ is a normal algebraic subgroup then G/H is canonically an algebraic group. The commutant $[H_1, H_2]$ of two algebraic subgroups $H_1, H_2 \subset G$ Is an algebraic subgroup of G, as are their intersection $H_1 \cap H_2$, their intersection, and the center $Z(G)$ are algebraic subgroups of G.

The basic example is $\mathrm{GL}(V)$, the group of invertible linear transformations of a vector space V. If an algebraic group G can be embedded into some $\mathrm{GL}(V)$ then G is called a *linear algebraic group.* G is linear if and only if G is an affine algebraic variety. The class of linear algebraic groups is closed with respect to taking subgroups, quotient groups and direct products. All algebraic groups in this book are linear.

Lie Algebras

Any algebraic group G is smooth, and therefore G is irreducible if and only if it is connected. The connected component of G containing the unity element is an algebraic subgroup $G^0 \subset G$. It is called the *connected component* of G. The commutant of two connected subgroups in G is connected.

The tangent space $T_{e,G}$ to G at unity has a natural structure of a Lie algebra. We denote Lie algebras of algebraic groups by the corresponding gothic letters, e.g. by $\mathfrak{g}$. Any homomorphism $f : G_1 \to G_2$ of algebraic groups induces the homomorphism of Lie algebras $df|_e : \mathfrak{g}_1 \to \mathfrak{g}_2$. If $G = \mathrm{GL}(V)$ then $\mathfrak{g} = \mathfrak{gl}(V)$ is the Lie algebra of all linear operators of V with the standard commutator

$$[X, Y] = XY - YX.$$

If $G \subset \mathrm{GL}(V)$ then $\mathfrak{g} \subset \mathfrak{gl}(V)$ and the commutator in $\mathfrak{g}$ is the restriction of the commutator in $\mathfrak{gl}(V)$.

A homomorphism of Lie algebras $f : \mathfrak{g}_1 \to \mathfrak{g}_2$ is called *algebraic* if there exists a homomorphism of algebraic groups $F : G_1 \to G_2$ such that f is its differential.

Algebraic Actions

An action of an algebraic group G on an algebraic variety X is called algebraic if the corresponding map

$$G \times X \to X, \ (g, x) \mapsto g \cdot x$$

is algebraic. All our actions will be algebraic.

If an algebraic group G acts on an algebraic variety X and $Y \subset X$ is an algebraic subvariety, then the *normalizer*

$$N_G(Y) = \{g \in G \,|\, gY \subset Y\}$$

is an algebraic subgroup of G. The normalizer of a point $x \in X$ is denoted by G_x and is called the *stabilizer* of x. For example, if G acts on itself by conjugation and $H \subset G$ is an algebraic subgroup (or just any subvariety), then the normalizer

$$N_G(H) = \{g \in G \,|\, gHg^{-1} \subset H\}$$

and the centralizer

$$Z_G(H) = \{g \in G \,|\, gh = hg \text{ for any } h \in H\}$$

are algebraic subgroups of G.

Representations

If we have a homomorphism $G \to \mathrm{GL}(V)$ then we also say that we have a representation of G in V, or that V is a G-module, or that G acts linearly on V. In the same way we define representations of Lie algebras. If V is a G-module then V is also a $\mathfrak{g}$-module, where $\mathfrak{g}$ is the Lie algebra of G.

For example, any algebraic group G has a natural representation in its Lie algebra $\mathfrak{g}$ called the *adjoint representation* $\mathrm{Ad} : G \to \mathrm{GL}(\mathfrak{g})$. If $G = \mathrm{GL}(V)$ then Ad is just the matrix conjugation

$$\mathrm{Ad}(X)A = XAX^{-1}.$$

If $G \subset \mathrm{GL}(V)$ then $\mathfrak{g} \subset \mathfrak{gl}(V)$ and the adjoint representation is again given by the conjugation (restricted on $\mathfrak{g}$). The differential of the adjoint representation is the adjoint representation of a Lie algebra $\mathfrak{g}$:

$$\mathrm{ad} : \mathfrak{g} \to \mathfrak{gl}(\mathfrak{g}), \quad \mathrm{ad}(x) = [x, \cdot].$$

If G acts linearly on V then G automatically acts on $V^\vee$ as well. This representation (or module, or action) is called *dual*.

The group of all projective transformations of $\mathbb{P}^N$ is an algebraic group denoted by PGL_{N+1}. It is a quotient of GL_{N+1} by its center. PGL_{N+1} also acts dually on $(\mathbb{P}^N)^\vee$. In fact any automorphism of $\mathbb{P}^N$ is given by a projective transformation, and therefore any action of an algebraic group G on $\mathbb{P}^N$ is represented by projective transformations and corresponds to some homomorphism $G \to \mathrm{PGL}_{N+1}$.

Invariants

If G acts on the affine variety X then the ring of regular functions $\mathbb{C}[X]$ is an infinite-dimensional G-module. The action is defined by the formula

$$(g \cdot f)(x) = f(g^{-1} \cdot x), \text{ for any } g \in G,\ x \in X.$$

This module is a union of finite-dimensional G-modules. The subring of G-invariant (constant along G-orbits) functions is denoted by $\mathbb{C}[X]^G$.

Connected Solvable Groups

The group B_n of upper-triangular $(n \times n)$-matrices is a typical example of a solvable algebraic group.

Any action of a connected solvable algebraic group G on a projective variety has a fixed point (Borel Theorem). In particular, for any representation $f : G \to \mathrm{GL}(V)$ there exists a basis such that all operators $f(x)$, $x \in G$, have an upper-triangular form, i.e. G preserves some flag of subspaces in V (Lie Theorem).

Any algebraic group G has a unique maximal normal connected solvable algebraic subgroup $\mathrm{Rad}(G)$. It is called the *radical* of G.

Jordan Decomposition

A linear operator $A \in \mathrm{GL}(V)$ is called unipotent (resp. semisimple) if in some basis A is represented by an upper-triangular matrix with 1's on the diagonal (resp. is represented by a diagonal matrix). Let G be an algebraic group. An element $x \in G$ is called *unipotent* (resp. *semisimple*) if for some (and hence or any) embedding $i : G \hookrightarrow \mathrm{GL}(V)$ the element $i(x)$ is a unipotent operator (resp. a semisimple operator). Any homomorphism of algebraic groups takes semisimple (resp. unipotent) elements to semisimple (resp. unipotent) elements. Any element $x \in G$ has a unique *Jordan decomposition*

$$x = x_s x_u = x_u x_s,$$

where x_s is semisimple and x_u is unipotent.

Unipotent Groups

An algebraic group is called *unipotent* if all its elements are unipotent. Any unipotent group is connected and solvable. For example, the group U_n of upper-triangular $(n \times n)$ matrices with 1's on the diagonal is unipotent. Another example is a vector space with respect to addition.

Any character, i.e. a homomorphism $G \to \mathbb{C}^*$, of a unipotent group G is trivial. It follows from the Lie Theorem that for any representation $f : G \to \mathrm{GL}(V)$ there exists a basis such that all operators $f(x)$, $x \in G$, have an upper-triangular form with 1's on the diagonal. Equivalently, any finite-dimensional G-module contains a non-zero fixed vector.

If a unipotent group G acts on the affine variety X then all orbits are closed. Indeed, let Y be a closure of a certain orbit. Then G has a dense open orbit in Y. If the complement to this orbit is not empty then its ideal of vanishing functions contains a non-trivial finite-dimensional G-module, and therefore a non-trivial invariant function $f \in \mathbb{C}[Y]^G$. This function is constant along the dense orbit, and therefore it is constant along Y. Hence it can not vanish on the complement to the dense open orbit. This is a contradiction.

Any algebraic group G has a unique maximal normal unipotent subgroup $R_u(G)$. It is called the *unipotent radical* of G.

Reductive Groups

An algebraic group G is called *reductive* if $R_u(G)$ is trivial. An algebraic group G is called *semisimple* if $\mathrm{Rad}(G)$ is trivial. Obviously, semisimple groups are reductive.

G is reductive if and only if any finite-dimensional G-module V is totally reducible, i.e. splits as a direct sum $\oplus V_i$ of irreducible submodules V_i.

If G is reductive then $[G, G]$ is semisimple. If G is also connected then

$$G = Z(G) \times_{loc} [G, G],$$

where the *locally direct product* is a quotient of the direct product by a finite central subgroup. The corresponding decomposition of $\mathfrak{g}$ is the direct product

$$\mathfrak{g} = \mathfrak{z}(\mathfrak{g}) \oplus [\mathfrak{g}, \mathfrak{g}].$$

For any semisimple Lie algebra $\mathfrak{g}$ there exists a unique connected simply-connected semisimple algebraic group G with a Lie algebra $\mathfrak{g}$. Any representation of $\mathfrak{g}$ is a differential of the unique representation of G. All other connected semisimple algebraic groups with a Lie algebra $\mathfrak{g}$ are quotients of G by finite central subgroups.

Standard Invariant Form

Let $f : \mathfrak{g} \to \mathfrak{gl}(V)$ be an algebraic representation of a Lie algebra $\mathfrak{g}$ of a connected algebraic group G. Then

$$(x, y)_V = \mathrm{Tr}\big(f(x)f(y)\big)$$

is an invariant symmetric bilinear form on $\mathfrak{g}$, i.e.

$$\big(\mathrm{Ad}(g)x, \mathrm{Ad}(g)y\big)_V = (x, y)_V \text{ for any } x, y \in \mathfrak{g} \text{ and } g \in G.$$

It is called the *standard invariant form*. G is reductive if and only if $(\cdot, \cdot)_V$ is non-degenerate for some (and hence for any) injective f.

If G is semisimple then its center is finite and the adjoint representation $\mathrm{Ad} : G \to \mathrm{GL}(\mathfrak{g})$ gives an embedding $\mathrm{ad} : \mathfrak{g} \hookrightarrow \mathfrak{gl}(\mathfrak{g})$. The image of $\mathfrak{g}$ in $\mathfrak{gl}(\mathfrak{g})$ is a Lie algebra of *derivations* $D \in \mathfrak{gl}(\mathfrak{g})$ such that

$$D([x, y]) = [D(x), y] + [x, D(y)] \text{ for any } x, y \in \mathfrak{g}.$$

The corresponding standard invariant form is called the *Killing form*,

$$(x, y) = \mathrm{Tr}[x, [y, \cdot]], \quad x, y \in \mathfrak{g}.$$

The Killing form is non-degenerate if and only if $\mathfrak{g}$ is semisimple.

Algebraic Tori

A connected algebraic group is called an *algebraic torus* if all its elements are semisimple. Any algebraic torus is isomorphic to $\mathbb{C}^* \times \ldots \times \mathbb{C}^*$, where $\mathbb{C}^*$ is the group of invertible elements in $\mathbb{C}$.

An algebraic torus T is obviously reductive. All irreducible representations of T are one-dimensional (any set of commuting semisimple operators can be taken into the diagonal form in some basis). Let $\mathcal{X}(T)$ be the group of *characters* of T, i.e. homomorphisms $T \to \mathbb{C}^*$. This is a free abelian group,

$$\operatorname{rank} \mathcal{X}(T) = \dim T.$$

$\mathcal{X}(T)$ admits the canonical embedding in $\mathfrak{t}^\vee$ by assigning to each character $\chi \in \mathcal{X}(T)$ its differential $d\chi$ at unity. We identify $\mathcal{X}(T)$ with its image in $\mathfrak{t}^\vee$ and we write $\mathcal{X}(T)$ additively. Let

$$\mathfrak{t}_{\mathbb{R}}^\vee = \mathcal{X}(T) \otimes \mathbb{R} \subset \mathfrak{t}^\vee.$$

Then

$$\mathfrak{t}^\vee = \mathfrak{t}_{\mathbb{R}}^\vee + i\mathfrak{t}_{\mathbb{R}}^\vee,$$

i.e. $\mathfrak{t}_{\mathbb{R}}^\vee$ is the real form of $\mathfrak{t}^\vee$ and $\mathfrak{t}_{\mathbb{R}} \subset \mathfrak{t}$ is the real form of $\mathfrak{t}$. The restriction of any standard invariant form of $\mathfrak{t}$ on $\mathfrak{t}_{\mathbb{R}}$ is a positive-definite scalar product.

Root Decomposition

Let G be a connected reductive group with a Lie algebra $\mathfrak{g}$. Any subtorus of G which is maximal with respect to inclusion is called a *maximal torus* of G. All maximal tori are conjugated to each other. Let T be one of them. Its dimension r is called the *rank* of G, and the Lie algebra $\mathfrak{t}$ of T is called a *Cartan subalgebra.*

We have the *root decomposition*

$$\mathfrak{g} = \mathfrak{t} \oplus \bigoplus_{\alpha \in \Delta} \mathfrak{g}_\alpha, \quad \alpha \neq 0, \quad \mathfrak{g}_\alpha \neq 0,$$

where

$$\mathfrak{g}_\alpha = \{x \in \mathfrak{g} \mid \operatorname{Ad}(t)x = \alpha(t)x \text{ for any } t \in T\}.$$

Elements of $\Delta \subset \mathcal{X}(T)$ are called *roots.* We have

$$\dim \mathfrak{g}_\alpha = 1 \text{ for any } \alpha \in \Delta;$$

$$\text{if } \alpha \in \Delta \text{ then } -\alpha \in \Delta;$$

$$\text{if } \alpha, k\alpha \in \Delta \text{ then } k = \pm 1.$$

Positive and Simple Roots

Choose any hyperplane in $\mathfrak{t}_{\mathbb{R}}^{\vee}$ not passing through Δ. Roots from one half-space are called *positive* and from another one *negative*. We denote the set of positive roots by Δ^+ and the set of negative roots by Δ^-. We have

$$\Delta = \Delta^+ \cup \Delta^-, \quad \Delta^- = -\Delta^+.$$

There exists a unique set of *simple roots* $\Pi \subset \Delta^+$,

$$\Pi = \{\alpha_1, \ldots, \alpha_r\},$$

such that any root $\alpha \in \Delta$ is an integral combination

$$\alpha = \sum_{i=1}^{r} n_i \alpha_i$$

with $n_i \geq 0$ for positive α and $n_i \leq 0$ for negative α. Simple roots are linearly independent.

Cartan Matrix and Dynkin Diagrams

We choose any standard invariant form on $\mathfrak{g}$. Its restriction on $\mathfrak{t}_{\mathbb{R}}$ is positive-definite. Using it, we define the scalar product on $\mathfrak{t}_{\mathbb{R}}^{\vee}$. We use the notation

$$\langle \lambda | \mu \rangle = \frac{2(\lambda, \mu)}{(\mu, \mu)} \text{ for } \lambda, \mu \in \mathfrak{t}_{R}^{\vee}, \ \mu \neq 0.$$

Then

$$\langle \alpha | \beta \rangle \in \mathbb{Z}$$

for any $\alpha, \beta \in \Delta$. Let $\Pi = \{\alpha_1, \ldots, \alpha_r\}$ be the set of simple roots. The *Cartan matrix*

$$A = \langle \alpha_i | \alpha_j \rangle_{i,j=1,\ldots,r}$$

has the properties

$$A_{ii} = 2, \quad A_{ij} \leq 0 \text{ for } i \neq j, \quad A_{ij=0} \text{ iff } A_{ji} = 0.$$

The *Dynkin diagram* is a graph with r vertices indexed by simple roots, where i-th and j-th vertices are connected by $|A_{ij} A_{ji}|$ edges. If $|A_{ij}| < |A_{ji}|$ then this edge has an arrow pointing from the j-th vertex to the i-th vertex.

Let us summarize some further properties of the root decomposition.

$$\text{If } \alpha, \beta \in \Delta \text{ then } \alpha + \beta \in \Delta \cup \{0\} \text{ iff } [\mathfrak{g}_\alpha, \mathfrak{g}_\beta] \neq 0;$$

$$\text{if } \alpha, \beta \in \Delta \text{ and } (\alpha, \beta) < 0 \text{ then } \alpha + \beta \in \Delta;$$

$$\text{if } \alpha, \beta, \alpha + \beta \in \Delta \text{ and } (\alpha, \beta) = 0 \text{ then } \alpha - \beta \in \Delta;$$

$$\text{if } x \in \mathfrak{g}_\alpha, \ y \in \mathfrak{g}_{-\alpha}, \ z \in \mathfrak{g}_\beta \text{ then } [[x, y], z] = (x, y)(\alpha, \beta) z.$$

Simple Groups

Any connected semisimple group is a locally direct product of *simple* algebraic groups, where a connected non-abelian algebraic group is called simple if all its normal subgroups are finite. G is simple if and only if $\mathfrak{g}$ is simple (i.e. $\mathfrak{g} \neq \mathbb{C}$ and $\mathfrak{g}$ has no proper ideals). A semisimple G is simple if and only if its Dynkin diagram is connected. Simple Lie algebras are parametrized by their Dynkin diagrams. There are four *classical series* A_n, B_n, C_n, D_n and five *exceptional cases* E_6, E_7, E_8, F_4, and G_2.

- A_n is isomorphic to $\mathfrak{sl}_{n+1}$, the Lie algebra of traceless endomorphisms of $\mathbb{C}^{n+1}$. The corresponding simply-connected simple group is SL_{n+1}, the group of all linear transformations of $\mathbb{C}^{n+1}$ with determinant equal to 1. The quotient of SL_{n+1} by its center is PGL_{n+1}. The Dynkin diagram is

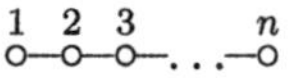

  Here we fix the numbering of simple roots that will be used afterwards.
- B_n, $n \geq 2$, is isomorphic to $\mathfrak{so}_{2n+1}$ and D_n, $n \geq 4$, is isomorphic to $\mathfrak{so}_{2n}$, where $\mathfrak{so}_n$ is the Lie algebra of skew-symmetric operators in $\mathbb{C}^n$ with respect to the given quadratic form Q. The group O_n of orthogonal transformations has $\mathfrak{so}_n$ as its Lie algebra. This group has two connected components — transformations with determinant equal to 1 and -1. $(\mathrm{O}_n)^0$ is called the special orthogonal group SO_n. This group is not simply-connected. The Dynkin diagram of B_n is

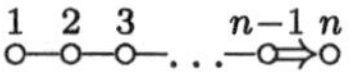

  The Dynkin diagram of D_n is

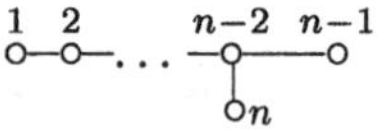

- C_n, $n \geq 2$, is isomorphic to $\mathfrak{sp}_{2n}$, the Lie algebra of skew-symmetric operators in $\mathbb{C}^{2n}$ with respect to the given symplectic form ω. The group Sp_{2n} of symplectic transformations has $\mathfrak{sp}_{2n}$ as its Lie algebra. This group is simply-connected. The Dynkin diagram of C_n is

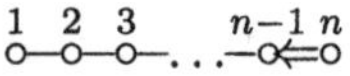

- Finally, exceptional Dynkin diagrams are

F_4 : 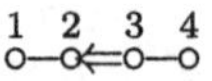

G_2 : 

E_6 : 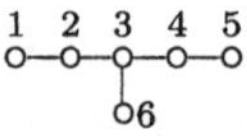

E_7 : 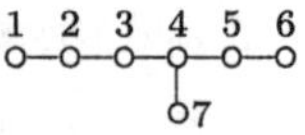

E_8 : 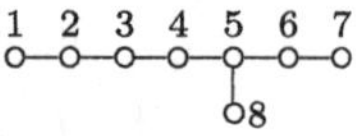

Weyl Group

For any $\mu \neq 0$ we denote by

$$r_\mu(\lambda) = \lambda - \langle \lambda, \mu \rangle \mu$$

the reflection with respect to the hyperplane orthogonal to μ. If $\alpha \in \Delta$ then $r_\alpha(\Delta) \subset \Delta$. The group W generated by reflection r_α, $\alpha \in \Delta$, is called the *Weyl group*. It is isomorphic to $N_G(T)/T$, where the latter group acts on $\mathfrak{t}$ by the adjoint representation (and dually on $\mathfrak{t}^\vee$).

We define the *length function*

$$l : W \to \mathbb{Z}_{\geq 0}, \quad l(w) = \left|\{\alpha \in \Delta_+ \,|\, w(\alpha) \in \Delta^-\}\right|.$$

There exists a unique element $w_0 \in W$ of the maximal length $|\Delta_+|$.

The hyperplanes in $\mathfrak{t}_\mathbb{R}^\vee$ orthogonal to roots divide $\mathfrak{t}_\mathbb{R}^*$ into cones called *Weyl chambers*. W acts transitively on the set of Weyl chambers, moreover, each Weyl chamber C is a fundamental domain for the action of W on $\mathfrak{t}_\mathbb{R}^\vee$, i.e. each W-orbit intersects C in one orbit. In particular, the number of Weyl chambers is equal to $|W|$. The choice of a Weyl chamber is equivalent to the choice of a set of positive roots. Namely, Δ^+ defines the Weyl chamber

$$C^+ = \{\lambda \in \mathfrak{t}_\mathbb{R}^* \,|\, (\lambda, \alpha) \geq 0 \text{ for any } \alpha \in \Delta^+\}.$$

Of course, in this formula Δ^+ can be substituted for the set of simple roots Π.

If G is simple and of types A, D, E then all roots are conjugated to each other by the Weyl group. If G is simple and of types B, C, F, G then W has two orbits on Δ – the set of *long roots* and the set of *short roots*. In $A-D-E$ cases we postulate that all roots are long.

Irreducible Representations

Take any representation V of a connected reductive group G. Then we have the weight decomposition

$$V = \bigoplus_{\lambda \in \mathcal{X}(T)} V^\lambda,$$

where

$$V^\lambda = \{v \in V \,|\, t \cdot v = \lambda(t)v \text{ for any } t \in T\}.$$

The subset

$$\mathcal{X}(T)_V = \{\lambda \in \mathcal{X}(T) \,|\, V^\lambda \neq 0\}$$

is called the *weight system* of V. If V is irreducible then there exists a unique *highest weight* $\lambda \in \mathcal{X}(T)_V$ such that $\lambda - \mu$ is an integral combination of simple roots with non-negative coefficients for any $\mu \in \mathcal{X}(T)_V$. Moreover,

$$\dim V^\lambda = 1 \text{ and } \lambda \in C^+.$$

Similarly, there exists a unique *lowest weight* $\hat{\lambda} \in \mathcal{X}(T)_V$ such that $\hat{\lambda} - \mu$ is an integral combination of simple roots with non-positive coefficients for any $\mu \in \mathcal{X}(T)_V$. The longest element in the Weyl group takes the highest weight to the lowest weight.

For any character $\lambda \in C^+ \cap \mathcal{X}(T)$ there exists a unique irreducible representation with the highest weight λ. Highest weights of irreducible representations are also called *dominant* weights. We denote the semigroup of dominant weights by $\mathcal{P}^+$.

If G is simple then Ad is irreducible, and we denote by γ the *highest root* (the highest weight of the adjoint representation) in Δ^+. The highest root is long.

Nilpotent Cone

Take any homomorphism with finite kernel $G \to \mathrm{GL}(V)$ and the corresponding embedding $i : \mathfrak{g} \hookrightarrow \mathfrak{gl}(V)$. An element $x \in \mathfrak{g}$ is called *semisimple* (resp. *nilpotent*) if $i(x)$ is a semisimple (resp. nilpotent) linear operator. This notion is independent of the choice of i. We have a version of the Jordan decomposition: any element $x \in \mathfrak{g}$ can be expressed uniquely as $x = x_n + x_s$, where x_n is nilpotent, x_s is semisimple, and $[x_s, x_n] = 0$.

Suppose that G is a connected reductive group. Nilpotent elements in $\mathfrak{g}$ form an irreducible conical algebraic subvariety $R(\mathfrak{g})$ called the *nilpotent cone.* Nilpotent elements $e \in \mathfrak{g}$ are characterized by the following equivalent properties:

- $0 \in \overline{\mathrm{Ad}(G)e}$.
- Any $\mathrm{Ad}(G)$-invariant polynomial on $\mathfrak{g}$ of positive degree vanishes at e.
- $e \in [\mathfrak{g}, \mathfrak{g}]$ and $[e, \cdot]$ is a nilpotent operator in $\mathfrak{gl}(\mathfrak{g})$.
- (Jacobson–Morozov Theorem) There exists an $\mathfrak{sl}_2$-triple $\langle e, h, f\rangle \subset \mathfrak{g}$, i.e. a triple of elements that satisfy the relations

$$[e, f] = h, \quad [h, e] = 2e, \quad [h, f] = -2f.$$

The number of $\mathrm{Ad}(G)$-orbits in $R(\mathfrak{g})$ is finite. For example, nilpotent orbits in $\mathfrak{gl}(V)$ are indexed by Jordan normal forms of nilpotent linear operators.

Homogeneous Spaces

If an algebraic group G acts transitively on an algebraic variety X then X is called a *homogeneous space.* If $H \subset G$ is an algebraic subgroup then the coset space G/H is a smooth algebraic variety and the action of G on G/H by left multiplication is algebraic. Moreover, G/H is quasiprojective (open subset of a projective variety): there exists an action of G on $\mathbb{P}^N$ such that H is the stabilizer of a point $p \in \mathbb{P}^N$ (Chevalley Theorem). In particular, G/H is identified with Gp.

Parabolic and Borel Subgroups

Let G be a connected reductive group. Consider any algebraic subgroup $H \subset G$. If G/H is a projective variety then H is called a *parabolic subgroup* and G/H is called a *flag variety.*

A maximal connected solvable subgroup of G is called a *Borel subgroup.* All Borel subgroups are conjugated to each other. For example, the group B_N of upper-triangular $(N \times N)$ matrices is a Borel subgroup in GL_N and G/B is the variety of complete flags in $\mathbb{C}^N$ (or in $\mathbb{P}^{N-1}$).

A subgroup $P \subset G$ is parabolic if and only if it contains some Borel subgroup. In particular, a Borel subgroup itself is parabolic.

The Lie subalgebra

$$\mathfrak{b} = \mathfrak{t} \oplus \bigoplus_{\alpha \in \Delta^+} \mathfrak{g}_\alpha$$

of $\mathfrak{g}$ is the Lie algebra of a Borel subgroup $B \subset G$. Any parabolic subgroup $P \supset B$ can be described as follows. Let

$$\Pi_P \subset \Pi$$

be some subset of simple roots. Let

$$\Delta_P^+ \subset \Delta^+$$

denote the positive roots that are linear combinations of the roots in Π_P. Then P is uniquely determined by its Lie algebra

$$\mathfrak{p} = \mathfrak{b} \oplus \bigoplus_{\alpha \in \Delta_P^+} \mathfrak{g}_{-\alpha}.$$

For example, if $\Pi_P = \Pi$ then $P = G$; if $\Pi_P = \emptyset$ then $P = B$. We let

$$\Pi_{G/P} = \Pi \setminus \Pi_P \quad \text{and} \quad \Delta_{G/P}^+ = \Delta^+ \setminus \Delta_P^+.$$

P is a a maximal proper parabolic subgroup if and only if $\Pi_{G/P}$ is a single simple root. In this case G/P is called the *minimal flag variety.*

If V is an irreducible G-module with the highest weight λ and $v \in V_\lambda$ is a non-zero vector, then B normalizes the line in V spanned by v_λ, and therefore the orbit $\mathcal{O}_V$ of this line in $\mathbb{P}(V)$ is the closed G-orbit isomorphic to some flag variety. $\mathcal{O}_V$ is called the *variety of the highest weight vectors.* This is a unique closed G-orbit in $\mathbb{P}(V)$.

If G/P is a minimal flag variety then $\operatorname{Pic} G/P = \mathbb{Z}$ is generated by a very ample line bundle. Let us describe minimal flag varieties of classical groups.

Projective Spaces and Grassmannians

Let $V = \mathbb{C}^N$ and $G = \mathrm{SL}(V)$. The projective space $\mathbb{P}(V)$ is identified with the flag variety G/P, where P is the normalizer in G of a line in V:

$$P = \{g \in G \,|\, g \cdot \langle e_1 \rangle \subset \langle e_1 \rangle\}.$$

All line bundles on $\mathbb{P}(V)$ have a form $\mathcal{O}(d)$ for $d \in \mathbb{Z}$. The line bundle $\mathcal{O}(d)$ is ample if and only if it is very ample if and only if $d > 0$. In this case

$$H^0(\mathbb{P}(V), \mathcal{O}(d)) = S^d V^\vee.$$

The embedding $\mathbb{P}(V) \subset \mathbb{P}(S^d V)$ is called the *Veronese embedding.* It assigns to a line spanned by $v \in V$ the line in $S^d V$ spanned by v^d. The zero scheme $Z(s)$ of a global section $s \in S^d(V^\vee)$ Is just the hypersurface in $\mathbb{P}(V)$ of degree d given by $s = 0$. Therefore the dual variety in this case parametrises singular hypersurfaces. In particular, the corresponding discriminant is the classical discriminant of a homogeneous form of degree d in N variables.

The Grassmannian $\mathrm{Gr}(k, V)$ of k-dimensional vector subspaces of V is the flag variety G/P, where P is the normalizer in $\mathrm{SL}(V)$ of a k-dimensional subspace in V:

$$P = \{g \in G \,|\, g \cdot \langle e_1, \ldots, e_k \rangle \subset \langle e_1, \ldots, e_k \rangle\}.$$

k-dimensional subspaces in V are in bijective correspondence with $(k-1)$-dimensional subspaces in $\mathbb{P}(V)$. Having this in mind we shall sometimes write $\mathrm{Gr}(k-1, \mathbb{P}(V))$ instead of $\mathrm{Gr}(k, V)$. The Grassmannian $\mathrm{Gr}(k, V)$ carries a natural equivariant vector bundle $\mathcal{S}$, called the *tautological vector bundle.* The fiber of $\mathcal{S}$ over a k-dimensional linear subspace $U \subset V$ is U itself. Therefore, $\mathcal{S}$ is a subbundle of the trivial bundle $\mathrm{Gr}(k, V) \times V$.

The line bundle $\Lambda^k \mathcal{S}^\vee$ is a very ample generator of $\mathrm{Pic}(\mathrm{Gr}(k, V)) = \mathbb{Z}$. The corresponding embedding is the Plücker embedding, $\mathrm{Gr}(k, V) \subset \mathbb{P}(\Lambda^k V)$. Namely, if $U \in \mathrm{Gr}(k, V)$ and $u_1, \ldots, u_k$ is any basis of U, then $U \mapsto [u_1 \wedge \ldots \wedge u_k]$.

The tangent space to $\mathrm{Gr}(k, V)$ at the point U is a vector space with an additional structure. Namely, it is easy to verify the natural isomorphism

$$T_{U, \mathrm{Gr}(k,V)} = \mathrm{Hom}(U, V/U) = U^\vee \otimes V/U.$$

In particular, the tangent bundle $T_{\mathrm{Gr}(k,V)}$ is canonically isomorphic to $\mathcal{S}^\vee \otimes (V/\mathcal{S})$. Notice also that

$$\mathrm{Gr}(k, V) \simeq \mathrm{Gr}(n-k, V^\vee).$$

Namely, any subspace $U \in \mathrm{Gr}(k, V)$ maps to $U^\perp \in \mathrm{Gr}(n-k, V^\vee)$.

Isotropic Grassmannians

Let $V = \mathbb{C}^N$ be a vector space equipped with a non-degenerate symmetric scalar product Q or a non-degenerate symplectic form ω (N should be even in this case). A subspace $U \subset V$ is called *isotropic* if for any $v_1, v_2 \in U$ we have $Q(v_1, v_2) = 0$ (resp. $\omega(v_1, v_2) = 0$). Let $G = \mathrm{SO}(V)$ or $G = \mathrm{Sp}(V)$, respectively.

For any $k \leq N/2$ let $\mathrm{Gr}_Q(k, V)$ (resp. $\mathrm{Gr}_\omega(k, V)$) denote the *isotropic Grassmannian* of k-dimensional isotropic subspaces. Then $\mathrm{Gr}_Q(k, V)$ and $\mathrm{Gr}_\omega(k, V)$ are projective G-equivariant varieties.

It follows from the Witt Theorem that $\mathrm{Gr}_Q(k, V)$ and $\mathrm{Gr}_\omega(k, V)$ are irreducible homogeneous varieties (hence flag varieties) in all cases except the symmetric case when $N = 2k$. In the latter case $\mathrm{Gr}_Q(k, \mathbb{C}^{2k})$ consists of two homogeneous components $\mathrm{Gr}_Q^+(k, \mathbb{C}^{2k})$ and $\mathrm{Gr}_Q^-(k, \mathbb{C}^{2k})$. These two varieties are in fact isomorphic to each other (as algebraic varieties, not as algebraic varieties with the action of G): if s is any orthogonal reflection then

$$s(\mathrm{Gr}_Q^{\pm}(k, \mathbb{C}^{2k})) = \mathrm{Gr}_Q^{\mp}(k, \mathbb{C}^{2k}).$$

Moreover, $\mathrm{Gr}_Q^{\pm}(k, \mathbb{C}^{2k})$ is isomorphic to $\mathrm{Gr}_Q(k-1, \mathbb{C}^{2k-1})$. To see this let $H = \mathbb{C}^{2k-1} \subset \mathbb{C}^{2k}$ be any hyperplane such that the restriction of Q on H is non-degenerate. Then we have obvious maps

$$\mathrm{Gr}_Q^{\pm}(k, \mathbb{C}^{2k}) \to \mathrm{Gr}_Q(k-1, \mathbb{C}^{2k-1}), \quad U \mapsto U \cap H.$$

It is easy to see that both maps are isomorphisms. The variety $\mathrm{Gr}_Q^{\pm}(k, \mathbb{C}^{2k})$ is called the *spinor variety*, denoted by $\mathbb{S}_k$.

The introduced varieties represent the complete list of minimal flag varieties of groups $\mathrm{SO}(V)$ and $\mathrm{Sp}(V)$. In particular, the Picard group of all these varieties is isomorphic to $\mathbb{Z}$. Let us describe the embedding corresponding to the ample generator of $\mathbb{Z}$.

Consider the symplectic case $G = \mathrm{Sp}(V)$ first. Then we have the Plücker embedding

$$\mathrm{Gr}_\omega(k, V) \subset \mathbb{P}(\Lambda^k V).$$

However, $\Lambda^k V$ is reducible as a $\mathrm{Sp}(V)$-module since it contains the submodule

$$\omega \wedge \Lambda^{k-2} V \subset \Lambda^k V.$$

The complement $\Lambda_0^k V$ of this submodule is an irreducible module and the Plücker embedding in fact gives the embedding of $\mathrm{Gr}_\omega(k, V)$ in $\mathbb{P}(\Lambda_0^k V)$. This embedding corresponds to the ample generator of the Picard group.

Now consider the orthogonal case $G = \mathrm{SO}(V)$. Then again we have the Plücker embedding $\mathrm{Gr}_Q(k, V) \subset \mathbb{P}(\Lambda^k V)$. Suppose first that

$$k < m \text{ if } N = 2m+1 \quad \text{or} \quad k < m-1 \text{ if } N = 2m.$$

Then $\Lambda^k V$ is irreducible as a G-module, and this embedding corresponds to the ample generator of the Picard group. The most interesting case is the minimal embedding of a spinor variety.

Clifford Algebra

Let $V = \mathbb{C}^{2m}$ be an even-dimensional vector space with a non-degenerate symmetric scalar product Q. Let $\mathrm{Cl}(V)$ be the *Clifford algebra* of V, i.e. the quotient of the tensor algebra T^*V by the ideal generated by elements

$$v \otimes v - Q(v,v)\cdot 1, \quad v \in V.$$

$\mathrm{Cl}(V)$ has the filtration

$$\mathrm{Cl}(V)_0 \subset \mathrm{Cl}(V)_1 \subset \ldots,$$

where $\mathrm{Cl}(V)_k$ is spanned by products

$$v_1 \cdot \ldots \cdot v_l, \; v_i \in V$$

with $l \leq k$. The associated graded algebra is the exterior algebra $\Lambda^\bullet V$. The Clifford algebra $\mathrm{Cl}(V)$ is in fact a superalgebra,

$$\mathrm{Cl}(V) = \mathrm{Cl}^0(V) \oplus \mathrm{Cl}^1(V),$$

where $\mathrm{Cl}^0(V)$ (resp. $\mathrm{Cl}^1(V)$) is a linear span of elements of the form $v_1 \cdot \ldots \cdot v_r$, $v_i \in V$, r is even (resp. r is odd). We define

$$\mathrm{Spin}(V) = \{a \in \mathrm{Cl}^0(V) \,|\, a = v_1 \cdot \ldots \cdot v_r, \; v_i \in V, \; Q(v_i, v_i) = 1\}.$$

This is a connected simply-connected algebraic group, called the *spinor group*.

Given $a \in \mathrm{Cl}(V)$, $a = v_1 \cdot \ldots \cdot v_r$, $v_i \in V$, let

$$\overline{a} = (-1)^r v_r \cdot \ldots \cdot v_1.$$

This is a well-defined involution of $\mathrm{Cl}(V)$. The formula

$$R(a)v = a \cdot v \cdot \overline{a}, \quad a \in \mathrm{Spin}(V), \; v \in V$$

gives the well-defined orthogonal action of $\mathrm{Spin}(V)$ on V. In fact, this gives the double covering

$$\mathrm{Spin}(V) \to \mathrm{SO}(V).$$

Let $U \subset V$ be a maximal isotropic subspace, and hence $\dim U = m$. Take also any maximal isotropic subspace U' such that $U \oplus U' = V$. For any $v \in U$ we define an operator $\rho(v) \in \mathrm{End}(\Lambda^\bullet U)$

$$\rho(v) \cdot u_1 \wedge \ldots \wedge u_r = v \wedge u_1 \wedge \ldots u_r.$$

For any $v \in U'$ we define an operator $\rho(v) \in \mathrm{End}(\Lambda^\bullet U)$

$$\rho(v) \cdot u_1 \wedge \ldots \wedge u_r = \sum_{i=1}^{r} (-1)^{i-1} Q(v, u_i) \wedge u_1 \wedge \ldots u_{i-1} \wedge u_{i+1} \wedge \ldots \wedge u_r.$$

These operators are well-defined and therefore give a linear map $\rho : V \to \mathrm{End}(\Lambda^\bullet U)$. It is easy to check that

$$\rho(v)^2 = Q(v,v)\,\mathrm{Id}$$

for any $v \in V$. Therefore we have a homomorphism (actually, an isomorphism) of associative algebras

$$\mathrm{Cl}(V) \to \mathrm{End}(\Lambda^\bullet U).$$

$\Lambda^\bullet U$ is an irreducible $\mathrm{Cl}(V)$-module but a reducible $\mathrm{Spin}(V)$-module. However, its even part $\Lambda^{ev}U$ is an irreducible $\mathrm{Spin}(V)$-module, called the *half-spinor module* $\mathcal{S}^+$.

Spinor varieties

Now let us describe the embedding

$$\mathbb{S}_m \subset \mathbb{P}(\mathcal{S}^+).$$

It will correspond to the ample generator of $\mathrm{Pic}(\mathbb{S}_m)$. We are going to show that $\mathbb{S}_m$ is isomorphic to the projectivization of the orbit of 1 in the half-spinor representation.

By definition, $\mathbb{S}_m$ is an $\mathrm{SO}(V)$-orbit of a fixed maximal isotropic subspace, say U'. We claim that the normalizer of U' in $\mathrm{Spin}(V)$ (acting on V via the representation R) is equal to the normalizer in $\mathrm{Spin}(V)$ (acting on $\Lambda^{ev}U$ via the representation ρ) of the line spanned by $1 \in \Lambda^{ev}U$. This will show that $\mathbb{S}_m$ is naturally isomorphic to the projectivization of the orbit of 1 in the half-spinor representation. Consider the map

$$\Psi : V \times \Lambda^{ev}U \to \Lambda^{odd}U, \quad (v,w) \mapsto \rho(v)w.$$

$\mathrm{Spin}\,V$ acts on V via the representation R and on Λ^{ev} and Λ^{odd} via the representation ρ. We claim that Ψ is $\mathrm{Spin}(V)$-equivariant. Indeed, for any $g \in \mathrm{Spin}(V)$ we have

$$\begin{aligned}\Psi(R(g)v, \rho(g)w) &= \rho(R(g)v)(\rho(g)w) = \rho(gv\overline{g})(\rho(g)w)\\ &= \rho(gv\overline{g}g)w = \rho(g)(\rho(v)w) = \rho(g)\Psi(v,w),\end{aligned}$$

since $g\overline{g} = 1$. It is easy to check that $\Psi(U',1) = 0$ and if for any $v \in U'$ we have $\Psi(v,w) = 0$ then $w = \lambda 1$ for some $\lambda \in \mathbb{C}$.

Suppose that $g \in \mathrm{Spin}(V)$ normalizes the line spanned by $1 \in \Lambda^{ev}U$. Then for any $v \in U'$ we have

$$\Psi(R(g)v, \rho(g)1) = \rho(g)\Psi(v,1) = 0,$$

and therefore g normalizes U'. Suppose now that g normalizes U'. Then for any $v \in U'$

$$0 = \Psi(R(g)(R(g^{-1})v), \rho(g)1) = \Psi(v, \rho(g)1),$$

and therefore g normalizes the line spanned by $1 \in \Lambda^{ev}U$.

2.2 Pyasetskii Pairing and Kashin Examples

In [Py] Pyasetskii has shown that if a connected algebraic group acts linearly on a vector space with finitely many orbits then the dual representation has the same property and, moreover, the number of orbits is the same. In this section we show how this result is related to projective duality.

Theorem 2.1 *Suppose that a connected algebraic group G acts on a projective space $\mathbb{P}^n$ with finitely many orbits. Then the dual action $G : (\mathbb{P}^n)^\vee$ has the same number of orbits. Let*

$$\mathbb{P}^n = \bigsqcup_{i=1}^{N} \mathcal{O}_i \quad \text{and} \quad (\mathbb{P}^n)^\vee = \bigsqcup_{i=1}^{N} \mathcal{O}'_i$$

be the orbit decompositions. Let $\mathcal{O}_0 = \mathcal{O}'_0 = \emptyset$. Then the bijection is defined as follows: $\mathcal{O}_i$ corresponds to $\mathcal{O}'_j$ if and only if $\overline{\mathcal{O}_i}$ is projectively dual to $\overline{\mathcal{O}'_j}$.

Proof. Indeed, take any G-orbit $\mathcal{O} \subset (\mathbb{P}^n)^\vee$. If $\overline{\mathcal{O}} \neq (\mathbb{P}^n)^\vee$ then the projectively dual variety $\overline{\mathcal{O}}^*$ is G-invariant, irreducible, and non-empty. Therefore it is the closure of some orbit $\mathcal{O}' \subset \mathbb{P}^n$, because there are finitely many of them. By the Reflexivity Theorem we have $\overline{\mathcal{O}} = \overline{\mathcal{O}'}^*$. □

Remark 2.2 If the number of orbits for the action $G : \mathbb{P}^n$ is infinite then, in general, there is no natural bijection between orbits in $\mathbb{P}^n$ and $(\mathbb{P}^n)^*$. However, it can be shown (see e.g. [Py]) that these actions have the same *modality*, i.e. the maximal number of parameters that the family of orbits can depend on. This can also be deduced from projective duality.

The linear action of an algebraic group on a vector space V is called *conical* if any G-orbit $\mathcal{O}$ is conical, i.e. preserved by homotheties. For example, if there are only finitely many orbits then the action is obviously conical. Non-zero orbits of the conical action are in a 1-1 correspondence with G-orbits in $\mathbb{P}(V)$. The action is conical if and only if any $v \in V$ belongs to the tangent space of its orbit $T_{v,Gv}$.

Proposition 2.3 *If the linear action $G : V$ is conical then the dual action $G : V^\vee$ is conical as well.*

Proof. We can embed $V = \mathbb{C}^n$ and $V^\vee = (\mathbb{C}^n)^\vee$ in $\mathbb{P}^n$ and $(\mathbb{P}^n)^\vee$ as affine charts. These embeddings are G-equivariant. Points of $V^\vee$ correspond to hyperplanes in $\mathbb{P}^n$ not passing through the origin $0 \in V \subset \mathbb{P}^n$. Suppose that $\mathcal{O} \subset V^\vee$ is a non-conical orbit. Then the dual variety $\overline{\mathcal{O}}^* \subset \mathbb{P}^n$ intersects with $\mathbb{C}^n$ non-trivially. Therefore $\overline{\mathcal{O}}^* \subset \mathbb{P}^n$ is the closure of a conical variety in $\mathbb{C}^n$. Therefore its dual variety $\overline{\mathcal{O}}^{**} \subset (\mathbb{P}^n)^\vee$ does not intersect $(\mathbb{C}^n)^\vee$. But this contradicts the Reflexivity Theorem. □

Corollary 2.4 ([**Py**]) *Suppose that a connected algebraic group G acts linearly on a vector space V with a finite number of orbits. Then the dual action $G : V^\vee$ has the same number of orbits. Let*

$$V = \bigsqcup_{i=1}^{N} \mathcal{O}_i \quad \text{and} \quad V^\vee = \bigsqcup_{i=1}^{N} \mathcal{O}'_i$$

be the orbit decompositions. Then the bijection is defined as follows: $\mathcal{O}_i$ corresponds to $\mathcal{O}'_j$ if and only if $\mathbb{P}(\overline{\mathcal{O}_i})$ is projectively dual to $\mathbb{P}(\overline{\mathcal{O}'_j})$.

Remark 2.5 Recall that the Reflexivity Theorem is equivalent to the formula (1.2). Varieties $\mathrm{Lag}(\mathcal{O}_i)$ have a nice interpretation. Namely, the representation of $\mathfrak{g}$ (the Lie algebra of G) in V is given by the element of

$$\mathrm{Hom}(\mathfrak{g}, \mathrm{End}(V)) = \mathfrak{g}^\vee \otimes V^\vee \otimes V = \mathrm{Hom}(V^\vee \otimes V, \mathfrak{g}^\vee).$$

The corresponding bilinear map

$$\mu : V \oplus V^\vee \to \mathfrak{g}^\vee$$

is called the *moment map*. Let

$$L = \mu^{-1}(0)$$

be its zero fiber. Then it is easy to see that if G acts on V with finitely many orbits $\mathcal{O}_i$ then

$$L = \cup_i \mathrm{Lag}(\mathcal{O}_i)$$

is the decomposition of L into irreducible components.

Fig. 2.1. B_2-orbits on $\mathfrak{u}_2$ and $\mathfrak{u}_2^\vee$

Let us consider some examples of Pyasetskii pairing when the orbit decomposition is really complicated. They were found by Kashin [Kas].

Let B_n be the group of upper-triangular $(n \times n)$ matrices. We denote by $\mathfrak{u}_n$ the Lie algebra of upper-triangular $(n \times n)$ matrices with zeros on the diagonal. B_n acts on $\mathfrak{u}_n$ by conjugation. This action has finitely many orbits if and only if $n \leq 5$ (see also 8.6.1).

Kashin has found all orbits in these cases, calculated Pyasetskii pairing and described Hasse diagrams for the poset of orbit closures. These results are summarised below. For $n = 2, 3, 4, 5$ we give the Hasse diagram for the set

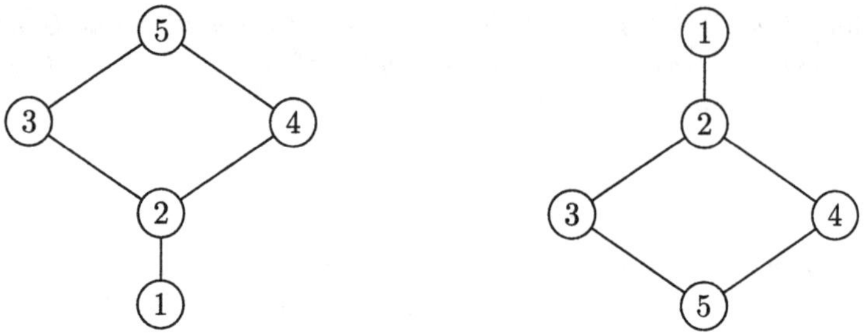

Fig. 2.2. B_3-orbits on $\mathfrak{u}_3$ and $\mathfrak{u}_3^\vee$

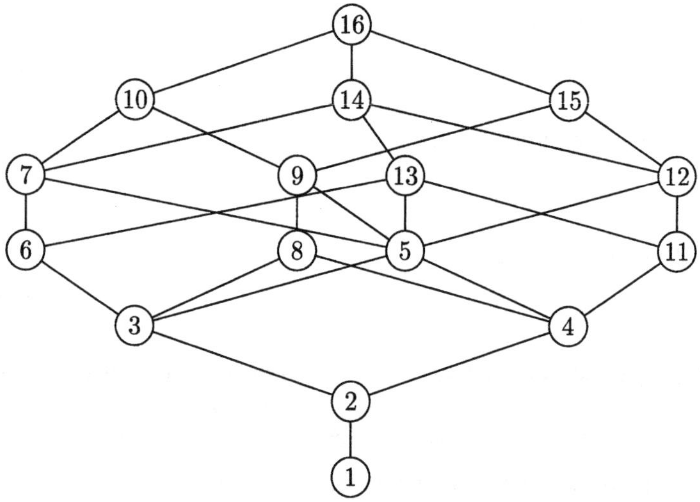

Fig. 2.3. B_4-orbits on $\mathfrak{u}_4$

of orbits in $\mathfrak{u}_n^\vee$ and $\mathfrak{u}_n$. We number orbits in such a way that an orbit $\mathcal{O}_i' \subset \mathfrak{u}_n^\vee$ corresponds to an orbit $\mathcal{O}_i'' \subset \mathfrak{u}_n$ via Pyasetskii pairing.

These diagrams have two interesting properties. Firstly, these graphs are networks, i.e. vertices are located on several levels and edges go from level $i+1$ to level i. This is easy to explain: the level of the orbit is just its dimension. The group B_n is solvable, but all orbits of solvable groups are affine varieties and therefore have divisorial boundaries.

The second property (discovered by Kashin) is much more mysterious. Namely, Hasse diagrams of $\mathfrak{u}_n^\vee$ and $\mathfrak{u}_n$ have the same number of edges! I don't know if it is a coincidence. This question is interesting only for non-reductive groups because if G is reductive then Hasse diagrams of orbit closures in any representation V and in its dual $V^\vee$ are equal.

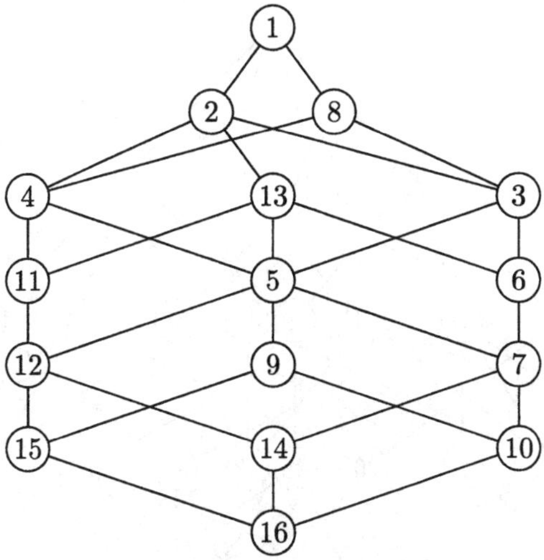

Fig. 2.4. B_4-orbits on $\mathfrak{u}_4^\vee$

2.3 Actions Related to Gradings

2.3.1 Construction

A vast number of linear actions of reductive groups with finitely many orbits is provided by the following construction. Let L be a connected reductive group with the Lie algebra $\mathfrak{l}$. Let $\mathfrak{z}$ be the center of $\mathfrak{l}$. Suppose that $\mathfrak{l}$ is *graded*, i.e.

$$\mathfrak{l} = \bigoplus_{k \in \mathbb{Z}} \mathfrak{l}_k, \quad [\mathfrak{l}_i, \mathfrak{l}_j] \subset \mathfrak{l}_{i+j}.$$

We also assume that $\mathfrak{z} \subset \mathfrak{l}_0$. The commutant $[\mathfrak{l}, \mathfrak{l}]$ is automatically graded, and we have

$$\mathfrak{l}_0 = [\mathfrak{l}, \mathfrak{l}]_0 \oplus \mathfrak{z}, \quad \mathfrak{l}_k = [\mathfrak{l}, \mathfrak{l}]_k \text{ for } k \neq 0.$$

It is easy to see that

$$D : [\mathfrak{l}, \mathfrak{l}] \to [\mathfrak{l}, \mathfrak{l}], \quad D|_{[\mathfrak{l}, \mathfrak{l}]_k} = k \cdot \mathrm{Id}$$

is a derivation of $[\mathfrak{l}, \mathfrak{l}]$. All derivations of a semisimple Lie algebra are inner, i.e. there exists a unique element $\xi \in [\mathfrak{l}, \mathfrak{l}]_0$ such that

$$\mathfrak{l}_k = \{x \in \mathfrak{l} \mid [\xi, x] = kx\}.$$

Let

$$L_\xi = \{x \in L \mid \mathrm{Ad}(x)\xi = \xi\}$$

be the centralizer of ξ in L. Let $G = (L_\xi)^0$. Then $\mathfrak{l}_0$ is the Lie algebra of G and G acts on each graded piece $\mathfrak{l}_k$. It is easy to see that ξ is semisimple and G is reductive.

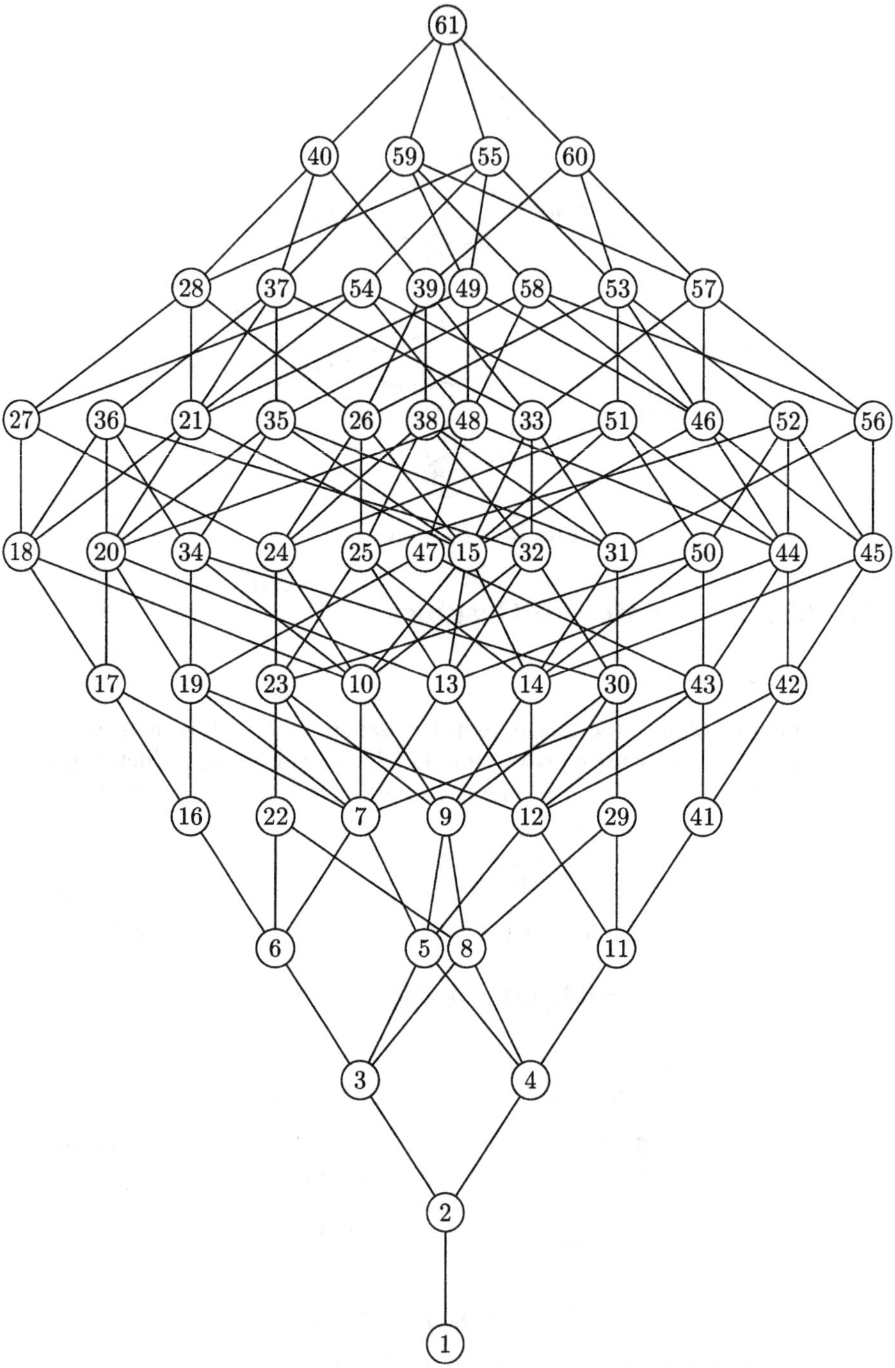

Fig. 2.5. B_5-orbits on $\mathfrak{u}_5$

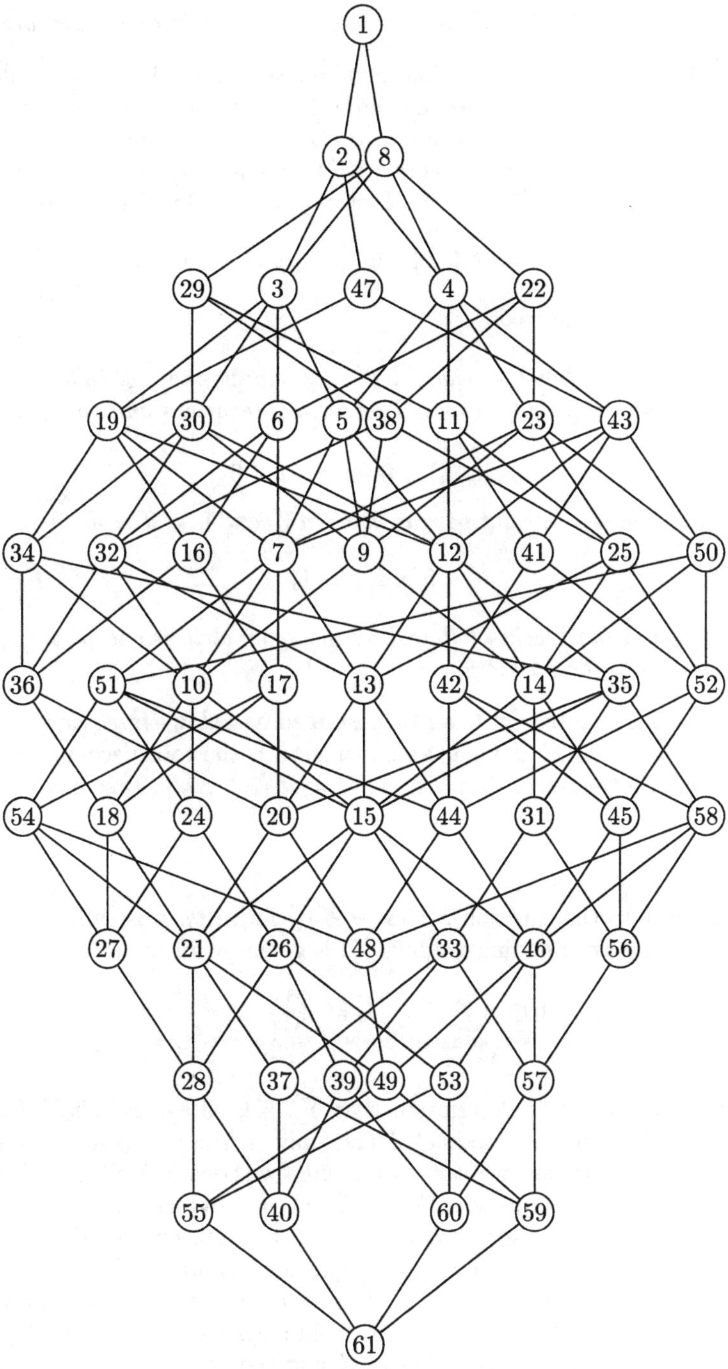

Fig. 2.6. B_5-orbits on $\mathfrak{u}_5^{\vee}$

Theorem 2.6 ([Ri, Vi]) *G acts on $\mathfrak{l}_k$ ($k \neq 0$) with finitely many orbits.*

Proof. All elements in $\mathfrak{l}_k$ are nilpotents. Since L has finitely many nilpotent orbits, it is sufficient to prove that any nilpotent L-orbit $\mathcal{O}$ intersects $\mathfrak{l}_k$ in finitely many G-orbits. It is enough to check that, for any $x \in \mathcal{O} \cap \mathfrak{l}_k$, the tangent space $[\mathfrak{l}_0, x]$ to the G-orbit $\mathrm{Ad}(G)x$ is equal to the intersection of $\mathfrak{l}_k$ with the tangent space $[\mathfrak{l}, x]$ to the L-orbit $\mathrm{Ad}(L)x$. But this is clear:

$$[\mathfrak{l}, x] \cap \mathfrak{l}_k = [\oplus_i \mathfrak{l}_i, x] \cap \mathfrak{l}_k = \oplus_i [\mathfrak{l}_i, x] \cap \mathfrak{l}_k = [\mathfrak{l}_0, x],$$

and the theorem is proved. □

Remark 2.7 *The dual G-module $(\mathfrak{l}_k)^\vee$ is isomorphic to $\mathfrak{l}_{-k}$. Indeed, take any standard invariant form $(\cdot,\cdot)$ on $\mathfrak{l}$. This form is invariant and non-degenerate, and therefore*

$$(\mathfrak{l}_k, \mathfrak{l}_l) = 0 \text{ for } k + l \neq 0$$

and it gives the G-invariant pairing between $\mathfrak{l}_k$ and $\mathfrak{l}_{-k}$. The moment map

$$\mathfrak{l}_k \times \mathfrak{l}_{-k} \to \mathfrak{l}_0^\vee$$

is given by the Lie bracket followed by the symmetric isomorphism $\mathfrak{l}_0 \to \mathfrak{l}_0^\vee$ corresponding to the restriction of $(\cdot,\cdot)$ on $\mathfrak{l}_0$.

All possible gradings of $\mathfrak{l}$ can be described as follows (see [Kac]). Fix the maximal torus $T \subset L$, the Cartan subalgebra $\mathfrak{t}$, the root decomposition $\mathfrak{l} = \mathfrak{t} \oplus \bigoplus_{\alpha \in \Delta} \mathfrak{l}_\alpha$, and the set of simple roots $\Pi = \{\alpha_1, \dots, \alpha_r\}$. Take any function

$$h : \Pi \to \mathbb{Z}_{\geq 0}$$

and extend it to the function $h : \Delta \to \mathbb{Z}$ by linearity. h is called the *height function.* The corresponding grading of $\mathfrak{l}$ is given by formula

$$\mathfrak{l}_0 = \mathfrak{h} \oplus \bigoplus_{\substack{\alpha \in \Delta \\ h(\alpha)=0}} \mathfrak{l}_\alpha, \quad \mathfrak{l}_k = \bigoplus_{\substack{\alpha \in \Delta \\ h(\alpha)=k}} \mathfrak{l}_\alpha \text{ if } k \neq 0.$$

It is clear from this description that $\sum_{k \geq 0} \mathfrak{l}_k$ is a parabolic subalgebra in $\mathfrak{l}$. This subalgebra is maximal if and only if $h(\alpha_i) = 0$ for all i except for some i_0. In this case we can also assume without loss of generality that $h(\alpha_{i_0}) = 1$. Then the representation of G in $\mathfrak{l}_1$ is irreducible, and hence there exists a unique closed G-orbit $X \subset \mathbb{P}(\mathfrak{l}_1)$. In the following tables we look at all possible gradings of this form for simple Lie algebras.

In the first column we give the type of L. Second column contains the unique i such that $h(i) = 1$. The third column is the description of X. The last two columns contain the defect of X and the degree of the discriminant. In all formulas we assume that $k, n \geq 1$. There are several possible cases.

By $\mathbb{P}^N$ we mean the projective space embedded into itself.

L	i	X	def	d
A_{n+1}	1 or $n+1$	$\mathbb{P}^n$	n	0
B_{n+1}	$n+1$	$\mathbb{P}^n$	n	0
C_{n+1}	1	$\mathbb{P}^{2n-1}$	$2n-1$	0
G_2	1	$\mathbb{P}^1$	1	0
A_{k+n+1} if $k \neq n$	$k+1$	$\mathbb{P}^k \times \mathbb{P}^n$	$\lvert k-n\rvert$	0
C_{k+n+1} if $k \neq 2n-1$	$k+1$	$\mathbb{P}^k \times \mathbb{P}^{2n-1}$	$\lvert 2n-1-k\rvert$	0
F_4	2	$\mathbb{P}^1 \times \mathbb{P}^2$	1	0
A_{2k+1}	$k+1$	$\mathbb{P}^k \times \mathbb{P}^k$	0	$k+1$
C_{3n}	$2n$	$\mathbb{P}^{2n-1} \times \mathbb{P}^{2n-1}$	0	$2n$
E_6	3	$\mathbb{P}^1 \times \mathbb{P}^2 \times \mathbb{P}^2$	0	12
E_7	4	$\mathbb{P}^1 \times \mathbb{P}^2 \times \mathbb{P}^3$	0	15
E_8	5	$\mathbb{P}^1 \times \mathbb{P}^2 \times \mathbb{P}^4$	1	0
B_{n+1}	1	Q^{2n-1}	0	2
D_{n+3}	1	Q^{2n+2}	0	2
F_4	1	Q^6	0	2
B_{k+n+1} if $k < 2n-1$	$k+1$	$\mathbb{P}^k \times Q^{2n-1}$	0	$2k+2$
D_{k+n+2} if $k < 2n$	$k+1$	$\mathbb{P}^k \times Q^{2n}$	0	$2k+2$
B_{3n}	$2n$	$\mathbb{P}^{2n-1} \times Q^{2n-1}$	0	$2n$
D_{3n+2}	$2n+1$	$\mathbb{P}^{2n} \times Q^{2n}$	0	$2n+1$
B_{k+n+1} if $k > 2n-1$	$k+1$	$\mathbb{P}^k \times Q^{2n-1}$	$k-2n+1$	0
D_{k+n+2} if $k > 2n$	$k+1$	$\mathbb{P}^k \times Q^{2n}$	$k-2n$	0

Table 2.1.

If $X_1 \subset \mathbb{P}^{N_1}$ and $X_2 \subset \mathbb{P}^{N_2}$, then let $X_1 \times X_2$ be the product of X_1 and X_2 in its *Segre embedding* in $\mathbb{P}^{(N_1+1)(N_2+1)-1}$. This is the composition of embeddings

$$X_1 \times X_2 \subset \mathbb{P}^{N_1} \times \mathbb{P}^{N_2} \subset \mathbb{P}^{(N_1+1)(N_2+1)-1},$$

where the last embedding corresponds to the line bundle

$$\mathrm{pr}_1^*(\mathcal{O}_{\mathbb{P}^{N_1}}(1)) \otimes \mathrm{pr}_2^*(\mathcal{O}_{\mathbb{P}^{N_2}}(1)).$$

L	i	X	def	d
C_{n+1}	$n+1$	$v_2(\mathbb{P}^n)$	0	$n+1$
G_2	2	$v_3(\mathbb{P}^1)$	0	4
F_4	3	$\mathbb{P}^1 \times v_2(\mathbb{P}^2)$	0	12
D_{n+3} if n is even	$n+2$ or $n+3$	$\mathrm{Gr}(1, \mathbb{P}^{n+2})$	2	0
D_{n+3} if n is odd	$n+2$ or $n+3$	$\mathrm{Gr}(1, \mathbb{P}^{n+2})$	0	$\frac{n+3}{2}$
E_6	2 or 4	$\mathbb{P}^1 \times \mathrm{Gr}(1, \mathbb{P}^4)$	1	0
E_7	3	$\mathbb{P}^2 \times \mathrm{Gr}(1, \mathbb{P}^4)$	0	12
E_8	4	$\mathbb{P}^3 \times \mathrm{Gr}(1, \mathbb{P}^4)$	0	40
E_7	5	$\mathbb{P}^1 \times \mathrm{Gr}(1, \mathbb{P}^5)$	0	12
E_8	6	$\mathbb{P}^1 \times \mathrm{Gr}(1, \mathbb{P}^6)$	1	0
E_6	6	$\mathrm{Gr}(2, \mathbb{P}^5)$	0	4
E_7	7	$\mathrm{Gr}(2, \mathbb{P}^6)$	0	7
E_8	8	$\mathrm{Gr}(2, \mathbb{P}^7)$	0	16
E_6	1 or 5	$\mathbb{S}_5$	4	0
E_7	2	$\mathbb{P}^1 \times \mathbb{S}_5$	3	0
E_8	3	$\mathbb{P}^2 \times \mathbb{S}_5$	2	0
E_7	6	$\mathbb{S}_6$	0	4
E_8	7	$\mathbb{S}_7$	0	8
F_4	4	$\mathrm{Gr}_\omega(3, \mathbb{C}^6)$	0	4
L	i	X	def	d
E_7	1	X_1	0	3
E_8	2	$\mathbb{P}^1 \times X_1$	0	12
E_8	1	X_2	0	4

Table 2.2.

In more prosaic terms, if $(z_0^1 : \ldots : z_{N_1}^1)$ are homogeneous coordinates on $\mathbb{P}^{N_1}$ and $(z_0^2 : \ldots : z_{N_2}^2)$ are homogeneous coordinates on $\mathbb{P}^{N_2}$, then the embedding $\mathbb{P}^{N_1} \times \mathbb{P}^{N_2} \subset \mathbb{P}^{(N_1+1)(N_2+1)-1}$ is given by formula

$$((z_0^1 : \ldots : z_{N_1}^1), (z_0^2 : \ldots : z_{N_2}^2)) \to (\ldots : z_i^1 z_j^2 : \ldots)_{ij}.$$

A smooth quadric hypersurface in $\mathbb{P}^{n+1}$ is denoted by Q^n. We denote by $v_k(\mathbb{P}^N)$ the k-th Veronese embedding of $\mathbb{P}^N$ in $\mathbb{P}^{\binom{N+k}{k}-1}$. We denote by X_1 the unique closed E_6-orbit in the projectivization of the 27-dimensional irreducible E_6-module and we denote by X_2 the unique closed E_7-orbit in the projectivization of the 56-dimensional irreducible E_7-module.

2.3.2 Short Gradings

Let L be a simple algebraic group with Lie algebra $\mathfrak{l}$.

Definition 2.8 A grading of $\mathfrak{l}$ with only three non-zero parts

$$\mathfrak{l} = \mathfrak{l}_{-1} \oplus \mathfrak{l}_0 \oplus \mathfrak{l}_1$$

is called *short*.

As in 2.3, there exists a unique semisimple element $\xi \in \mathfrak{l}_0$ such that

$$l_k = \{x \in \mathfrak{l} \,|\, [\xi, x] = kx\}.$$

We denote by G the connected component of the centralizer of ξ in L. By Theorem 2.6, G acts on $\mathfrak{l}_{\pm 1}$ with finitely many orbits. We are going to describe these G-orbits and the Pyasetskii pairing explicitly. These results were rediscovered several times, either using case-by-case considerations or in the invariant setting. We follow [MRS] and [Pan].

Let T be the maximal torus in G and hence in L. Then we have the partition

$$\Delta = \Delta_{-1} \cup \Delta_0 \cup \Delta_1,$$

where

$$\Delta_k = \{\alpha \in \Delta \,|\, \mathfrak{l}_\alpha \in \mathfrak{l}_k\}.$$

G is reductive and Δ_0 is its root system. We fix positive roots $\Delta_0^+ \subset \Delta_0$. Then

$$\Delta^+ = \Delta_0^+ \cup \Delta_1$$

is the set of positive roots of Δ.

Let Π be the set of simple roots in Δ^+, then $\Pi_0 = \Pi \cap \Delta_0$ is the set of simple roots in Δ_0^+. Let $h : \Delta \to \{-1, 0, 1\}$ be the height function. Consider the highest root

$$\gamma = \sum_{\alpha \in \Pi} n_\alpha \alpha, \quad h(\gamma) = \sum_{\alpha \in \Pi} n_\alpha h(\alpha).$$

Since γ obviously belongs to Δ_1 and all $n_\alpha > 0$ (we shall write them down explicitly later) we see that there is a unique simple root $\beta(\Delta_1)$ in Δ_1 and $n_{\beta(\Delta_1)} = 1$. In particular, the representation of G in $\mathfrak{l}_1$ is irreducible.

Moreover, if L is a simple group with the set of roots Δ, simple roots Π, the highest root γ, and for some simple root β we have

$$\gamma = \beta + \sum_{\alpha \in \Pi \setminus \{\beta\}} n_\alpha \alpha,$$

then we can define a short grading of L with $\beta = \beta(\Delta_1)$ using the height function $h(\beta) = 1$, $h(\alpha) = 0$ for $\alpha \in \Pi \setminus \{\beta\}$.

Now we can describe all short gradings. It suffices to draw all Dynkin diagrams marking the vertex corresponding to the simple root β by the number n_β. If $n_\beta = 1$ then this simple root defines the short grading as above. In this case we mark this vertex by a black circle.

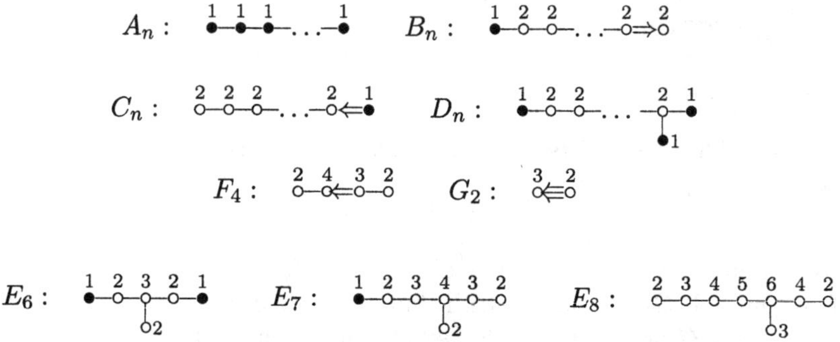

Most results about shortly graded simple Lie algebras can be proved by induction using the following procedure. Let

$$\widehat{\Delta} = \{\alpha \in \Delta \,|\, (\alpha, \beta(\Delta_1)) = 0\}.$$

Then $\widehat{\Delta} = \widehat{\Delta}_{-1} \cup \widehat{\Delta}_0 \cup \widehat{\Delta}_1$ is the root system of the graded reductive subalgebra

$$\widehat{\mathfrak{l}} = \mathfrak{t} \oplus \bigoplus_{\alpha \in \widehat{\Delta}} \mathfrak{l}_\alpha \subset \mathfrak{l}.$$

Proposition 2.9 *Suppose that $\widehat{\Delta}_1 \neq \emptyset$. Then $\gamma \in \widehat{\Delta}_1$ and there exists a unique decomposition of graded reductive subalgebras $\widehat{\mathfrak{l}} = \mathfrak{l}' \oplus \mathfrak{m}$ such that $\mathfrak{m} = \mathfrak{m}_0$, $\mathfrak{l}'$ is simple, and $\mathfrak{l}'_{\pm 1} = \widehat{\mathfrak{l}}_{\pm 1}$. We denote by Δ' the set of roots of $\mathfrak{l}'$.*

We postpone the proof until the end of this section. This proposition allows us to define the chain of shortly graded simple Lie algebras

$$\mathfrak{l} = \mathfrak{l}^{(1)} \supset \mathfrak{l}^{(2)} \supset \mathfrak{l}^{(3)} \supset \ldots,$$

where $\mathfrak{l}^{(k+1)} = (\mathfrak{l}^{(k)})'$. In a similar way we define $\Delta^{(k)}$, etc. We also define the *canonical string* of elements in Δ_1:

$$\beta_1 = \beta(\Delta_1^{(1)}), \ldots, \beta_r = \beta(\Delta_1^{(r)}).$$

The process terminates when $\Delta_1^{(r+1)}$ is empty.

Theorem 2.10 *For any $\alpha \in \Delta_1^+$, let $e_\alpha \in \mathfrak{g}_\alpha$ be any non-zero vector and $e_{-\alpha} \in \mathfrak{g}_{-\alpha}$ be a vector such that $(e_\alpha, e_{-\alpha}) = 1$. We set*

$$e_k = e_{\beta_1} + \ldots + e_{\beta_k}, \quad f_k = e_{-\beta_1} + \ldots + e_{-\beta_k} \quad \textit{for } k = 0, \ldots, r,$$

and so $e_0 = f_0 = 0$. Let $\mathcal{O}_k = G \cdot e_k \subset \mathfrak{l}_1$ and $\mathcal{O}'_k = G \cdot f_k \subset \mathfrak{l}_{-1}$. Then

(i) $\mathfrak{l}_1 = \bigsqcup_{i=0}^{r} \mathcal{O}_i$, $\mathfrak{l}_{-1} = \bigsqcup_{i=0}^{r} \mathcal{O}'_i$.
(ii) $\mathcal{O}_i \subset \overline{\mathcal{O}_j}$ *if and only if* $i \leq j$.
(iii) $\mathcal{O}'_i \subset \overline{\mathcal{O}'_j}$ *if and only if* $i \leq j$.
(iv) $\mathcal{O}_k$ *corresponds to* $\mathcal{O}'_{r-k}$ *in the Pyasetskii pairing.*

The proof is given at the end of this section. First let us look at the examples. The following results easily follow from Theorem 2.10 by explicit calculations.

Example 2.11 Consider the short gradings of $\mathfrak{l} = \mathfrak{sl}_{n+m}$. Then

$$G = \{(A, B) \in \mathrm{GL}_n \times \mathrm{GL}_m \mid \det(A)\det(B) = 1\}.$$

$\mathfrak{l}_1$ can be identified with $\mathbb{C}^n \times \mathbb{C}^m$. There are $r = \min(n, m)$ non-zero G-orbits $\mathcal{O}_i$, $i = 1, \ldots, r$. $\mathcal{O}_i$ is the variety of matrices of rank i. The projectivization of $\mathcal{O}_1$ is identified with $X = \mathbb{P}^{n-1} \times \mathbb{P}^{m-1}$ in the Segre embedding. Therefore, the dual variety X^* is equal to the projectivization of the closure $\overline{\mathcal{O}_{r-1}}$, the variety of matrices of rank less than or equal to $r-1$. X^* is a hypersurface if and only if $n = m$, in which case Δ_X is the ordinary determinant of a square matrix. Another interesting case is $n = 2$, $m \geq 2$: we see that the Segre embedding of $\mathbb{P}^1 \times \mathbb{P}^k$ is self-dual.

Example 2.12 Consider the short grading of $\mathfrak{l} = \mathfrak{so}_{n+2}$ that corresponds to $\beta = \alpha_1$. Then $G = \mathbb{C}^* \times \mathrm{SO}_n$ and $l_1 = \mathbb{C}^n$ with the simplest action. There are two non-zero orbits: the dense one and the self-dual quadric hypersurface $Q \subset \mathbb{C}^n$ preserved by G.

Example 2.13 Consider the short grading of $\mathfrak{l} = \mathfrak{so}_{2n} = D_n$ that corresponds to $\beta = \alpha_n$ or $\beta = \alpha_{n-1}$. Then $G = \mathbb{C}^* \times \mathrm{SL}_n$ acts naturally on $\mathfrak{l}_1 = \Lambda^2\mathbb{C}^n$. There are $r = [n/2]$ non-zero orbits $\mathcal{O}_i$, $i = 1, \ldots, r$. $\mathcal{O}_i$ is the variety of skew–symmetric matrices of rank $2r$. The projectivization of $\mathcal{O}_1$ is identified with $X = \mathrm{Gr}(2, \mathbb{C}^n)$ in the Plücker embedding. Therefore the dual variety X^* is equal to the projectivization of the closure $\overline{\mathcal{O}_{r-1}}$, the variety of matrices of rank less than or equal to $2r-2$. X^* is a hypersurface if and only if n is even, in which case Δ_X is the Pfaffian of a skew-symmetric matrix whose degree is equal to $n/2$. If n is odd then $\operatorname{def} X = \operatorname{codim} \mathcal{O}_{r-1} = 2$.

Example 2.14 Consider the short grading of $\mathfrak{l} = \mathfrak{sp}_{2n}$. Then $G = \mathbb{C}^* \times \mathrm{SL}_n$ acts naturally on $\mathfrak{l}_1 = S^2\mathbb{C}^n$. There are $r = n$ non-zero orbits $\mathcal{O}_i$, $i = 1, \ldots, r$. $\mathcal{O}_i$ is the variety of symmetric matrices of rank i. The projectivization of $\mathcal{O}_1$ is identified with $X = \mathbb{P}^{n-1}$ in the second Veronese embedding. Therefore the dual variety X^* is equal to the projectivization of the closure $\overline{\mathcal{O}_{r-1}}$, the variety of matrices of rank less than r. X^* is a hypersurface and Δ_X is the determinant a symmetric matrix.

Example 2.15 The short grading of E_6 gives the following action. $G = \mathbb{C}^* \times \mathrm{Spin}_{10}$ and $\mathfrak{l}_1$ is the half-spinor representation. Except for the zero orbit and the dense orbit there is only one orbit $\mathcal{O}$. In particular, the projectivization of $\mathcal{O}$ is smooth and self-dual. It is a spinor variety $\mathbb{S}_5$. It could also be described via Cayley numbers. Let $\mathbb{C}\mathrm{a}$ be the algebra of split Cayley numbers (therefore $\mathbb{C}\mathrm{a} = \mathbb{O} \otimes \mathbb{C}$, where $\mathcal{O}$ is the real division algebra of octonions). Let $u \mapsto \overline{u}$ be the canonical involution in $\mathbb{C}\mathrm{a}$. Let $\mathbb{C}^{16} = \mathbb{C}\mathrm{a} \oplus \mathbb{C}\mathrm{a}$ have octonionic coordinates u, v. Then the spinor variety $\mathcal{S}_5 \subset \mathbb{P}(\mathbb{C}^{16})$ is defined by homogeneous equations

$$u\overline{u} = 0, \quad v\overline{v} = 0, u\overline{v} = 0,$$

where the last equation is equivalent to 8 complex equations.

Example 2.16 The final example appears from the short grading of E_7. Here $G = \mathbb{C}^* \times E_6$, and $\mathfrak{l}_1 = \mathbb{C}^{27}$ can be identified with the exceptional simple Jordan algebra (the Albert algebra); see [J] and also the next Example. Then E_6 is the group of norm similarities and $\mathbb{C}^*$ acts by homotheties. There are 3 non-zero orbits: the dense one, the cubic hypersurface (defined by the norm in the Jordan algebra), and the closed conical variety with smooth projectivization consisting of elements of rank one.

Example 2.17 (Severi varieties) Some of the varieties introduced previously are known as Severi varieties. They have the following algebraic description. Let $\mathbb{D}_{\mathbb{R}}$ denote one of four division algebras over $\mathbb{R}$: real numbers $\mathbb{R}$, complex numbers $\mathbb{C}$, quaternions $\mathbb{H}$, or octonions $\mathbb{O}$. Let $\mathbb{D} = \mathbb{D}_{\mathbb{R}} \otimes_{\mathbb{R}} \mathbb{C}$ be its complexification, and therefore $\mathbb{D}$ is either $\mathbb{C}$, $\mathbb{C} \oplus \mathbb{C}$, $\mathrm{Mat}_2(\mathbb{C})$, or $\mathbb{C}\mathrm{a}$ (the algebra of complex Cayley numbers). All these algebras have the standard involution. Let $\mathcal{H}_3(\mathbb{D})$ denote the 3×3 Hermitian matrices over $\mathbb{D}$. If $x \in \mathcal{H}_3(\mathbb{D})$ then we may write

$$x = \begin{pmatrix} c_1 & u_1 & u_2 \\ \overline{u_1} & c_2 & u_3 \\ \overline{u_2} & \overline{u_3} & c_3 \end{pmatrix}, \quad c_i \in \mathbb{C}, \; u_i \in \mathbb{D}.$$

$\mathcal{H}_3(\mathbb{D})$ is a Jordan algebra with respect to the operation $x \circ y = \dfrac{xy + yx}{2}$, that is, for any a, b we have

$$a \circ b = b \circ a \quad \text{and} \quad a \circ ((a \circ a) \circ b) = (a \circ a) \circ (a \circ b),$$

see [J]. In case $\mathbb{D} = \mathbb{C}\mathrm{a}$ this is the *Albert algebra* . The group of automorphisms $G = \mathrm{Aut}(\mathcal{H}_3)$ is a semisimple algebraic group of type SO_3, SL_3, Sp_6, or F_4, respectively. The action of G on $\mathcal{H}_3$ has two irreducible components: scalar matrices and traceless matrices. It is possible to enlarge the group of automorphisms by considering the group of norm similarities. Namely, there exists a cubic form det on $\mathcal{H}_3(\mathbb{D})$ defined by the formula

$$\det(x) = \frac{1}{6}\left((\mathrm{Tr}\, x)^3 + 2\,\mathrm{Tr}(x^3) - 3(\mathrm{Tr}\, x)\,\mathrm{Tr}(x^2)\right).$$

Consider the subgroup in $\mathrm{SL}(\mathcal{H}_3)$ preserving det. Its connected component is called the *group of norm similarities* $\tilde{G}$. It is a semisimple algebraic group of type SL_3, $\mathrm{SL}_3 \times \mathrm{SL}_3$, SL_6, or E_6, respectively. The representation of $\tilde{G}$ on $\mathcal{H}_3$ is irreducible. The projectivization of the cone of the highest weight vectors is called the Severi variety corresponding to $\mathbb{D}$.

The Severi variety corresponding to $\mathbb{C}\mathrm{a}$ is the minimal flag variety of E_6 embedded into $\mathbb{P}^{26}$ by the ample generator of its Picard group. The corresponding elements of the Albert algebra are known as elements of rank 1, see [J]. Other Severi varieties have the following description.

If $\mathbb{D}_{\mathbb{R}} = \mathbb{R}$ then $\mathcal{H}_3$ is equal to the Jordan algebra of symmetric 3×3 complex matrices and det is the standard determinant. $\tilde{G} = \mathrm{SL}_3$ acts on symmetric matrices as on quadratic forms, the Severi variety is the projectivization of the cone of rank 1 symmetric matrices. Therefore this Severi variety is isomorphic to $\mathbb{P}^2$ in the second Veronese embedding.

If $\mathbb{D}_{\mathbb{R}} = \mathbb{C}$ then $\mathcal{H}_3$ is equal to the Jordan algebra of 3×3 complex matrices and det is the standard determinant. $\tilde{G} = \mathrm{SL}_3 \times \mathrm{SL}_3$ acts on matrices by left and right multiplication, and the Severi variety is the projectivization of the cone of rank 1 matrices. Therefore this Severi variety is isomorphic to $\mathbb{P}^2 \times \mathbb{P}^2$ in the Segre embedding.

If $\mathbb{D}_{\mathbb{R}} = \mathbb{H}$ then $\mathcal{H}_3$ is equal to the Jordan algebra $\Lambda^2\mathbb{C}^6$ with the multiplication given by $\alpha_1 \circ \alpha_2 = \alpha_1 \wedge \alpha_2$, where we identify $\Lambda^2\mathbb{C}^6$ and $\Lambda^4\mathbb{C}^6$ by means of a standard symplectic form ω (it is easy to see that $\Lambda^4\mathbb{C}^6 = \omega \wedge \Lambda^2\mathbb{C}^6$). The function det is defined as $\det(\alpha) = \alpha \wedge \alpha \wedge \alpha$ (in the matrix form det is equal to the Pfaffian of the skew-symmetric matrix corresponding to α). The Severi variety is the projectivization of the variety of decomposable bivectors of the form $u \wedge v$, $u, v \in \mathbb{C}^6$. Therefore this Severi variety is isomorphic to $\mathrm{Gr}(2, \mathbb{C}^6)$ in the Plücker embedding.

For $\mathbb{D}_{\mathbb{R}}$ equal to $\mathbb{R}$, $\mathbb{C}$, or $\mathbb{H}$ it is possible to define higher dimensional Severi varieties in the same way (isomorphic to $\mathbb{P}^n$ in the second Veronese embedding, $\mathbb{P}^n \times \mathbb{P}^n$ in the Segre embedding, or $\mathrm{Gr}(2, \mathbb{C}^{2n+2})$ in the Plücker embedding). However, for $\mathbb{D}_{\mathbb{R}} = \mathbb{O}$ only the case of 3×3 matrices makes sense.

The last theorem of this section gives a geometric interpretation of the inductive process based on Proposition 2.9:

Theorem 2.18 *Consider the minimal nontrivial orbit* $X = \mathbb{P}(\mathcal{O}_1) \subset \mathbb{P}(\mathfrak{l}_1)$ *and let* $x \in X$. *Then the quotient projective space* $\mathbb{P}(\mathfrak{l}_1)/\hat{T}_{x,X}$ *is isomorphic to* $\mathbb{P}(\mathfrak{l}'_1)$ *and the projection with center* $\hat{T}_{x,X}$ *takes* $\mathbb{P}(\mathcal{O}_k)$ *to* $\mathbb{P}(\mathcal{O}'_k)$ *for* $k < r$.

The proof is given below.

Proof of Proposition 2.9. Suppose that $(\beta(\Delta_1), \delta) = 0$ for some

$$\delta = \beta(\Delta_1) + \sum_{\alpha \in \Pi_0} m_\alpha \alpha \in \Delta_1.$$

Since γ is the highest root, $0 \le m_\alpha \le n_\alpha$ for any $\alpha \in \Pi_0$. Then

$$(\beta(\Delta_1), \gamma) = (\beta(\Delta_1), \delta) + \left(\beta(\Delta_1), \sum_{\alpha \in \Pi_0} (n_\alpha - m_\alpha)\alpha \right).$$

Since $(\beta(\Delta_1), \alpha) \leq 0$ for any $\alpha \in \Pi_0$, we get $(\beta(\Delta_1), \gamma) \leq 0$. If $(\beta(\Delta_1), \gamma) < 0$ then $\beta(\Delta_1) + \gamma$ is a root and this is a contradiction, because γ is the highest root. Therefore, $(\beta(\Delta_1), \gamma) = 0$ and $\gamma \in \Delta_1'$.

We take the canonical decomposition

$$\widehat{\mathfrak{l}} = \mathfrak{z}(\widehat{\mathfrak{l}}) \oplus \widehat{\mathfrak{l}^1} \oplus \ldots \oplus \widehat{\mathfrak{l}^k},$$

where $\widehat{\mathfrak{l}^i}$ is simple for $i = 1, \ldots, k$. Each summand is graded, $\mathfrak{z}(\widehat{\mathfrak{l}}) = \mathfrak{z}(\widehat{\mathfrak{l}})_0$ and, for any $i = 1, \ldots, k$, the subalgebra $\widehat{\mathfrak{l}^i}$ is either shortly graded or $\widehat{\mathfrak{l}^i} = (\widehat{\mathfrak{l}^i})_0$. Therefore, in order to finish the proof of the proposition, it remains to prove that the representation of $\widehat{\mathfrak{l}_0}$ on $\widehat{\mathfrak{l}_1}$ is irreducible.

Suppose that this representation is reducible, let $V_1, V_2 \subset \widehat{\mathfrak{l}_1}$ be two nontrivial irreducible submodules. Obviously, $\widehat{\Delta}_0^+ = \widehat{\Delta}_0 \cap \Delta_0^+$ is the set of positive roots in $\widehat{\Delta}_0$. Moreover, $\widehat{\Pi}_0 = \Pi_0 \cap \widehat{\Delta}_0$ is the set of simple roots in $\widehat{\Delta}_0$. Indeed, any root $\delta \in \widehat{\Delta}_0^+$ is equal to

$$\delta = \sum_{\alpha' \in \widehat{\Pi}_0} k_{\alpha'} \alpha' + \sum_{\alpha'' \in \Pi_0 \setminus \widehat{\Pi}_0} k_{\alpha''} \alpha'',$$

where $k_\alpha \geq 0$ for any $\alpha \in \Pi_0$. Since $(\delta, \beta(\Delta_1)) = 0$ and

$$(\alpha, \beta(\Delta_1)) = 0 \text{ for } \alpha \in \widehat{\Pi}_0, \quad (\alpha, \beta(\Delta_1)) < 0 \text{ for } \alpha \in \Pi_0 \setminus \widehat{\Pi}_0,$$

we have $k_\alpha = 0$ for any $\alpha \in \Pi_0 \setminus \widehat{\Pi}_0$.

Let γ_i be the highest weight of V_i. Then $\gamma_i \in \widehat{\Delta}_1$ and $\gamma_i + \alpha$ is not a root for any $\alpha \in \widehat{\Pi}_0$. We may assume that $\gamma_1 \neq \gamma$. Then there exists $\alpha \in \Pi_0 \setminus \widehat{\Pi}_0$ such that $\gamma_1 + \alpha \in \Delta_1$. Since $(\beta(\Delta_1), \gamma_1 + \alpha) < 0$, we have $\gamma_1 + \alpha \neq \gamma$. Therefore there exists $\alpha' \in \Pi_0$ such that $\gamma_1 + \alpha + \alpha' \in \Delta_1$. Since $(\beta(\Delta_1), \gamma_1 + \alpha + \alpha') < 0$, we have $\gamma_1 + \alpha + \alpha' \neq \gamma$, etc. This process can be continued infinitely many times and we get the contradiction. □

Proof of Theorem 2.10. (i) We shall prove the first equality $\mathfrak{l}_1 = \bigsqcup_{i=0}^{r} \mathcal{O}_i$, the second equality being similar. Let us show that any element in $\mathfrak{l}_1$ is G-conjugate to one of the e_k. We argue by the induction on r. Let $x \in \mathfrak{l}_1$, $x \neq 0$. Then

$$x = \sum_{\alpha \in \Delta_1} x_\alpha e_\alpha, \quad x_\alpha \in \mathbb{C}.$$

Since $\mathfrak{l}_1$ is an irreducible G-module, after conjugating x we may assume that $x_{\beta_1} \neq 0$. Since the G-orbit of x is conical, we may assume that $x_{\beta_1} = 1$. Let

$$\widetilde{\Delta}_0 = \{\alpha \in \Delta_0 \mid (\alpha, \beta_1) < 0\}.$$

Clearly, $\widetilde{\Delta}_0 \subset \Delta_0^+$. Then

$$\mathfrak{u} = \bigoplus_{\alpha \in \widetilde{\Delta}_0} \mathfrak{g}_\alpha \subset \mathfrak{g}$$

is a unipotent subalgebra. Let $U \subset G$ be a unipotent subgroup with Lie algebra $\mathfrak{u}$. Then U acts on $\mathfrak{l}_1$.

Lemma 2.19 *U acts on $\mathfrak{l}'_1$ trivially.*

Proof. For any roots $\alpha \in \widetilde{\Delta}_0$, $\alpha' \in \Delta'_1$, we have $\alpha + \alpha' \notin \Delta$. Indeed, if $\alpha + \alpha' \in \Delta$ then, since $(\alpha + \alpha', \beta_1) < 0$, we have $\alpha + \alpha' + \beta_1 \in \Delta$. But $h(\alpha + \alpha' + \beta_1) = 2$. This means that U acts on $\mathfrak{l}'_1$ trivially.
□ Consider the action of U on $\mathfrak{l}_1/\mathfrak{l}'_1$. Clearly, the function x_{β_1} is U-invariant, and therefore each U-orbit in $\mathfrak{l}_1/\mathfrak{l}'_1$ lies in a hyperplane $x_{\beta_1} = \text{const}$. In particular,

$$U \cdot (x \mod \mathfrak{l}'_1) \subset \langle x_{\beta_1} = 1 \rangle.$$

Let us show that in fact we have an equality. Since any orbit of a unipotent group is closed, it is sufficient to show the equality of tangent spaces

$$\mathfrak{u} \cdot (x \mod \mathfrak{l}'_1) = \langle x_{\beta_1} = 0 \rangle.$$

But this is clear: for any root $\alpha \in \Delta_1$ we have $(\alpha, \beta_1) \geq 0$ and if $(\alpha, \beta_1) > 0$ then $\alpha = \beta_1 + \widetilde{\alpha}$, where $\widetilde{\alpha} \in \widetilde{\Delta}_0$.

In particular, x is U-conjugate to an element $e_{\beta_1} + x'$, where $x' \in \mathfrak{l}'_1$. By induction, x' is G'-conjugate to one of $e'_0, \ldots, e'_{r-1}$, where $\operatorname{Lie} G' = \mathfrak{l}'_0$. We now need the following Lemma:

Lemma 2.20 $L' \cdot e_{\beta_1} = e_{\beta_1}$.

Proof. We notice that $\mathfrak{l}'$ is generated by subalgebras $\mathfrak{l}'_{-1}$ and $\mathfrak{l}'_1$. Indeed, take any root $\alpha \in \Delta'_0$. Since $\mathfrak{l}'$ is simple, $\mathfrak{l}'_0$ acts on $\mathfrak{l}'_1$ faithfully (otherwise $\{x \in \mathfrak{l}'_0 \mid x \cdot \mathfrak{l}'_1 = 0\}$ is the ideal in $\mathfrak{l}'$). It follows that there exists a root $\alpha' \in \Delta'_1$ such that $\alpha + \alpha' \in \Delta'_1$. Then $\mathfrak{l}'_\alpha = [\mathfrak{l}'_{\alpha+\alpha'}, \mathfrak{l}_{-\alpha'}]$. Therefore the subalgebra generated by $\mathfrak{l}'_{-1}$ and $\mathfrak{l}'_1$ contains all root spaces $\mathfrak{l}'_\alpha$, $\alpha \in \Delta'$, and hence coincides with $\mathfrak{l}'$.

Obviously, $\mathfrak{l}'_1 \cdot e_{\beta_1} = 0$. Let us show that $\mathfrak{l}'_{-1} \cdot e_{\beta_1} = 0$. Indeed, take any root $\alpha \in \Delta'_{-1}$. Then $(\alpha, \beta_1) = 0$ and if $\alpha + \beta_1$ is a root then $\beta_1 - \alpha$ is also a root, but this is impossible because $h(\beta_1 - \alpha) = 2$.

We see that $\mathfrak{l}' \cdot e_{\beta_1} = 0$ and therefore $L' \cdot e_{\beta_1} = e_{\beta_1}$. □

Since $G' \cdot e_{\beta_1} = e_{\beta_1}$, we see that any nonzero element of $\mathfrak{l}_1$ is G-conjugate to $e_{\beta_1} + e'_k = e_{k+1}$ for some $k = 0, \ldots, r-1$.

To finish the proof of (i) we need to show that all orbits $\mathcal{O}_i$ are distinct. It is sufficient to check that

$$\dim \mathfrak{g} \cdot e_i = \dim \mathcal{O}_i < \dim \mathcal{O}_j = \dim \mathfrak{g} \cdot e_j$$

for $i < j$. This will follow from the following Lemma:

Lemma 2.21 $\mathrm{Ann}(\mathfrak{g}\cdot e_k) = \bigoplus_{\alpha\in\Delta_+^{(k)}} \mathfrak{l}_{-\alpha}$.

Proof. Indeed, $\mathrm{Ann}(\mathfrak{g}\cdot e_k) \subset \mathfrak{l}_{-1}$ and $x \in \mathrm{Ann}(\mathfrak{g}\cdot e_k)$ iff $(x, [z, e_k]) = 0$ for any $z \in \mathfrak{g}$. But $(x, [z, e_k]) = ([e_k, x], z)$ by the invariance of the Killing form. Since it is non-degenerate, we see that $x \in \mathrm{Ann}(\mathfrak{g}\cdot e_k)$ if and only if $[x, e_k] = 0$. If $\alpha \in \Delta^{(k)}$ then $-\alpha + \beta_i$ is not a root for $i \leq k$ because $-\alpha - \beta_i$ is not a root and $(\alpha, \beta_i) = 0$. Therefore $e_k \cdot l_{-\alpha} = 0$ and

$$\bigoplus_{\alpha\in\Delta_+^{(k)}} \mathfrak{l}_{-\alpha} \subset \mathrm{Ann}(\mathfrak{g}\cdot e_k).$$

Let $x \in \mathfrak{l}_{-1}$ and $[x, e_k] = 0$. Since $[x, f_k] = 0$, we have $[h_k, x] = 0$, where $h_k = [e_k, f_k]$. Since $-\beta_i + \beta_j$ is not a root for $i \neq j$, we have

$$h_k = [e_{\beta_1}, e_{-\beta_1}] + \ldots + [e_{\beta_k}, e_{-\beta_k}].$$

Let

$$x = \sum_{\alpha\in\Delta_{-1}} x_\alpha e_\alpha.$$

Then

$$[h_k, x] = \sum_{\alpha\in\Delta_{-1}} \left(\sum_{i=1}^{k} (\beta_i, \alpha)\right) x_\alpha e_\alpha.$$

But $(\beta_i, \alpha) \leq 0$ for any $i = 1, \ldots, k$ and, therefore, $x_\alpha = 0$ for $\alpha \notin \Delta^{(k)}$. Finally,

$$x \in \bigoplus_{\alpha\in\Delta_+^{(k)}} \mathfrak{l}_{-\alpha}.$$

Lemma 2.21 is proved. □

To prove (ii) we need only show that $\mathcal{O}_{i-1} \subset \overline{\mathcal{O}_i}$ for $i > 0$. But

$$\lim_{t\to-\infty} \exp^{t[e_{\beta_i}, e_{-\beta_i}]} e_{\beta i} = 0, \quad \lim_{t\to-\infty} \exp^{t[e_{\beta_i}, e_{-\beta_i}]} e_{\beta j} = e_{\beta_j} \text{ for } i \neq j,$$

and hence

$$\lim_{t\to-\infty} \exp^{t[e_{\beta_i}, e_{-\beta_i}]} e_i = e_{i-1}.$$

In order to prove (iii) we need to check that $\mathrm{Ann}(\mathfrak{g}\cdot e_k) \cap G\cdot f_{r-k}$ is open in $\mathrm{Ann}(\mathfrak{g}\cdot e_k)$. Let

$$f'_{r-k} = e_{-\beta_{k+1}} + \ldots + e_{-\beta_r}.$$

By the previous Lemma, $\mathrm{Ann}(\mathfrak{g}\cdot e_k) = \bigoplus_{\alpha\in\Delta_+^{(k)}} \mathfrak{l}_{-\alpha}$. Now we apply (i) to $\mathfrak{l}^{(k)}$ and see that $f'_{r-k} \in \mathrm{Ann}(\mathfrak{g}\cdot e_k)$ and $G^{(k)} \cdot f'_{r-k}$ is open in $\mathrm{Ann}(\mathfrak{g}\cdot e_k)$. It remains to prove that f'_{r-k} and f_{r-k} belong to the same orbit. We shall prove a more general result.

Lemma 2.22 *β_k is a long root for $k = 1, \dots, r$. Let W_0 be the Weyl group of $\mathfrak{g}$. If $\delta_1, \dots, \delta_{r'}$ is any sequence of pairwise orthogonal long roots in Δ_1, then $r' \le r$ and there exists $w \in W_0$ such that $\delta_i = w(\beta_i)$ for $i \le r'$.*

Proof. Indeed, since $\mathfrak{l}_1^{(k)}$ is an irreducible $\mathfrak{g}^{(k)}$-module, β_k is the unique lowest weight. The longest element in $W_0^{(k)}$ takes β_k to the highest weight of $\mathfrak{l}_1^{(k)}$, i.e. to γ. Hence β_k is long.

The rest of the proof goes by induction. If $(\delta_1, \alpha) < 0$ for some $\alpha \in \Pi_0$ then $r_\alpha(\delta_1) > \delta_1$ and we may finish by induction. Suppose that $(\delta_1, \alpha) \ge 0$ for all $\alpha \in \Pi_0$. Since $(\delta_1, \beta_1) \ge 0$ as well and δ_1 is long, $\delta_1 = \gamma$ (this is a general fact about root systems). But γ is W_0-conjugate to β_1. Therefore, up to W_0-action, we may assume that $\delta_1 = \beta_1$. Then both $\delta_2, \dots, \delta'_{r'}$ and $\beta_2, \dots, \beta_r$ belong to Δ'_1 and hence there exists $w \in W'_0$ (the Weyl group of $\mathfrak{l}'_0$) such that $\beta'_i = w(\beta_i)$ for $i \le r'$. It remains to notice that $w(\beta_1) = \beta_1$ by Lemma 2.20. Lemma 2.22 and therefore Theorem 2.10 are now proved. □

Proof of Theorem 2.18. Since $(\alpha, \beta) \ge 0$ for any $\alpha, \beta \in \Delta_1$, $\mathfrak{l}_1 = \mathfrak{m} \oplus \mathfrak{l}'_1$, where

$$\mathfrak{m} = \bigoplus_{\alpha \in \Delta_1,\ (\alpha, \beta_1) > 0} \mathfrak{l}_\alpha, \qquad \mathfrak{l}'_1 = \bigoplus_{\alpha \in \Delta_1,\ (\alpha, \beta_1) = 0} \mathfrak{l}_\alpha.$$

We take $x = e_{\beta_1}$. Since β_1 is long, if $\alpha \in \Delta_1$ then $(\alpha, \beta_1) > 0$ if and only if $\alpha - \beta$ is a root. Therefore $\mathfrak{m}$ is the cone over $\hat{T}_{x,X}$. It follows that the quotient projective space $\mathbb{P}(\mathfrak{l}_1)/\hat{T}_{x,X}$ is equal to $\mathbb{P}(\mathfrak{l}'_1)$. By Theorem 2.10 and Lemma 2.22,

$$y = e_{\beta_2} + e_{\beta_3} + \ldots + e_{\beta_{k+1}} \in \mathcal{O}_k.$$

Since $y \in \mathfrak{l}'_1$, we see that $\pi(\mathbb{P}(\mathcal{O}_k)) \supset \mathbb{P}(\mathcal{O}'_k)$, where π is the projection with center $\hat{T}_{x,X}$. Suppose that this inclusion is strict. Since the image of π is G'-invariant, it follows that

$$z = e_{\beta_2} + e_{\beta_3} + \ldots + e_{\beta_{k+2}} \in \mathrm{Cone}(\pi(\mathbb{P}(\mathcal{O}_k))),$$

i.e. there exists $m \in \mathfrak{m}$ such that $m + z \in \mathcal{O}_k$. Let $h = [e_{\beta_1}, e_{-\beta_1}] \in \mathfrak{t}$. Then

$$\lim_{t \to -\infty} \exp(th) \cdot m = 0, \qquad \lim_{t \to -\infty} \exp(th) \cdot z = z.$$

Therefore $z \in \overline{\mathcal{O}_k}$, but this contradicts Theorem 2.10. □

2.3.3 Multisegment Duality

Apart from actions associated with $\mathbb{Z}$-graded semisimple Lie algebras, another class of actions with finitely many orbits is provided by the theory of representations of quivers (see [Ga]). These two classes overlap: the representations of quivers of type A give the same class of actions as the standard gradings of $\mathrm{GL}(V)$. The Pyasetskii pairing in this case was studied under the name of the multisegment duality, or the Zelevinsky involution. Here we give an overview of some of these results.

We fix a positive integer r and consider the set $S = S_r$ of pairs of integers (i, j) such that $1 \leq i \leq j \leq r$. Let $\mathbb{Z}_+^S$ denote the semigroup of families $m = (m_{ij})_{(i,j)\in S}$ of non-negative integers indexed by S. We regard a pair $(i, j) \in S$ as a segment $[i, j] = \{i, i + 1, \ldots, j\}$ in $\mathbb{Z}$. A family $m = m_{ij} \in \mathbb{Z}_+^S$ can be regarded as a collection of segments, containing m_{ij} copies of each $[i, j]$. Thus, elements of $\mathbb{Z}_+^S$ can be called *multisegments*. The *weight* $|m|$ of a multisegment m is defined as a sequence $\gamma = \{d_1, \ldots, d_r) \in \mathbb{Z}_+^r$ given by

$$d_i = \sum_{i\in[k,l]} m_{kl} \quad \text{for } i = 1, \ldots, r.$$

In other words, $|m|$ records how many segments of m contain any given number $i \in [1, r]$. For any $\gamma \in \mathbb{Z}_+^r$ we set $\mathbb{Z}_+^S(\gamma) = \{m \in \mathbb{Z}_+^S \mid |m| = \gamma\}$.

Another important interpretation of $\mathbb{Z}_S^+(\gamma)$ is that it parametrizes isomorphism classes of representations of quivers of type A. Let A_r be the quiver equal to the Dynkin diagram of type A_r, where all edges are oriented from left to right. Let $A_r^\vee$ be the dual quiver with all orientations reversed. A representation of A_r with the dimension vector $\gamma = \{d_1, \ldots, d_r) \in \mathbb{Z}_+^r$ is a collection of vector spaces $\mathbb{C}^{d_1}, \ldots, \mathbb{C}^{d_r}$ and linear maps

$$\varphi_1 : \mathbb{C}^{d_1} \to \mathbb{C}^{d_2}, \ldots, \varphi_{r-1} : \mathbb{C}^{d_{r-1}} \to \mathbb{C}^{d_r}.$$

A representation of $A_r^\vee$ with the dimension vector $\gamma = \{d_1, \ldots, d_r) \in \mathbb{Z}_+^r$ is a collection of vector spaces $\mathbb{C}^{d_1}, \ldots, \mathbb{C}^{d_r}$ and linear maps

$$\varphi_1 : \mathbb{C}^{d_2} \to \mathbb{C}^{d_1}, \ldots, \varphi_{r-1} : \mathbb{C}^{d_r} \to \mathbb{C}^{d_{r-1}}.$$

Therefore representations of A_r with dimension vector γ are parametrized by points of a vector space

$$V(\gamma) = \bigoplus_{i=1}^{r-1} \mathrm{Hom}(\mathbb{C}^{d_i}, \mathbb{C}^{d_{i+1}}).$$

Representations of $A_r^\vee$ with dimension vector γ are parametrized by points of a vector space

$$V(\gamma)^\vee = \bigoplus_{i=1}^{r-1} \mathrm{Hom}(\mathbb{C}^{d_{i+1}}, \mathbb{C}^{d_i}).$$

Notice that the vector spaces $V(\gamma)$ and $V(\gamma)^\vee$ are naturally dual to each other as vector spaces and as modules of the group

$$G(\gamma) = \mathrm{GL}_{d_1} \times \ldots \times \mathrm{GL}_{d_r}$$

with respect to the natural action. These actions could also be described in terms of graded Lie algebras. Namely, consider the Lie algebra $\mathfrak{g} = \mathfrak{gl}_{d_1+\ldots+d_r}$.

Elements of $\mathfrak{g}$ can be represented as block matrices with diagonal blocks of shapes $d_1 \times d_1, \dots, d_r \times d_r$. This determines $\mathbb{Z}$-grading by diagonals: block diagonal matrices have grade 0, etc. Then $\mathfrak{g}_0$ is a Lie algebra of $G(\gamma)$ and the action of $G(\gamma)$ on $V(\gamma)$ and $V(\gamma)^\vee$ is equivalent to the action of $G(\gamma)$ on $\mathfrak{g}_{-1}$ and $\mathfrak{g}_1$.

Orbits of $G(\gamma)$ on $V(\gamma)$ (resp. $V(\gamma)^\vee$) correspond to isoclasses of representations of A_r (resp. $A_r^\vee$) with the dimension vector γ. These orbits are parametrized by elements of $\mathbb{Z}_S^+(\gamma)$. Elements of S parametrize indecomposable A_r-modules (or $A_r^\vee$-modules). Namely, each $(i,j) \in S$ corresponds to an indecomposable module R_{ij} with dimension vector

$$|(i,j)| = (0^{i-1}, 1^{j-i+1}, 0^{r-j})$$

and all maps are isomorphisms, if that is possible, or zero maps otherwise. Then any family $(m_{ij}) \in \mathbb{Z}_S^+$ corresponds to an A_r-module (or $A_r^\vee$-module) $\oplus_S R_{ij}^{m_{ij}}$.

By the Pyasetskii Theorem 2.4, there is a natural bijection of $G(\gamma)$-orbits in $V(\gamma)$ and $V(\gamma)^\vee$. Therefore there exists a natural involution ζ of $\mathbb{Z}_+^S(\gamma)$, which can be extended to a weight-preserving involution of $\mathbb{Z}_+^S$ called the *multisegment duality.*

The inductive description of ζ was given in [MW]. Let $m = (m_{ij})$ be a multisegment of weight $\gamma = (d_1, \dots, d_r)$. We set $i_1 = \min\{i \mid d_i \neq 0\}$ and define the sequence of indices $j_1, \dots, j_p$ as follows:

$$j_1 = \min\{j \mid m_{i_1 j} \neq 0\}, \dots, j_{t+1} = \min\{j \mid j > j_t,\ m_{i_1+t,j} \neq 0\},$$

where $t = 1, \dots, p-1$. The sequence terminates when j_{p+1} does not exist. Let $i_t = i_1 + t - 1$ for $t = 1, \dots, p+1$. We associate to m the multisegment m' given by

$$m' = m - (i_1, j_1) - (i_2, j_2) - \dots - (i_p, j_p) + (i_2, j_1) + (i_3, j_2) + \dots + (i_{p+1}, j_p),$$

where we use the convention that $(i,j) = 0$ unless $1 \le i \le j \le r$.

Theorem 2.23 ([MW]) *If the multisegment m' is associated to m then*

$$\zeta(m) = \zeta(m') + (i_1, i_p).$$

The involution ζ can also be described in terms of irreducible finite-dimensional representations of affine Hecke algebras and in terms of canonical bases for quantum groups; see [KZ].

An explicit description of ζ was found in [KZ]. To formulate it, we need the following definition. For any multisegment $m \in \mathbb{Z}_+^S$ the ranks $r_{ij}(m)$ are given by

$$r_{ij}(m) = \sum_{[i,j] \subset [k,l]} m_{kl}.$$

It is easy to see that the multisegment m can be recovered from its ranks by formula

$$m_{ij} = r_{ij}(m) - r_{i-1,j}(m) + r_{i-1,j+1}(m).$$

If the multisegment corresponds to the representation of A_r given by

$$(\varphi_1, \ldots, \varphi_{r-1}) \in V(\gamma)$$

then r_{ij} is equal to the rank of the map $\varphi_{j-1} \circ \ldots \circ \varphi_{i+1} \circ \varphi_i$. In particular, $r_{ii} = d_i$.

For any $(i,j) \in S$, let T_{ij} denote the set of all maps $\nu : [1,i] \times [j,r] \to [i,j]$ such that $\nu(k,l) \leq \nu(k',l')$ whenever $k \leq k'$, $l \leq l'$ (in other words, ν is a morphism of partially ordered sets, where $[1,i] \times [j,r]$ is supplied with the product order).

Theorem 2.24 ([KZ]) *For every $m = (m_{ij}) \in \mathbb{Z}_+^S$ we have*

$$r_{ij}(\zeta(m)) = \min_{\nu \in T_{ij}} \sum_{(k,l) \in [1,i] \times [j,r]} m_{\nu(k,l)+k-i, \nu(k,l)+l-j}.$$

The proof occupies the rest of this section. We follow [KZ].

Definition 2.25 A finite oriented graph without multiple edges (but possibly with loops) is called *a digraph.* By a *p-network* we mean a digraph such that the set of vertices I is decomposed into the disjoint union of subsets (levels) $I_0, \ldots, I_p$, and arrows from the level I_k can go only to the next level I_{k+1}. A *p-path* in a p-network is a sequence $\alpha_0 \in I_0$, …, $\alpha_p \in I_p$ such that there is an arrow from α_k to α_{k+1} for each k. A *flow* in a p-network is a collection of mutually disjoint p-paths. The maximal possible number of p-paths in a flow is called the *capacity* of a p-network. Finally, *a cut* of a p-network is a subset of vertices that meets every p-path.

The following result is a special case of Menger's theorem [FF].

Theorem 2.26 *The capacity of a network is equal to the minimal cardinality of its cuts.*

Now we state Poljak's theorem describing the maximal rank of the p-th power of a matrix with a given pattern of zeros. Consider a digraph D with a set of vertices I. We say that a square matrix $A = (a_{ij})_{i,j \in I}$ is supported on D if $a_{ij} = 0$ in case there is no arrow from i to j. We associate to a digraph D a p-network $D[p]$ such that $I_0 = \ldots = I_p = I$ and there is an arrow from $\alpha_k \in I_k$ to $\alpha_{k+1} \in I_{k+1}$ if and only if there is an arrow from α_k to α_{k+1} in D.

Theorem 2.27 ([Po])

(i) *Let D be a digraph and let $p \geq 0$. The maximal value of* $\operatorname{rk} A^p$ *over all matrices supported on D is equal to the capacity of $D[p]$.*

(ii) *Suppose a digraph D has no oriented cycles and no loops. Then for each $p \geq 0$ there exist mutually disjoint paths $\alpha_0^{(1)} \to \alpha_1^{(1)} \to \ldots \to \alpha_{p_1}^{(1)}$, $\alpha_0^{(2)} \to \alpha_1^{(2)} \to \ldots \to \alpha_{p_2}^{(2)}, \ldots, \alpha_0^{(m)} \to \alpha_1^{(m)} \to \ldots \to \alpha_{p_m}^{(m)}$ in D such that $p_k \geq p$ for all k and the capacity of $D[p]$ is equal to $\sum_{h=1}^{m}(p_h - p + 1)$. Let A be a square $(0,1)$-matrix supported on D and such that $a_{ij} = 1$ if and only if $(i,j) = (\alpha_q^{(h)}, \alpha_{q+1}^{(h)})$ for some h, q. Then $\operatorname{rk} A^p$ is equal to the capacity of $D[p]$.*

We also need an explicit description of cuts for some special networks associated with posets (partially ordered sets). Suppose that a digraph D is a poset (w.r.t. the relation $\alpha \leq \beta$ iff there is an arrow from α to β). For each poset morphism $\nu : D \to [0,p]$ we denote by $C(\nu) \in D \times [0,p] = D(p)$ the graph of ν.

Theorem 2.28 ([KZ]) *The sets $C(\nu)$ for all poset morphisms $\nu : D \to [0,p]$ are exactly all minimal cuts (w.r.t. inclusion) of a p-network $D(p)$.*

Now we can start to prove Theorem 2.24. We fix a positive integer r, a weight $\gamma = (d_1, \ldots, d_r) \in \mathbb{Z}_+^r$, a multisegment $m = (m_{kl}) \in \mathbb{Z}_+^S(\gamma)$, a graded vector space $V = V_1 \oplus \ldots \oplus V_r$ with dimension vector γ, and a linear map $X : V \to V$ of degree 1 corresponding to m. By definition of Pyasetskii pairing, the multisegment $\zeta(m)$ corresponds to a linear map $Y : V \to V$ of degree -1 such that $Y \in Z(X)$, where $Z(X)$ consists of all linear maps of degree -1 commuting with X, and such that the intersection of $Z(X)$ with $G(\gamma)$-orbit of Y is dense in $Z(X)$. Functions r_{ij} are lower-semicontinuous on $Z(X)$, and therefore $Y \in Z(X)$ corresponds to $\zeta(m)$ if and only if $r_{ij}(Y) = \max_{Y' \in Z(X)} r_{ij}(Y')$ for any $(i,j) \in S$. We see that Theorem 2.24 reduces to the following equation:

$$\max_{Y \in Z(X)} r_{ij}(Y) = \min_{\nu \in T_{ij}} \sum_{(k,l) \in [1,i] \times [j,r]} m_{\nu(k,l)+k-i, \nu(k,l)+l-j}. \qquad (*)$$

We set

$$I^1 = \{(k,l,u) \mid (k,l) \in S,\ 1 \leq u \leq m_{kl}\}$$

and

$$I^2 = \{(\alpha, s) \mid \alpha \in I^1,\ k \leq s \leq l\}.$$

Then I^1 parametrizes Jordan cells occurring in the decomposition of E and I^2 is the index set for a corresponding Jordan basis. Thus, we represent V as $V = \oplus_{\alpha \in I^1} L_\alpha$, where L_α for $\alpha = (k,l,u)$ is a Jordan cell of type (k,l), i.e. L_α is an X-invariant subspace of the form $L_\alpha = \oplus_{s \in [k,l]} L_{\alpha,s}$, where each $L_{\alpha,s}$ is a one-dimensional subspace of V_s and $X(L_{\alpha,s}) = L_{\alpha,s+1}$ for $s < l$. We choose non-zero vectors (Jordan basis) $v_{\alpha,s} \in L_{\alpha,s}$ for all $(\alpha, s) \in I^2$ so that $X(v_{\alpha,s}) = v_{\alpha,s+1}$ for $s < l$ and $X(v_{\alpha,l}) = 0$.

We represent each $Y \in \operatorname{End}(V)$ by its matrix in the Jordan basis $v_{\alpha,s}$, so that

$$Y(v_{\alpha,s}) = \sum_{\alpha',s'} y_{\alpha,s}^{\alpha',s'} v_{\alpha',s'}.$$

We shall describe the pattern of zeros of this matrix for $Y \in Z(X)$. We use the following terminology: for $(k,l) \in S$, $(k',l') \in S$ we say that $(k,l) \in S$ *precedes* $(k',l') \in S$ and write $(k,l) \prec (k',l')$ if $k'+1 \leq k \leq l'+1 \leq l$.

The following lemma is a routine computation.

Lemma 2.29 *A linear map Y belongs to $Z(X)$ if and only if its matrix in the Jordan basis satisfies the following two conditions:*

- $y_{\alpha,s}^{\alpha',s'} = 0$ *unless $(k',l') \prec (k,l)$ and $s' = s-1$.*
- *If $s' = s-1$ then $y_{\alpha,s}^{\alpha',s'}$ depends only on $\alpha, \alpha' \in I^1$.*

From now on until the end of this section we fix a pair $(i,j) \in S$ and set $p = j - i$. We consider the digraph D with the set of vertices given by $I = \{(\alpha,s) \in I^2 \mid s \in [i,j]\}$ and there is an arrow from (α,s) to (α',s') if and only if $(k',l') \prec (k,l)$ and $s' = s-1$. Clearly, D is a p-network.

Lemma 2.30 $\max\limits_{Y \in Z(X)} r_{ij}(Y)$ *is equal to the capacity $c(D)$ of D.*

Proof. First we show that c is the maximal possible value of r_{ij} over all $Y : V \to V$ satisfying the first condition in Lemma 2.29. This is an immediate consequence of the first part of Theorem 2.27 and the obvious fact that the capacity of a p-network N is equal to the capacity of $N[p]$. It remains to construct Y satisfying both conditions of Lemma 2.29 and having $r_{ij}(Y) = c$. For this we shall use the second part of Theorem 2.27, but not directly. First, we shall embed our network D into a network of type $D'[p]$ for some digraph D'. We define the set of vertices as $I' = \{(k,l,u) \in I^1 \mid [k,l] \cap [i,j] \neq \emptyset\}$ (Jordan cells intersecting $[i,j]$) and there is an arrow in D' from (k,l,u) to (k',l',u') if and only if $(k',l') \prec (k,l)$. By definition, $I'[p] = \{(\alpha,t) \mid \alpha = (k,l,u) \in I',\ t \in [0,p]\}$. Clearly, the embedding $I \to I'[p]$ sending each (α,s) to $(\alpha, j-s)$ allows us to identify I with the subset $\{(k,l,u,t) \in I'[p] \mid j-t \in [k,l]\}$. Furthermore, under this identification the embedding does not add new arrows. It is easy to verify that every p-path on $D'[p]$ has all its vertices lying on I. Therefore every flow on $D'[p]$ is in fact a flow on D. In particular, the capacity of D equals the capacity of $D'[p]$. Now we apply the second part of Theorem 2.27 to the network D'. Thus, there exist mutually disjoint paths $\alpha_0^{(1)} \to \alpha_1^{(1)} \to \dots \to \alpha_{p_1}^{(1)}$, $\alpha_0^{(2)} \to \alpha_1^{(2)} \to \dots \to \alpha_{p_2}^{(2)}$, $\dots$, $\alpha_0^{(m)} \to \alpha_1^{(m)} \to \dots \to \alpha_{p_m}^{(m)}$ in D' such that $p_h \geq p$ for all $h = 1,\dots,m$ and $\sum_{h=1}^{m}(p_h - p + 1) = c$. We associate to this family of paths a linear map $Y : V \to V$ with the matrix defined as follows. For $(\alpha,s), (\alpha',s') \in I^2$ we set $y_{\alpha,s}^{\alpha',s'} = 1$ whenever $s' = s-1$ and $\alpha \to \alpha'$ is an arrow belonging to one of our paths. By construction, Y satisfies both conditions of Lemma 2.29, and hence $Y \in Z(X)$. The image of $Y^p : V_j \to V_i$ is the linear span of the vectors $v_{\alpha,i}$ for $\alpha = \alpha_q^{(h)}$, $h = 1,\dots,m$, $q = p, p+1, \dots, p_h$. Therefore $r_{ij}(Y) = c$, which completes the proof of lemma. □

Now we can rewrite the equality $(*)$ in the following purely combinatorial form:

$$c(D) = \min_{\nu \in T_{ij}} \sum_{(k,l)\in[1,i]\times[j,r]} m_{\nu(k,l)+k-i,\nu(k,l)+l-j}. \qquad (**)$$

Consider now the digraph $\tilde{D}$ with the set of vertices

$$\tilde{I} = \{(k,l,s) \mid (k,l) \in S,\ s \in [k,l] \cap [i,j]\}$$

and there is an arrow from (k,l,s) to (k',l',s') iff $(k',l') \prec (k,l)$ and $s' = s-1$. Then, clearly, $\tilde{D}$ is a p-network and there is a natural surjective map of p-networks $D \to \tilde{D}$ such that two vertices in D are connected by an arrow if and only if their images in $\tilde{D}$ are connected by an arrow. Now the following lemma is a simple consequence of Theorem 2.26.

Lemma 2.31 *We have*

$$c(D) = \min_{C} \sum_{(k,l,s)\in C} m_{k,l},$$

where C runs over all cuts of $\tilde{D}$.

Now $(**)$ can be rewritten as

$$\min_{C} \sum_{(k,l,s)\in C} m_{k,l} = \min_{\nu\in T_{ij}} \sum_{(k,l)\in[1,i]\times[j,r]} m_{\nu(k,l)+k-i,\nu(k,l)+l-j}, \qquad (***)$$

where C runs over all cuts of $\tilde{D}$, and T_{ij} is the set of all morphisms of posets $[1,i] \times [j,r] \to [i,j]$. Let us give an explicit description of minimal cuts C. To every $\nu \in T_{ij}$ we associate a map $\tilde{\nu} : [1,i] \times [j,r] \to \mathbb{Z}^3$ given by

$$\tilde{\nu}(k,l) = (\nu(k,l) + k - i, \nu(k,l) + l - j, \nu(k,l)).$$

It is clear that $\tilde{\nu}$ is injective. Now the equality $(***)$ and Theorem 2.24 follow from the next lemma.

Lemma 2.32 *For each $\nu \in T_{ij}$ we have* $\operatorname{Im}\tilde{\nu} \subset \tilde{I}$, *and the sets* $\operatorname{Im}(\tilde{\nu})$ *are exactly all the minimal cuts of $\tilde{D}$.*

Proof. We define the subset $I^{ess} \subset \tilde{I}$ of essential points by

$$I^{ess} = \{(k',l',s') \in \tilde{I} \mid s' - i \le k' - 1,\ j - s' \le r - l'\}.$$

This terminology is explained by the following observation: every p-path on $\tilde{D}$ has all its vertices lying in I^{ess} (this follows at once from the definition of $\tilde{D}$). Thus, setting D^{ess} to be the digraph with the set of vertices I^{ess} and the set of arrows inherited from $\tilde{D}$, we see that the minimal cuts of $\tilde{D}$ are the same as the minimal cuts of the p-network D^{ess}. Now we consider two mutually inverse bijections $\varphi : \mathbb{Z}^3 \to \mathbb{Z}^3$ and $\psi : \mathbb{Z}^3 \to \mathbb{Z}^3$ given by

$$\varphi(k', l', s') = (k' - s' + i, l' - s' + j, s'),$$

$$\psi(k, l, s) = (s + k - i, s + l - j, s).$$

A routine check shows that

$$\varphi(I^{ess}) = [1, i] \times [j, r] \times [i, j]$$

and that φ transforms the conditions

$$k'_2 + 1 \leq k'_1 \leq l'_2 + 1 \leq l'_1, \quad s'_2 + 1 = s'_1$$

defining arrows in D^{ess} into the conditions

$$k_2 \leq k_1, \quad l_2 \leq l_1, \quad s_2 + 1 = s_1.$$

Let Π be the product $[1, i] \times [j, r]$ supplied with the product partial order. We see that φ transforms D^{ess} into the network which is essentially isomorphic to the network $\Pi[p]$. It follows from Theorem 2.28 that the minimal cuts of D^{ess} are obtained from the sets $C(\nu)$ by the transformation ψ. But clearly $\psi(C(\nu)) = \mathrm{Im}(\tilde{\nu})$. □

This finishes the proof of Theorem 2.24.

Poljak's Theorem 2.27 has another important consequence which can be viewed as a far-reaching generalization of the fact that the maximum possible value of the rank of a linear map $V \to W$ is $\min(\dim V, \dim W)$. Let D be a p-network with $I = I_0 \cup I_1 \cup \ldots \cup I_p$. A *representation* of D is an assignment of a finite-dimensional vector space W_α to each vertex $\alpha \in I$ along with a linear map $f_{\beta\alpha} : W_\alpha \to W_\beta$ to each arrow $\alpha \to \beta$. Let $W_i = \oplus_{\alpha \in I_i} W_\alpha$ and $W = \oplus_{i=0}^{p} W_i$. Then the maps $f_{\beta\alpha}$ give rise to a linear map $f : W \to W$ of degree 1.

Theorem 2.33 ([KZ]) *Suppose that we are given the dimensions d_α of the spaces W_α. Then the maximal possible value of the rank of the map $f^p : W_0 \to W_p$ is equal to $\min_C \sum_{\alpha \in C} d_\alpha$, where C runs over all cuts of D.*

3

Local Calculations

Here we study projectively dual varieties using calculations in local coordinates. In Sect. 3.1 we prove a formula due to Katz that expresses the dimension of a dual variety in terms of the hessian of local equations of a variety. We use it to prove a formula of Weyman and Zelevinsky that expresses the defect of a Segre embedding of a product of two varieties. In Sect. 3.2 we introduce a gadget called the second fundamental form that incorporates these calculations. We prove some results of Griffiths and Harris about the second fundamental form. We finish this section with the description of higher fundamental forms of flag varieties obtained by Landsberg.

3.1 Calculations in Coordinates

3.1.1 Katz Dimension Formula

Let $X \subset \mathbb{P}^N = \mathbb{P}(V)$ be an irreducible n-dimensional projective variety and let $X^* \subset (\mathbb{P}^N)^\vee = \mathbb{P}(V^\vee)$ be the projectively dual variety. The Katz dimension formula expresses $\operatorname{def}(X)$ as a corank of a certain Hessian matrix.

For $x \in \mathbb{P}(V)$ let $x^\perp \subset V^\vee$ denote the annihilator of the line in V corresponding to x. Let $x_0 \in X_{sm}$. Then one can choose linear functionals

$$T_0 \in V^* \setminus x_0^\perp, \quad T_1, \ldots, T_n \in x_0^\perp$$

so that

$$t_1 = T_1/T_0,\ t_2 = T_2/T_0, \ldots, t_n = T_n/T_0$$

are local coordinates on X in the neighborhood of x_0. For $U \in x_0^\perp$, the function $u = U/T^0$ on X near x_0 is an analytic function of $t_1, \ldots, t_n$ such that $u(0, \ldots, 0) = 0$. Consider the Hessian matrix

$$\operatorname{Hes}(u) = \left(\frac{\partial^2 u}{\partial t_i \partial t_j}(0, \ldots, 0)\right)_{i,j=1,\ldots,n}$$

Theorem 3.1 ([Ka]) $\operatorname{def}(X) = \min\{\operatorname{corank}\operatorname{Hes}(u)\}$, *the minimum over all possible choices of x_0 and U. In particular, X^* is a hypersurface iff for some choice of x_0 and u the matrix* $\operatorname{Hes}(u)$ *is invertible.*

Proof. Let $I_X^0 \subset \mathbb{P}(V) \times \mathbb{P}(V^\vee)$ be a "smooth part" of the conormal variety consisting of all pairs (x, H) such that $x \in X_{sm}$ and H is tangent to X at x. Let $\pi : I_X^0 \to \mathbb{P}(V^\vee)$ be the second projection. Then X^* is the closure of $\pi(I_X^0)$. We shall compute the Jacobian matrix of π in appropriate local coordinates and relate it to the Hessian matrices appearing in the theorem. Then we shall use the well-known fact that, for any regular map $\pi : Z \to S$ of algebraic varieties, the dimension of the image is equal to the maximal rank of the Jacobian matrix of π at smooth points of Z.

We extend $T_1, \dots, T_n$ to a basis

$$\{T_1, \dots, T_n, U_1, \dots, U_{N-n}\}$$

of $x_0^\perp$. Then each of functions $u_i = U_i/T_0$ on X is an analytic function of $t_1, \dots, t_n$ near $x_0 = (0, \dots, 0)$, and we have

$$u_i(0, \dots, 0) = 0 \quad \text{for} \quad i = 1, \dots, N-n.$$

Let $x \in X$ be a point close to x_0 with local coordinates $(t_1, \dots, t_n)$, and so in homogeneous coordinates

$$x = (1 : t_1 : \dots : t_n : u_1 : \dots : u_{N-n}).$$

Let $H \subset \mathbb{P}(V^\vee)$ be a hyperplane with an equation

$$\sum_{j=0}^{n} \tau_j T_j + \sum_{i=1}^{N-n} \eta_i U_i = 0.$$

Then H is tangent to X at x if and only if

$$\tau_0 + \sum_{j=1}^{n} \tau_j t_j + \sum_{i=1}^{N-n} \eta_i u_i$$

vanishes at x together with all its first derivatives. Therefore, for a given x, the hyperplanes tangent to X at x form a projective space of dimension $N - n - 1$ with homogeneous coordinates $(\eta_1 : \dots : \eta_{N-n})$, and the remaining coordinates are given by

$$\tau_j = -\sum_{i=1}^{N-n} \eta_i \frac{\partial u_i}{\partial t_j}, \quad j = 1, \dots, n, \tag{3.1}$$

$$\tau_0 = -\left(\sum_{j=1}^{n} \tau_j t_j + \sum_{i=1}^{N-n} \eta_i u_i \right) = -\sum_{i=1}^{N-n} \eta_i u_i + \sum_{j=1}^{n} \sum_{i=1}^{N-n} \eta_i t_j \frac{\partial u_i}{\partial t_j}. \tag{3.2}$$

Let $(x_0, H_0) \in I_X^0$. Without loss of generality we may assume that the hyperplane H_0 has coordinates

$$(\eta_1 : \ldots : \eta_{N-n}) = (0 : \ldots : 0 : 1).$$

It follows that we can set $\eta_{N-n} = 1$ in the above formulas, and use

$$(t_1, \ldots, t_n, \eta_1, \ldots, \eta_{N-n-1})$$

as local coordinates on I_X^0 near (x_0, H_0). In these coordinates the projection $\pi : I_X^0 \to \mathbb{P}(V^\vee)$ has the form

$$(t_1, \ldots, t_n, \eta_1, \ldots, \eta_{N-n-1}) \mapsto (\tau_0, \tau_1, \ldots, \tau_n, \eta_1, \ldots, \eta_{N-n-1}),$$

where τ_i are given by formulas (3.1) and (3.2) with $\eta_{N-n} = 1$. Therefore the Jacobian matrix of π at the origin is equal to

$$\frac{\partial(\tau_0, \tau_1, \ldots, \tau_n, \eta_1, \ldots, \eta_{N-n-1})}{\partial(t_1, \ldots, t_n, \eta_1, \ldots, \eta_{N-n-1})}(0, \ldots, 0) =$$

$$= \begin{pmatrix} 0 & 0 \\ -\operatorname{Hes}(u_{N-n}) & -\left(\frac{\partial(u_1, \ldots, u_{N-n})}{\partial(t_1, \ldots, t_n)}(0, \ldots, 0)\right)^t \\ 0 & \mathrm{Id}_{N-n-1} \end{pmatrix}.$$

Here the vertical sizes of the blocks are $1, n, N-n-1$, and the horizontal sizes are n, $N-n-1$. Clearly, this matrix has rank

$$N - n - 1 + \operatorname{rk}\operatorname{Hes}(u_{N-n}).$$

It follows that

$$\dim X^* = N - n - 1 + \max \operatorname{rk}\operatorname{Hes}(u),$$

and the theorem is proved. □

If $X \subset \mathbb{P}^N$ is a hypersurface (or a complete intersection) then it is possible to rewrite Hessian matrices in homogeneous coordinates. In the case of hypersurfaces the corresponding result was first formulated by B. Segre [Se]:

Theorem 3.2 *Let $f(x_0, \ldots, x_N)$ be an irreducible homogeneous polynomial and let $X \subset \mathbb{P}^N$ be the hypersurface $\{f = 0\}$. Let m be the largest number with the following property: there exists an $(m \times m)$-minor of the Hessian matrix $(f_{x_i x_j})$ that is not divisible by f. Then* $\dim X^* = m - 2$.

Example 3.3 An irreducible surface in $\mathbb{P}^3$ with equation $f(x_0, \ldots, x_3) = 0$ is projectively dual to a space curve if and only if the Hessian $\det(f_{x_i x_j})$ is divisible by f.

The Segre Theorem is applicable in the most common situation: the projectively dual variety is usually a hypersurface and therefore by taking duals of hypersurfaces we can get a considerable number of varieties that are not hypersurfaces.

Example 3.4 The following remark belongs to Zak [Z7]. In notation of Theorem 3.2, suppose that the Hessian $H(f) = |\partial^2 f/\partial x_i \partial x_j|$ is not trivial. Then $H(f)$ is divisible by f^{N+1-m} (this result can be easily verified by induction using the well-known fact that if any $(p \times p)$-minor of a $(p+1) \times (p+1)$-matrix A over the factorial domain D is divisible by f^l, where $f \in D$ is prime, then $\det A$ is divisible by f^{l+1}). Let $\deg f = d$. We can compare degrees of $H(f)$ and f^{N+1-m} and obtain the formula

$$(N+1)(d-2) \geq d(N+1-m), \quad \text{i.e.} \quad d \geq \frac{2(N+1)}{m}.$$

In other words, if Y is a projective variety in $\mathbb{P}^N$ such that Y^* is a hypersurface with the non-trivial Hessian, then

$$\deg Y^* \geq \frac{2(N+1)}{\dim Y + 2}.$$

In fact, in this inequality the non-triviality of the Hessian can be substituted for the non-degeneracy of Y. Conjecturally, this inequality becomes an equality if and only if Y is the projectivization of the variety of rank 1 elements in a simple Jordan algebra.

3.1.2 Defect of a Product

Let $X_1 \subset \mathbb{P}^{N_1}$ and $X_2 \subset \mathbb{P}^{N_2}$ be irreducible projective varieties. Then $X_1 \times X_2$ is naturally embedded in $\mathbb{P}^{N_1} \times \mathbb{P}^{N_2}$ and the latter is embedded in $\mathbb{P}^N$ via the Segre embedding, where

$$N+1 = (N_1+1)(N_2+1).$$

This gives an embedding $X_1 \times X_2 \subset \mathbb{P}^N$. Let $(X_1 \times X_2)^* \subset (\mathbb{P}^N)^\vee$ be the dual variety. It turns out that it is quite easy to calculate $\operatorname{def}(X_1 \times X_2)$.

Theorem 3.5 ([WZ])

$$\operatorname{def}(X_1 \times X_2) = \max\{0, \operatorname{def} X_1 - \dim X_2, \operatorname{def} X_2 - \dim X_1\}.$$

Proof. We apply Theorem 3.1. Let $n_1 = \dim X_1$, $n_2 = \dim X_2$. Let $x_0 = (x_{01}, x_{02})$ be a smooth point of $X_1 \times X_2$. We choose local coordinates

$$t_{11}, t_{21}, \ldots, t_{n_1 1}$$

of X_1 near x_{01} and

$$t_{12}, t_{22}, \ldots, t_{n_2 2}$$

of X_2 near x_{02}. So for $\nu = 1, 2$, a point x_ν close to $x_{0\nu}$ has homogeneous coordinates

such that

$$\operatorname{corank}(\varphi') = \max(0, c_2 - k_1).$$

We define the form φ on

$$\mathbb{C}^{k_1} \oplus \mathbb{C}^{k_2} = \left(\mathbb{C}^{k_1-1} \oplus \langle e\rangle\right) \oplus \left(\mathbb{C}^{k_2-1} \oplus \langle f\rangle\right)$$

such that the restriction of φ on $\mathbb{C}^{k_1-1} \oplus \mathbb{C}^{k_2-1}$ coincides with φ',

$$\left(\mathbb{C}^{k_1-1} \oplus \mathbb{C}^{k_2-1}\right) \perp \langle e, f\rangle,$$

and

$$\varphi(e,e) = \varphi(e,f) = 1, \quad \varphi(f,f) = 0.$$

It is clear then that $\varphi \in \mathrm{Bil}(0, k_1; c_2, k_2)$ and

$$\operatorname{corank}(\varphi) = \operatorname{corank}(\varphi') = \max(0, c_2 - k_1).$$

It remains to consider only the cases when either $k_1 = 0$ or $c_1 = c_2 = 0$. In both cases lemma is obvious. □

It follows immediately from Theorem 3.5 that the same result holds for the product with any number of factors. Let $X_k \subset \mathbb{P}^{N_k}$, $k = 1 \ldots, r$, be irreducible projective varieties. Then $X = X_1 \times \ldots \times X_r$ is embedded in $\mathbb{P}^{N_1} \times \ldots \times P^{N_r}$ and the latter is embedded in $\mathbb{P}^N$ via the Segre embedding, where $N + 1 = (N_1 + 1) \times \ldots \times (N_r + 1)$. This gives an embedding $X \subset \mathbb{P}^N$. Let $X^* \subset (\mathbb{P}^N)^\vee$ be the dual variety.

Corollary 3.7

$$\operatorname{def} X + \dim X = \max\{\dim X, \operatorname{def} X_1 + \dim X_1, \ldots, \operatorname{def} X_r + \dim X_r\}.$$

In particular, X^* is a hypersurface if and only if

$$\operatorname{def} X_k + \dim X_k \leq \dim X, \quad k = 1 \ldots, r.$$

Example 3.8 The Segre embedding

$$\mathbb{P}^{l_1} \times \ldots \times \mathbb{P}^{l_r} \subset \mathbb{P}^l, \quad \mathbb{P}^l = \mathbb{P}(\mathbb{C}^{l_i+1} \otimes \ldots \otimes \mathbb{C}^{l_r+1}),$$

identifies $\mathrm{Cone}(\mathbb{P}^{l_1} \times \ldots \times \mathbb{P}^{l_r})$ with the variety of decomposable tensors in $\mathbb{C}^{l_i+1} \otimes \ldots \otimes \mathbb{C}^{l_r+1}$. By Theorem 3.5, $(\mathbb{P}^{l_1} \times \ldots \times \mathbb{P}^{l_r})^*$ is a hypersurface if and only if

$$2l_k \leq l_1 + \ldots + l_r \quad \text{for} \quad k = 1, \ldots, r.$$

The corresponding discriminant is called a *hyperdeterminant* for the matrix format $(l_1 + 1) \times \ldots \times (l_r + 1)$.

3.2 Fundamental Forms

3.2.1 Second Fundamental Form

To measure how $\hat{T}_{x,X}$ moves to the first order, one calculates

$$(d\gamma)_x : T_{x,X} \to T_{\hat{T}_{x,M},\mathrm{Gr}(n,\mathbb{P}(V))} = \mathrm{Hom}(U, V/U), \quad U = \mathrm{Cone}(T_{x,X}.$$

It is easy to see that $\langle x \rangle \subset \mathrm{Ker}(d\gamma)_x(v)$ for any $v \in T_{x,X}$, and therefore $d(\gamma)_x$ factors to a map

$$\mathbb{FF}^2_x : T_{x,X} \to \mathrm{Hom}(T_{x,X}, N_{X,\mathbb{P}(V)}|_x).$$

Example 3.9 Let us give a useful coordinate description of $\mathbb{FF}^2$. It will show that $\mathbb{FF}^2$ was implicitly used in the proof of Theorem 3.1. For $x \in X$, we choose a homogeneous coordinate system $T_0, \dots, T_N$ for $\mathbb{P}^N$ so that $x = (1 : 0 : \dots : 0)$ and $\hat{T}_{x,X}$ is spanned by the first $n+1$ coordinate vectors. Let $x_1 = T_1/T_0, \dots$, $x_N = T_N/T_0$ be the corresponding affine coordinate system for the affine chart of $\mathbb{P}^N$. Then $x_1, \dots, x_n$ is a holomorphic coordinate system for X centered at x, $dx_1, \dots, dx_n$ is a basis of $T^\vee_{x,X}$, and $\partial/\partial x_{n+1}, \dots, \partial/\partial x_N$ is a basis of $N_{X,\mathbb{P}(V)}|_x$. For $\mu = n+1, \dots, N$ we have in the neighborhood of x that $x_\mu|_X$ vanishes to the second order at x. Therefore

$$x_\mu|_X = \sum_{i,j=1}^{n} q_{ij\mu} x_i x_j + (\text{higher order terms}).$$

Then

$$\mathbb{FF}^2_x = \sum_{\mu=n+1}^{N} \left(\sum_{i,j=1}^{n} q_{ij\mu} dx_i \otimes dx_j \right) \otimes \partial/\partial x_\mu.$$

$\mathbb{FF}^2_x$ is symmetric, essentially because the Gauss map is already the derivative of a map and mixed partial derivatives commute.

Definition 3.10 The section $\mathbb{FF}^2$ of the vector bundle $S^2T^\vee_X \otimes N_{X,\mathbb{P}(V)}$ constructed above is called the *projective second fundamental form* of X. If $X \subset \mathbb{P}(V)$ is singular then $\mathbb{FF}^2$ is defined over X_{sm}.

It is convenient to consider $\mathbb{FF}^2$ as a map $\mathbb{FF}^2 : N^\vee_{X,\mathbb{P}(V)} \to S^2T^\vee_X$ and to set $|\mathbb{FF}^2| = \mathbb{P}(\mathbb{FF}^2(N^\vee_{X,\mathbb{P}(V)}))$. One can think of $|\mathbb{FF}^2|_x$ as a linear system of quadric hypersurfaces in $\mathbb{P}(T_{x,X})$. The space $\mathbb{P}(N^\vee_{X,\mathbb{P}(V)}|_x)$ has the geometric interpretation as the space of hyperplanes tangent to X at x, i.e., the hyperplanes H such that $X \cap H$ is singular at x. Then $\mathbb{FF}^2_x$ can be viewed as a map that sends a hyperplane to the quadratic part of the singularity of $X \cap H$ at x.

Let $\mathrm{Base}\,|\mathbb{FF}^2|_x \subset \mathbb{P}(T_{x,X})$ be the base locus of a linear system $|\mathbb{FF}^2|_x$, i.e. the variety of directions tangent to lines osculating to order 2 at x (that is, X appears to contain these lines to the second order). So we have

$$\begin{aligned}\operatorname{Base}|\mathbb{FF}^2|_x &= \mathbb{P}\{v \in T_{x,X} \mid \mathbb{FF}^2_x(v,v) = 0\} \\ &= \{p \in \mathbb{P}(T_{x,X}) \mid p \in Q \ \forall Q \in |\mathbb{FF}^2|_x\}.\end{aligned} \tag{3.3}$$

Let $\operatorname{Sing}|\mathbb{FF}^2|_x$ be the set of tangent directions such that the embedded tangent space does not move to the first order in these directions. Thus

$$\begin{aligned}\operatorname{Sing}|\mathbb{FF}^2|_x &= \mathbb{P}\{v \in T_{x,X} \mid \mathbb{FF}^2(v,w) = 0 \ \forall w \in T_{x,X}\} \\ &= \{p \in \mathbb{P}(T_{x,X}) \mid p \in Q_{sing} \ \forall Q \in |\mathbb{FF}^2|\}.\end{aligned}$$

Clearly this is just the kernel of the differential of the Gauss map. And so by Theorem 4.2, if X is smooth, then $\operatorname{Sing}|\mathbb{FF}^2|_x = \emptyset$ for any x. A weaker result that $\operatorname{Sing}|\mathbb{FF}^2|_x = \emptyset$ for generic $x \in X$ was proved in [GH]. However, these results do not imply that generic quadrics from the second fundamental form are not degenerate. In fact, the following theorem is just a reformulation of Theorem 3.1:

Theorem 3.11 *If X is any projective variety and $x \in X_{sm}$, then the projective second fundamental form $|\mathbb{FF}^2|_x$ of X at x is a system of quadrics of rank bounded above by* $\dim X - \operatorname{def} X$. *Moreover, if $x \in X_{sm}$ is a generic point and $Q \subset |\mathbb{FF}^2|_x$ is a generic quadric, then* $\operatorname{rank} Q = \dim X - \operatorname{def} X$.

A useful formalism for dealing with second (and higher) fundamental forms is provided by moving frames. We recall it briefly; more details and applications can be found in [GH].

In $\mathbb{P}^N$ a frame is denoted by $\{A_0, \dots, A_N\}$. It is given by a basis $A_0, \dots, A_N$ for $\mathbb{C}^{N+1}$. The set of all frames forms an algebraic variety $\mathcal{F}(\mathbb{P}^N)$. The general linear group GL_{N+1} acts on $\mathcal{F}(\mathbb{P}^N)$ transitively with trivial stabilizer (it is the principal homogeneous space of GL_{N+1}). Each of the vectors A_i may be viewed as a mapping $v : \mathcal{F}(\mathbb{P}^N) \to \mathbb{C}^{N+1}$. Expressing the exterior derivative dv in terms of the basis $\{A_i\}$ gives

$$dA_i = \sum_j \omega_{ij} A_j.$$

The $(N+1)^2$ differential forms ω_{ij} are called the *Maurer–Cartan forms* (on the group GL_{N+1}). Taking the exterior derivative of this formula we obtain *the Maurer–Cartan equations*

$$d\omega_{ij} = \sum_k \omega_{ik} \wedge \omega_{kj}.$$

Geometrically we may think of a frame $\{A_0, \dots, A_N\}$ as defining a coordinate simplex in $\mathbb{P}^N$, and then ω_{ij} gives the rotation matrix when this coordinate simplex is infinitesimally displaced.

There is a fibering $\pi : \mathcal{F}(\mathbb{P}^N) \to \mathbb{P}^N$ given by

$$\pi\{A_0, \dots, A_N\} \to A_0.$$

1-forms $\omega_i = \omega_{0i}$ are horizontal for π, i.e. they vanish on fibers of π. We may think of π as defined by the foliation

$$\omega_1 = \ldots = \omega_N = 0.$$

The equation

$$dA_0 = \sum_{i=0}^{N} \omega_i A_i = \sum_{i=1}^{N} \omega_i A_i \text{ mod } A_0$$

has the following geometric interpretation. For each choice of the frame $\{A_0, \ldots, A_N\}$ lying over $p \in \mathbb{P}^N$, the horizontal 1-forms $\omega_1, \ldots, \omega_N$ give a basis for the cotangent space $T^{\vee}_{p,\mathbb{P}^N}$. The corresponding basis $v_1, \ldots, v_N \in T_{p,\mathbb{P}^N}$ for the tangent space has the property that v_i is tangent to the line $\overline{A_0 A_i}$.

Assume now that we are given a connected n-dimensional smooth variety $M \subset \mathbb{P}^N$, not necessarily closed. Associated to M is the submanifold $\mathcal{F}(M) \subset \mathcal{F}(\mathbb{P}^N)$ of *Darboux frames*

$$\{A_0; A_1, \ldots, A_n; A_{n+1}, \ldots, A_N\}$$

defined by the conditions that A_0 lies over some $p \in M$ and $A_0, A_1, \ldots, A_n$ span $\hat{T}_{p,M}$, the embedded tangent space of M at p. On $\mathcal{F}(M)$ we have the condition that differential 1-forms $\omega_1, \ldots, \omega_n$ give a basis of $T^*_p M$ and $\omega_i = 0$ for $i > n$. Therefore it is natural to think of the $\omega_1, \ldots, \omega_n$ as homogeneous coordinates in the projectivized tangent spaces $\mathbb{P}(T_{p,M})$. For example, a quadric in $\mathbb{P}(T_{p,M})$ is defined by an equation $\sum_{i,j=1}^{n} q_{ij}\omega_i\omega_j = 0$ with $q_{ij} = q_{ji}$.

Since $\omega_i = 0$ for $i > n$, we also have $d\omega_i = 0$ on $\mathcal{F}(M)$. On the other hand, by the structure equations we have $d\omega_i = \sum_{j=1}^{N} \omega_j \wedge \omega_{ji}$. Therefore, for any $i > n$, we have

$$\sum_{j=1}^{n} \omega_j \wedge \omega_{ji} = 0.$$

Since $\omega_1, \ldots, \omega_n$ are linearly independent, an easy calculation (called the Cartan Lemma) shows that, for any $\mu > n$ and any $i = 1, \ldots, n$, we have

$$\omega_{i\mu} = \sum_{j=1}^{n} q_{ij\mu}\omega_j, \quad q_{ij\mu} = q_{ji\mu}.$$

Then for any $\mu = n+1, \ldots, N$ we set

$$Q_\mu = \sum_{i,j=1}^{n} q_{ij\mu}\omega_i\omega_j.$$

An easy local calculation (see [GH1]) shows:

Proposition 3.12 *The linear system of quadrics in $\mathbb{P}(T_{p,M})$ spanned by Q_μ, $\mu = n+1, \ldots, N$, is the projective second fundamental form $|\mathbb{FF}^2|$.*

If we think of $\mathbb{FF}^2$ as the map $S^2T_{p,M} \to (N_{M,\mathbb{P}^N})_p$ and identify $(N_{M,\mathbb{P}^N})_p$ with $\mathbb{C}^{N+1}/\operatorname{Cone}(\hat{T}_{p,M})$. Then for any vector $v \in T_{p,M}$ the vector $\mathbb{FF}^2(v)$ is given in coordinates by

$$\mathbb{FF}^2(v) = \sum_{i,j=1}^{n} \sum_{\mu=n+1}^{N} \omega_i(v)\omega_j(v)A_\mu,$$

where $v = \sum_{i=1}^n \omega_i(v)v_i \in T_{p,M}$.

3.2.2 Higher Fundamental Forms

Definition 3.13 If $\{p(t)\}$ is any holomorphic arc in $\mathbb{P}^N$ described by a vector-valued function $A_0(t)$, then the *osculating sequence* is the sequence of linear spaces spanned by the following collections of vectors:

$$\langle A_0(t)\rangle,\ \langle A_0(t), A_0'(t)\rangle,\ \langle A_0(t), A_0'(t), A_0''(t)\rangle,\ \ldots$$

We then have the Euler–Meusnier Theorem

Theorem 3.14 ([GH1]) *For $v \in T_{p,M}$, the vector $\mathbb{FF}^2(v,v) \in (N_{M,\mathbb{P}^N})_p$ gives the projection in $\mathbb{C}^{n+1}/\operatorname{Cone}(\hat{T}_{p,M})$ of the second osculating space to any curve $p(t)$, $p(0) = p$, with tangent v at $t = 0$*

Proof. We choose an arbitrary field of Darboux frames $\{A_i(t)\}$ along $p(t)$ and write

$$\frac{dA_0}{dt} = \sum_{i=1}^{n} \left(\frac{\omega_i}{dt}\right) A_i \mod A_0.$$

Recall that ω_i is a 1-form on the frame manifold. Hence the expression $\frac{\omega_i}{dt}$ makes sense and is equal to the value of this 1-form on the tangent vector to the curve $\{A_i(t)\}$ on the frame manifold. If we interpret $\{\omega_1, \ldots, \omega_n\}$ as a basis in the cotangent space $T_{p,M}^\vee$ then this equation shows that $\frac{\omega_i}{dt}$ is equal to the i-th coordinate of $v \in T_{p,M}$ in the dual basis $\{v_1, \ldots, v_n\}$ of $T_{p,M}$ (in fact, $v_i = A_i \mod A_0$).

Differentiating further, we get

$$\frac{d^2A_0}{dt^2} = \sum_{i=1}^{n} \left(\frac{\omega_i}{dt}\right)\left(\frac{dA_i}{dt}\right) \mod A_0, \ldots, A_n$$
$$= \sum_{i,j=1}^{n} \sum_{\mu=n+1}^{N} q_{ij\mu} \left(\frac{\omega_i}{dt}\right)\left(\frac{\omega_j}{dt}\right) \mod \operatorname{Cone}(\hat{T}_pM).$$

But we also have

$$\mathbb{FF}^2(v,v) = \sum_{i,j=1}^{n} \sum_{\mu=n+1}^{N} q_{ij\mu} \left(\frac{\omega_i}{dt}\right)\left(\frac{\omega_j}{dt}\right).$$

This finishes the proof. □

The higher fundamental forms are defined similar to the second fundamental form $\mathbb{FF}^2$. The most geometric way is to utilize the Euler–Meusnier Theorem 3.14 as the definition.

Suppose that $M \subset \mathbb{P}^N = \mathbb{P}(V)$ is a smooth variety, $p \in M$, and $\tilde{T}_{p,M} \subset V$ is the cone over $\hat{T}_{p,M}$. Then the sequence of osculating spaces

$$T^{(0)}_{p,M} = \mathbb{C}p, \quad T^{(1)}_{p,M} = \tilde{T}_{p,M}, \quad T^{(2)}_{p,M}, \quad \ldots$$

is defined as follows. The osculating space $T^{(k)}_{p,M}$ is the linear span of k-th osculating spaces for smooth curves $p(t) \subset M$, $p(0) = p$. The k-th normal space $N^{(k)}_{p,M}$ is defined as $V/T^{(k)}_{p,M}$, so we have $N^{(1)}_{p,M} = N_{M,\mathbb{P}^N}|_p$. Suppose that $v \in T_{p,M} = T^{(1)}_{p,M}/T^{(0)}_{p,M}$ is any tangent vector and $\{p(t)\}$ is any holomorphic arc in $\mathbb{P}^N$ with tangent vector v at $t = 0$. Let $A(t) \subset V$ be a curve projected to $p(t)$. Then it is easy to see that the vector

$$\mathbb{FF}^k(v) = \frac{d^k A(t)}{dt^k} \mod T^{(k-1)}_{p,M} \in N^{(k-1)}_{p,M}$$

depends only on v. This construction gives a map

$$\mathbb{FF}^k : S^k T_{p,M} \to N^{(k-1)}_{p,M}$$

called the k-th projective fundamental form. Of course, it can also be described via moving frames, in coordinates, by differentiating the Gauss map, etc., see e.g. [GH1, Lan4]. The sequence of dimensions of osculating spaces $T^{(k)}_{p,M}$ can, of course, depend on $p \in M$. One solution is to restrict to an open subset of M where these dimensions are fixed.

Example 3.15 ([GH1]) Suppose that $\mathbb{P}^m \times \mathbb{P}^n \subset \mathbb{P}^{nm+n+m}$ is the Segre embedding. Recall that if $\mathbb{P}^m = \mathbb{P}(V)$ and $\mathbb{P}^n = \mathbb{P}(W)$, then the image M of the Segre embedding is the image of the natural inclusion $\mathbb{P}(V) \times \mathbb{P}(W) \subset \mathbb{P}(V \otimes W)$. Choosing coordinates $\{X_\alpha\}$ in V and $\{Y_\mu\}$ in W, this inclusion is given by the mapping $(\{X_\alpha\}, \{Y_\mu\}) \to \{X_\alpha Y_\mu\}$. We denote by $\{A_0, \ldots, A_m\}$ and $\{B_0, \ldots, B_n\}$ frames for $\mathbb{P}(V)$ and $\mathbb{P}(W)$. We shall use the range of indices $1 \le i, j \le m$, $1 \le \alpha, \beta \le n$. If A_0 lies over $v_0 \in \mathbb{P}(V)$ and B_0 over $w_0 \in \mathbb{P}(W)$, then the frame

$$\{A_0 \otimes B_0, A_0 \otimes B_\alpha, A_i \otimes B_0, A_i \otimes B_\alpha\} \tag{3.4}$$

in $\mathbb{P}(V \otimes W)$ lies over $v_0 \otimes w_0$. If we write

$$dA_0 = \sum_i \phi_i A_i \mod A_0, \quad dB_0 = \sum_\alpha \psi_\alpha B_\alpha \mod B_0,$$

then

$$d(A_0 \otimes B_0) = \sum_\alpha \psi_\alpha A_0 \otimes B_\alpha + \sum_i \phi_i A_i \otimes B_0 \mod A_0 \otimes B_0.$$

Therefore (3.4) is a Darboux frame for M and

$$\hat{T}_{v_0 \otimes w_0, M} = \mathbb{P}(V \otimes w_0 + v_0 \otimes W),$$

so that

$$(N_M)_{v_0 \otimes w_0} = (V \otimes W)/(V \otimes w_0 + v_0 \otimes W).$$

Differentiating further, we get

$$d^2(A_0 \otimes B_0) = 2 \sum_{i,\alpha} \phi_i \psi_\alpha A_i \otimes B_\alpha \mod \{A_0 \otimes B_0, A_i \otimes B_0, A_0 \otimes B_\alpha\}.$$

Therefore the second fundamental form is given by

$$\mathbb{FF}^2(v_0 \otimes w + v \otimes w_0, v_0 \otimes w + v \otimes w_0) = 2v \otimes w \mod V \otimes w_0 + v_0 \otimes W.$$

In terms of homogeneous coordinates $\{\phi_i\}$ for $T_{v_0, \mathbb{P}^m}$ and $\{\psi_\alpha\}$ for $T_{w_0, \mathbb{P}^n}$, $|\mathbb{FF}^2|$ is the linear system of quadrics

$$\sum_{i,\alpha} q_{i\alpha} \phi_i \psi_\alpha$$

for any matrix $q_{i\alpha}$. In other words, matrices of quadrics from this linear system have the form

$$\begin{pmatrix} 0 & Q \\ Q^t & 0 \end{pmatrix},$$

where Q is an arbitrary matrix.

Summarizing, we see that the projectivization of the tangent space at any point $x \in M$ is $\mathbb{P}^{n+m-1}$ and $|\mathbb{FF}^2|_x$ is the complete linear system of quadrics having as a base locus the union $\mathbb{P}^{m-1} \cup \mathbb{P}^{n-1}$ of two skew linear subspaces.

Example 3.16 ([GH1]) Suppose that $\mathbb{P}^n = \mathbb{P}(V) \subset \mathbb{P}(S^d V)$ is the d-th Veronese embedding. In terms of homogeneous coordinates $\{X_0, \ldots, X_n\}$ for $\mathbb{P}(V)$, the embedding is given by

$$(\ldots : X_\alpha : \ldots) \mapsto (X_0^d : X_0^{d-1} X_1 : \ldots : X_0^{d-1} X_n : \ldots : F_\lambda(X) : \ldots),$$

where F_λ varies over a basis for the homogeneous forms of degree d. Given a point $v_0 \in \mathbb{P}^n$, e.g., $v_0 = (1 : \ldots : 0)$, v_0 maps to $(1 : \ldots : 0)$; the projectivization of the tangent space at v_0 is $\mathbb{P}^{n-1}$ spanned by first n coordinate vectors in $\mathbb{P}(S^d V)$. The affine coordinate system in the affine chart containing v_0 is given by $F_\lambda(X)/X_0^d = F_\lambda(X_1/X_0, \ldots, X_n/X_0)$. Therefore $|\mathbb{FF}^2|_{v_0}$ is the linear system of all quadrics.

3.2.3 Fundamental Forms of Flag Varieties

Let G be a connected semisimple complex algebraic group, $\mathfrak{g}$ its Lie algebra, and $U\mathfrak{g}$ the universal enveloping algebra of $\mathfrak{g}$, i.e. the quotient of the tensor algebra $T^\bullet\mathfrak{g}$ by the ideal generated by elements

$$x \otimes y - y \otimes x - [x, y] \quad \text{for} \quad x, y \in \mathfrak{g}.$$

$U\mathfrak{g}$ has a natural filtration $U\mathfrak{g}^{(0)} \subset U\mathfrak{g}^{(1)} \subset \ldots$ such that $U\mathfrak{g}^{(k)}$ is spanned by products $x_1 \ldots x_l$, where $x_i \in \mathfrak{g}$ and $l \leq k$. The associated graded algebra is the symmetric algebra $\operatorname{Sym}^\bullet \mathfrak{g}$ of $\mathfrak{g}$. There exists a G-equivariant linear isomorphism $\sigma : \operatorname{Sym}^k \mathfrak{g} \to U\mathfrak{g}^{(k)}/U\mathfrak{g}^{(k-1)}$ given by

$$\sigma(x^k) = x^k \bmod U\mathfrak{g}^{(k-1)} \quad \text{for} \quad x \in \mathfrak{g}.$$

Let V_λ be an irreducible G-module with the highest weight λ, and $v_\lambda \in V_\lambda$ a highest weight vector. The action of $U\mathfrak{g}$ on V_λ induces a filtration of V_λ whose k-th term is $V_\lambda^{(k)} = U\mathfrak{g}^{(k)} v_\lambda$.

Let $x_\lambda = [v_\lambda] \in \mathbb{P}(V_\lambda)$ be the line in V_λ generated by v_λ, and let $X = G/P \subset \mathbb{P}(V_\lambda)$ be the projectivization of the orbit of the highest weight vector. Here $P \subset G$ is a parabolic subgroup, namely, the stabilizer of x. Let $\mathfrak{p}$ be the Lie algebra of P. We can canonically identify $T_{x,X}$ with $\mathfrak{g}/\mathfrak{p}$ as a P-module. Then it is easy to verify that the osculating spaces and the fundamental forms of X have the following simple representation-theoretic interpretation.

Theorem 3.17 *Let $X = G/P \subset \mathbb{P}(V_\lambda)$ be a polarized flag variety and let $x = eP \in X$. Then the k-th osculating space $T_{x,X}^{(k)} = V_\lambda^{(k)}$ and $N_{x,X}^{(k)} = V_\lambda/V_\lambda^{(k)}$. There is a commutative diagram*

$$\begin{array}{ccc} \operatorname{Sym}^k \mathfrak{g} & \xrightarrow{\ \sigma\ } & U\mathfrak{g}^{(k)}/U\mathfrak{g}^{(k-1)} \\ {\scriptstyle \pi_1}\downarrow & & \downarrow{\scriptstyle \pi_2} \\ \operatorname{Sym}^k(T_{x,X}) = \operatorname{Sym}^k(\mathfrak{g}/\mathfrak{p}) & \xrightarrow[\mathbb{FF}^k]{} & V_\lambda/V_\lambda^{(k)} = N_{x,X}^{(k-1)} \end{array}$$

where π_1 is induced by the projection $\mathfrak{g} \to \mathfrak{g}/\mathfrak{p}$ and π_2 is the map $u \mapsto u \cdot v_\lambda$.

Proof. Indeed, let $\bar{\xi} \in \mathfrak{g}/\mathfrak{p}$ be a tangent vector. We take any $\xi \in \mathfrak{g}$ projecting on $\bar{\xi}$. Then $A_\xi = \exp(t\xi) \cdot v_\lambda \subset V$, $t \in \mathbb{C}$, is a curve projecting to a curve $\exp(t\xi) \cdot x_\lambda \subset X$ with a tangent vector $\bar{\xi}$. It follows that $T_{x,X}^{(k)}$ is spanned by k-th osculating spaces to curves A_ξ. Taking derivatives of the exponent, we see that

$$\langle A_\xi(t), A'_\xi(t), \ldots A_\xi^{(k)}(t) \rangle = \langle v_\lambda, x \cdot v_\lambda, \ldots, x^k \cdot v_\lambda \rangle.$$

Therefore $T_{x,X}^{(k)} = V_\lambda^{(k)}$ and $N_{x,X}^{(k)} = V_\lambda/V_\lambda^{(k)}$. □

Example 3.18 ([GH1]) Let $X = \mathrm{Gr}(k, V) \subset \mathbb{P}(\Lambda^k V)$ be the Grassmannian in the Plücker embedding with $\dim V = n$. Let us apply Theorem 3.17 and find fundamental forms of X. We choose the basis $\{e_1, \dots, e_n\}$ of V. Let $G = \mathrm{SL}(V)$. The maximal torus $T \subset G$ consists of diagonal matrices, and the Borel subgroup $B \subset G$ is the subgroup of upper triangular matrices. Then $v = e_1 \wedge \dots \wedge e_k$ is the highest weight vector. Its G-orbit Y is the variety of non-trivial decomposable tensors. Let $U = \langle e_1, \dots, e_k \rangle \subset V$. The projectivization $X = \mathbb{P}(Y) = G \cdot x$ is $\mathrm{Gr}(k, V) = G \cdot U$ in the Plücker embedding, where $x = [v]$ is the line spanned by v. The stabilizer $P_x = P_U$ is the parabolic subgroup of block matrices

$$P = \begin{pmatrix} A & B \\ 0 & C \end{pmatrix}, \quad \text{where } A \in \mathrm{Mat}_{k,k},\ B \in \mathrm{Mat}_{k,n-k},\ C \in \mathrm{Mat}_{n-k,n-k}\,.$$

$T_{x,X}$ is identified with $\mathfrak{g}/\mathfrak{p} = \mathrm{Hom}(U, V/U) = \mathrm{Mat}_{n-k,k}$. Applying Theorem 3.17, we see that $T^{(l)}_{x,X}$ is spanned by elements $(x_1 \dots x_m) \cdot v$, where $m \le l$ and x_i is any matrix of the form

$$\begin{pmatrix} 0 & 0 \\ X & 0 \end{pmatrix}, \quad \text{where } X \in \mathrm{Mat}_{n-k,k}\,.$$

Therefore

$$T^{(l)}_{x,X} = W_0 \oplus W_1 \oplus \dots \oplus W_l,$$

where

$$W_l = \langle e_{i_1} \wedge \dots \wedge e_{i_{k-l}} \wedge e_{i_{k-l+1}} \wedge \dots \wedge e_{i_k} \rangle$$

for

$$1 \le i_1 < \dots < i_{k-l} \le k < i_{k-1+1} < \dots < i_k \le n \rangle.$$

The same calculation shows that $|\mathbb{FF}^k|_x$ is generated by $(k \times k)$-minors of the matrix X. For instance, $|\mathbb{FF}^2|_x$ consists of all quadrics vanishing at the image of the Segre embedding $\mathbb{P}^{k-1} \times \mathbb{P}^{n-k-1} \subset \mathbb{P}(\mathrm{Mat}_{n-k,k})$,

Using this description, it is quite straightforward to find the defect of the Grassmannian in the Plücker embedding. All we need to do is find the maximal possible rank of a quadric generated by (2×2)-minors of a rectangular matrix. The answer is as follows. We may suppose that $k \le n/2$. Then the dual variety is always a hypersurface except except the case $k = 1$ (the dual variety is empty) and the case $k = 2$, n is odd (the defect is equal to 2).

Example 3.19 ([Lan4]) Let $\mathbb{S}_m \subset \mathbb{P}(\Lambda^{ev}\mathbb{C}^m)$ be the spinor variety. Then calculating the tangent space at the point $v_0 = 1$, it is easy to see that $T_{v_0,\mathbb{S}_m} = \Lambda^2\mathbb{C}^m$. Let $\mathrm{Gr}(2, \mathbb{C}^m) \subset \mathbb{P}(\Lambda^2\mathbb{C}^m)$ be the Plücker embedding. Then the linear system $|\mathbb{FF}^2|$ consists of all quadrics containing $\mathrm{Gr}(2, \mathbb{C}^m)$ and is generated by Pfaffians of principal (4×4)-minors of a skewsymmetric matrix (Plücker relations).

Example 3.20 (**[Lan4]**) Consider the Severi variety, c.f. Example 2.17. Let $\mathbb{D}_{\mathbb{R}}$ denote one of four division algebras over $\mathbb{R}$: real numbers $\mathbb{R}$, complex numbers $\mathbb{C}$, quaternions $\mathbb{H}$, or octonions $\mathbb{O}$. Let $\mathbb{D} = \mathbb{D}_{\mathbb{R}} \otimes_{\mathbb{R}} \mathbb{C}$ denotes its complexification. Therefore $\mathbb{D}$ is either $\mathbb{C}$, $\mathbb{C}\oplus\mathbb{C}$, $\mathrm{Mat}_2(\mathbb{C})$, or $\mathbb{C}\mathrm{a}$ (the algebra of Cayley numbers). All these algebras have the standard involution. Let $\mathcal{H}_3(\mathbb{D})$ denote the 3×3 Hermitian matrices over $\mathbb{D}$. If $x \in \mathcal{H}_3(\mathbb{D})$ then we may write

$$x = \begin{pmatrix} c_1 & u_1 & u_2 \\ \overline{u_1} & c_2 & u_3 \\ \overline{u_2} & \overline{u_3} & c_3 \end{pmatrix}, \quad c_i \in \mathbb{C},\ u_i \in \mathbb{D}.$$

The projectivization of the highest weight vector orbit with respect to the group of norm similarities is called the Severi variety corresponding to $\mathcal{D}$. One may think of Severi varieties as complexifications of projective planes over $\mathbb{D}_{\mathbb{R}}$. Take a point p of the Severi variety of the form

$$p = \begin{pmatrix} 1 & 0 & 0 \\ 0 & 0 & 0 \\ 0 & 0 & 0 \end{pmatrix}.$$

Choose the affine coordinates $u_1, u_2, u_3 \in \mathbb{D}$, $c_2, c_3 \in \mathbb{C}$ of $\mathcal{H}_3(\mathbb{D})$ centered at p. Then the tangent space at p is $\{u_1, u_2\}$ and the second fundamental form is generated by quadrics

$$u_1\overline{u_1} = 0,\ u_2\overline{u_2} = 0,\ \overline{u_2}u_1 = 0.$$

For example, if $\mathbb{D} = \mathbb{C}\mathrm{a}$ then the base locus of this linear system is the spinor variety $\mathbb{S}_5$ (cf. Example 2.15).

4

Projective Constructions

In this Chapter we study projective constructions related to projective duality but also having a merit of their own. We prove a theorem of Zak and Ran that the Gauss map of a smooth variety is a normalization. We introduce secant and tangent varieties in Sect. 4.2, prove the Terracini Lemma, give a method for the calculation of multisecant varieties of homogeneous spaces, discuss the relationship discovered by Zak between the degree of a dual variety and the order of the variety, and give an overview of old and new results related to the Waring problem for forms. In Sect. 4.3 we discuss theorems of Zak related to the Hartshorne conjecture – theorems on tangencies, on linear normality and on Severi varieties. The main tool is a connectedness theorem of Fulton and Hansen. We finish in Sect. 4.4 by explaining the Cayley trick for Chow forms.

4.1 Gauss Map

Suppose that $X \subset \mathbb{P}(V)$ is an irreducible n-dimensional variety.

Definition 4.1 The *Gauss map* is defined as

$$\gamma : X_{sm} \to \mathrm{Gr}(n, \mathbb{P}(V)), \quad x \mapsto \hat{T}_{x,X}.$$

$\overline{\gamma(X)} \subset \mathrm{Gr}(n, \mathbb{P}(V))$ is called the *Gauss image* of X.

The following important theorem shows that the Gauss image of X is almost isomorphic to X if X is smooth.

Theorem 4.2 ([Z3, E4, Ra3]) *If X is smooth then γ is finite and birational, i.e. X is the normalization of $\gamma(X)$.*

The proof will be given in Sect. 4.3.

One can define more general Gauss images in a completely similar fashion:

Definition 4.3 For any $k \geq n$ the variety

$$\gamma_k(X) = \overline{\cup_{x \in X_{sm}} \{L \subset \mathrm{Gr}(k, \mathbb{P}^N) \mid L \supset \hat{T}_x X\}} \subset \mathrm{Gr}(k, \mathbb{P}^N)$$

is called the k-th *Gauss image* of X.

It is easy to see that Gauss images are irreducible varieties. For example, $\gamma_{N-1}(X) \subset \mathrm{Gr}(N-1, \mathbb{P}^N) = (\mathbb{P}^N)^\vee$ is equal to the dual variety X^*.

4.2 Tangents and Secants

4.2.1 Terracini Lemma

Definition 4.4 Let $X, Y \subset \mathbb{P}^N$. Their *join* $S(X,Y)$ is defined as

$$S(X,Y) = \overline{\bigcup_{x \in X,\ y \in Y,\ x \neq y} \langle x, y \rangle},$$

where $\langle x, y \rangle$ is a line connecting x and y (and more generally, $\langle S \rangle$ denotes the minimal projective subspace containing S). The closure is not necessary if X and Y do not intersect.

Let

$$\tilde{X} = \mathrm{Cone}(X), \quad \tilde{Y} = \mathrm{Cone}(Y)$$

and consider the map

$$\Sigma : \mathbb{C}^{N+1} \times \mathbb{C}^{N+1} \to \mathbb{C}^{N+1}, \quad (x, y) \mapsto x + y.$$

Let $\tilde{S}(X,Y)$ be the closure of $\Sigma(\tilde{X} \times \tilde{Y})$. Then obviously

$$\tilde{S}(X,Y) = \mathrm{Cone}(S(X,Y)).$$

Let $\pi : \tilde{X} \times \tilde{Y} \to \tilde{S}(X,Y)$ be the map induced by Σ. By Sard's lemma for algebraic varieties (see [Mum1]), $d\pi|_z$ is surjective for a generic point $v \in \tilde{S}(X,Y)$ and any $z \in \pi^{-1}(v) \cap (\tilde{X} \times \tilde{Y})_{sm}$. This proves the Terracini Lemma:

Lemma 4.5 **([Te])** *Let $X, Y \subset \mathbb{P}(V)$ be irreducible varieties and let*

$$x \in X, \quad y \in Y, \quad z \in \langle x, y \rangle.$$

Then

$$\hat{T}_{z, S(X,Y)} \supset \langle \hat{T}_{x,X}, \hat{T}_{y,Y} \rangle.$$

Moreover, if x, y, z are generic points then the equality holds.

Definition 4.6 The *secant variety* $\operatorname{Sec} X = S(X, X)$ is the join of X with itself, $\operatorname{Sec}(X) = S(X, X)$. *Multisecant varieties* $\operatorname{Sec}^k X$ are defined inductively as

$$\operatorname{Sec}^0 X = X, \quad \operatorname{Sec}^1 X = \operatorname{Sec} X, \ \ldots, \quad \operatorname{Sec}^k X = S(\operatorname{Sec}^{k-1} X, X).$$

In other words, $\operatorname{Sec}^k X$ is the closure of the subvariety of $\mathbb{P}^N$ swept out by the linear spans of general collections of $k+1$ points of X.

Assume that X is non-degenerate. By the Terracini Lemma 4.5, if $\operatorname{Sec}^k X = \operatorname{Sec}^{k+1} X$ for some k, then $\operatorname{Sec}^k X = \mathbb{P}^N$. It follows that there is a strictly ascending filtration

$$X = \operatorname{Sec}^0 X \subset \operatorname{Sec}^1 X \subset \operatorname{Sec}^2 X \subset \ldots \subset \operatorname{Sec}^i X \subset \operatorname{Sec}^{i+1} X = \mathbb{P}^N.$$

$i+2$ is called the *order* of X, denoted by $\operatorname{ord} X$. In other words, $\operatorname{ord} X$ is the minimal integer k such that $\mathbb{P}^N$ is swept out by the linear spans of general collections of k points of X.

By the Terracini Lemma, $(\operatorname{Sec}^k X)^* \supset (\operatorname{Sec}^{k+1} X)^*$ and by the Reflexivity Theorem the following filtration descends strictly:

$$X^* \supset (\operatorname{Sec}^1 X)^* \supset \cdots \supset (\operatorname{Sec}^{\operatorname{ord} X - 2} X)^*.$$

4.2.2 Multisecant Varieties of Homogeneous Spaces

Example 4.7 Let $V = \mathbb{C}^{k+1} \otimes \mathbb{C}^{l+1}$, where $k \geq l$. Let $X \subset \mathbb{P}(V)$ be the projectivization of the variety of rank 1 matrices. Then X is isomorphic to the Segre embedding of $\mathbb{P}^k \times \mathbb{P}^l$. It is easy to see that $\operatorname{Sec}^{n-1} X$ is the projectivization of the variety of matrices of rank at most n. Indeed, if A and B are two matrices and $\operatorname{rk} A = n$, $\operatorname{rk} B = 1$ then $\operatorname{rk}(A+B) \leq n+1$, and for generic A and B we have

$$\operatorname{rk}(A+B) = \min\{n+1, k+1, l+1\}.$$

The degree of $\operatorname{Sec}^n X$ is given by Giambelli's formula [Ful]:

$$\deg \operatorname{Sec}^n(X) = \prod_{i=0}^{l-n-1} \frac{(k+i+1)!}{(l-i)!(k-n+i)!} \cdot 1! \cdot \ldots \cdot (l-n-1)!$$

Since

$$\operatorname{ord}(X) = \min\{k+1, l+1\},$$

if $k = l$ then $\operatorname{ord}(X)$ is equal to $\deg X^*$, i.e. to the degree of the determinant. If $k > l$ then X^* is not a hypersurface and $\deg X^* = \binom{k+1}{l}$.

If an algebraic group G acts on $\mathbb{P}^N$ with finitely many orbits, then for any two orbits $\mathcal{O}_1$ and $\mathcal{O}_2$ the join $S(\overline{\mathcal{O}_1}, \overline{\mathcal{O}_2})$ is G-invariant and therefore is equal to $\overline{\mathcal{O}}$ for some orbit $\mathcal{O}$. This gives an interesting associative product on the set of G-orbits. For example, the following theorem generalizes the previous example.

Let L be a simple algebraic group and P a parabolic subgroup with an abelian unipotent radical. In this case $\mathfrak{l} = \operatorname{Lie} L$ admits a short grading with only three non-zero parts: $\mathfrak{l} = \mathfrak{l}_{-1} \oplus \mathfrak{l}_0 \oplus \mathfrak{l}_1$. Here $\mathfrak{l}_0 \oplus \mathfrak{l}_1 = \operatorname{Lie} P$ and $\exp(\mathfrak{l}_1)$ is the abelian unipotent radical of P. Let $G \subset L$ be a reductive subgroup with Lie algebra $\mathfrak{l}_0$. By Theorem 2.10 G has finitely many orbits in $\mathfrak{l}_1$ naturally labeled by integers from the segment $[0, r]$. Let $X_i = \overline{\mathbb{P}(\mathcal{O}_i)}$, $i = 0, \ldots, r$.

Theorem 4.8 $\operatorname{Sec}^k X_1 = X_{k+1}$.

Proof. We use the notation of Theorem 2.10 and argue by induction on r and k. By induction, $\operatorname{Sec}^k X_1 = S(X_1, X_k)$. Moreover, $\operatorname{Sec}^k X_1$ is obviously a union of G-orbits, and hence $\operatorname{Sec}^k X_1 \supset X_{k+1}$. Suppose that this inclusion is strict, i.e. $\operatorname{Sec}^k X_1 \supset X_{k+2}$.

Let $k+2 < r$. By Theorem 2.18, if $x \in X_1$ then the quotient projective space $\mathbb{P}(\mathfrak{l}_1)/\hat{T}_{x,X_1}$ is isomorphic to $\mathbb{P}(\mathfrak{l}_1')$ and the projection with center $\hat{T}_{x,X_1}$ takes X_i to $X_i' = \mathbb{P}(\mathcal{O}_i')$ for $i < r$. Therefore $\operatorname{Sec}^k X_1' \supset X_{k+2}'$, but this contradicts the induction hypothesis.

Let $k+2 = r$. In this case our assumption means that

$$S(X_1, X_{r-2}) = \mathbb{P}(\mathfrak{l}_1).$$

By the Terracini Lemma 4.5, if $x \in \mathcal{O}_1$ and $y \in \mathcal{O}_{r-2}$ are generic then $T_{x,\mathcal{O}_1} + T_{y,\mathcal{O}_{r-2}} = \mathfrak{l}_1$. By Theorem 2.18, the quotient projective space $\mathbb{P}(\mathfrak{l}_1)/\hat{T}_{x,X_1}$ is isomorphic to $\mathbb{P}(\mathfrak{l}_1')$ and the projection with center $\hat{T}_{x,X_1}$ takes X_{r-2} to X_{r-2}'. It follows that $\hat{T}_{\pi(y),X_{r-2}'} = \mathbb{P}(\mathfrak{l}_1')$, and this is the contradiction. □

Example 4.9 Consider the Plücker embedding

$$X = \operatorname{Gr}(2, \mathbb{C}^n) \subset \mathbb{P}(\Lambda^2\mathbb{C}^n).$$

Then $\operatorname{Sec}^k X$ is the projectivization of the variety of skew-symmetric matrices of rank at most $2k+2$. Therefore

$$\operatorname{ord} X = \left[\frac{n}{2}\right].$$

If n is even then $\operatorname{ord} X = \deg X^* = n/2$ is the degree of the Pfaffian. If n is odd then $\operatorname{codim} X^* = 3$ and

$$\deg X^* = \frac{1}{4}\binom{r+1}{3}$$

(see [HT1]).

Example 4.10 Consider the second Veronese embedding

$$X = \mathbb{P}^{n-1} \subset \mathbb{P}(S^2\mathbb{C}^n).$$

$\operatorname{Sec}^k X$ is the projectivization of the variety of symmetric matrices of rank at most $k+1$. Therefore

$$\operatorname{ord}(X) = n = \deg X^*.$$

Example 4.11 Consider the spinor variety

$$\mathbb{S}_5 \subset \mathbb{P}^{15}.$$

Then $\operatorname{Sec} X = \mathbb{P}^{15}$.

Example 4.12 Let $\mathbb{C}^{27}$ be the exceptional simple Jordan algebra, and let

$$X \subset \mathbb{P}^{26}$$

be the projectivization of the variety of elements of rank one. Then $\operatorname{Sec} X$ is the cubic hypersurface (defined by the norm in the Jordan algebra) and $\operatorname{Sec}^2 X = \mathbb{P}^{26}$. Notice that

$$\operatorname{ord}(X) = 3$$

is again equal to $\deg X^*$.

4.2.3 $\deg X^*$ and $\operatorname{ord} X$

The previous section gives many examples of varieties with $\deg X^* = \operatorname{ord} X$. Here we study the relation between these invariants more closely.

Definition 4.13 Let $Z \subset \mathbb{P}^N$ be a projective variety, and let $z \in Z$. We denote by $\operatorname{mult}_z Z$ the *multiplicity* of Z at z. If L is a general $(\operatorname{codim} Z)$-dimensional subspace of $\mathbb{P}^N$ passing through z, then L meets Z at z and $\deg Z - \operatorname{mult}_z Z$ other points. Furthermore, if U is a small neighborhood of z in Z and L' is a general $(\operatorname{codim} Z)$-dimensional subspace of $\mathbb{P}^N$ sufficiently close to L, then L' meets U in $\operatorname{mult}_z Z$ nonsingular points. We denote by $\operatorname{Sing}^k Z$ those points of Z with multiplicity at least $k+1$.

Multiplicity gives the stratification

$$Z = \operatorname{Sing}^0 Z \supset \operatorname{Sing}^1 Z \supset \cdots.$$

Theorem 4.14 **([Z7])** $(\operatorname{Sec}^k X)^* \subset \operatorname{Sing}^k X^*$.

Proof. We argue by induction. Assume for simplicity that X^* is a hypersurface, the general case being similar. We will show that, for a generic point $H_k \in (\operatorname{Sec}^{k-1} X)^*$ and its arbitrary neighbourhood U in X^*, a general line close to H_k intersects U in at least k points.

Let $(x_1, \ldots, x_k)$ be a general collection of k points of X, let

$$P_{x_i} = \{H \,|\, H \supset \hat{T}_{x_i,X}\} \subset X^*, \quad i = 1, \dots, k,$$

and let

$$H_k \in P_{x_1} \cap \dots \cap P_{x_k}.$$

Then $H_k \in (\operatorname{Sec}^{k-1} X)^*$ by the Terracini Lemma. It suffices to show that $\operatorname{mult}_{H_k} X^* \geq k$. By induction, for a general point

$$H_{k-1} \in P_{x_1} \cap \dots \cap P_{x_{k-1}}$$

in a neighborhood of H_k we know that $\operatorname{mult}_{H_{k-1}} X^* \geq k-1$.

Now we join H_{k-1} to a general point $H \in P_{x_k}$ close to H_k. This line meets X^* in finitely many points, since otherwise X^* is the cone over H_{k-1} and X is degenerate. Perturbing this line a little we see that

$$\operatorname{mult}_{H_k} X^* \geq \operatorname{mult}_{H_{k-1}} X^* + \operatorname{mult}_H X^* \geq k.$$

□

Theorem 4.15 ([**Z7**])

(i) $\deg X \geq \operatorname{codim} X + 1$.
(ii) $\deg X^* \geq \operatorname{ord} X + \operatorname{def} X$.

Proof. (i) We argue by induction on $\operatorname{codim} X$. Let $x \in X$ be a smooth point and let $\pi : \mathbb{P}^N \dashrightarrow \mathbb{P}^{N-1}$ be the projection with center x. Then

$$\deg X = \deg \pi(X) + 1 \geq \operatorname{codim} \pi(X) + 2 = \operatorname{codim} X + 1.$$

(ii) We are going to show that

$$\deg X^* \geq k + \operatorname{def} X + 1, \tag{4.1}$$

where k is the minimal integer such that $\operatorname{Sing}^k X^* = \varnothing$. Then the claim will follow from Theorem 4.14. If $\operatorname{def} X = 0$, then there is nothing to prove. Suppose that $\operatorname{def} X > 0$, and let $H \in \operatorname{Sing}^{k-1} X^*$ be a general point. Let $\pi : (\mathbb{P}^N)^\vee \dashrightarrow (\mathbb{P}^{N-1})^\vee$ be the projection with center at H, and let $X^{*\prime} = \pi(X^*)$. Since X is nondegenerate, X^* is not a cone, and so $\dim X^{*\prime} = \dim X^*$ and

$$\deg X^{*\prime} = \deg X^* - \operatorname{mult}_H X^* = \deg X^* - k.$$

Moreover,

$$\operatorname{codim}_{(\mathbb{P}^{N-1})^\vee} X^{*\prime} = \operatorname{codim}_{(\mathbb{P}^N)^\vee} X^* - 1 = \operatorname{def} X,$$

and so $\deg X^{*\prime} \geq \operatorname{def} X + 1$ by (i). Thus we have (4.1). □

It is conjectured that the inequality in this Theorem becomes an equality if and only if X is a variety of rank 1 elements in a simple Jordan algebra.

4.2.4 Waring Problem for Forms

We finish this section with an overview of several results about multisecant varieties of Veronese embeddings. Let

$$X_{d,n} = v_d(\mathbb{P}^n) \subset \mathbb{P}^N$$

be the image of the d-fold Veronese embedding, where $N = \binom{n+d}{d} - 1$. Multisecant varieties of $X_{d,n}$ were studied since the middle of the 19-th century most notably due to the following observation. If $l_1, \ldots, l_s$ are linear forms then the sum of powers

$$f = l_1^d + \ldots + l_s^d$$

belongs to the cone over $\operatorname{Sec}^{s-1} X_{d,n}$ and the general point of $\operatorname{Sec}^{s-1} X_{d,n}$ has this form. For instance, if we want to find the order of $X_{d,n}$, we have to solve the following problem: to find the minimal s such that the general form of degree d in $n+1$ variables can be decomposed as a sum of s powers of linear forms.

Theorem 4.16 ([AH]) $\operatorname{ord} X_{d,n} = \left\lceil \frac{1}{n+1}\binom{n+d}{d} \right\rceil$ *except*

- $d = 2$, *where* $\operatorname{ord} X_{d,n} = n+1$ *instead of* $\lceil \frac{n+2}{2} \rceil$;
- $d = 3$ *and* $n = 4$, *where* $\operatorname{ord} X_{d,n} = 8$ *instead of* 7;
- $d = 4$ *and* $n = 2$, *where* $\operatorname{ord} X_{d,n} = 6$ *instead of* 5;
- $d = 4$ *and* $n = 3$, *where* $\operatorname{ord} X_{d,n} = 10$ *instead of* 9;
- $d = 4$ *and* $n = 4$, *where* $\operatorname{ord} X_{d,n} = 15$ *instead of* 14;

All exceptions were classically known [C, Re1, Re2, Re3, Ric, Pa, Dix].

Suppose now that $s = \frac{1}{n+1}\binom{n+d}{d}$ is an integer and $s = \operatorname{ord} X_{d,n}$. Then a generic form of degree d in $n+1$ variables has finitely many presentations as a sum of s powers of linear forms. In several classical cases this presentation is unique:

- $n = 1$, $d = 2k-1$, $s = k$, see [Syl1].
- $n = 2$, $d = 5$, $s = 7$, see [H, Ric, Pa].
- $n = 3$, $d = 3$, $s = 5$, see [Syl1, Re1, Re2, Re3].

It is not known if there are any further examples.

Let us explain these results in the most simple case of binary forms, i.e. when $n = 1$. The reader may wish to look at [RaS] for further results and references.

Theorem 4.17 *Let f be a binary form of degree d in variables x, y. Then*

$$f = \lambda_1(a_1x + b_1y)^d + \ldots + \lambda_s(a_sx + b_sy)^d,$$

where linear forms $a_ix + b_iy$ are distinct and non-trivial, if and only if

$$D \cdot f = 0, \text{ where } D = \left(b_1\frac{\partial}{\partial x} - a_1\frac{\partial}{\partial y}\right)\cdots\left(b_s\frac{\partial}{\partial x} - a_s\frac{\partial}{\partial y}\right).$$

Proof. Since

$$\left(b_i\frac{\partial}{\partial x} - a_i\frac{\partial}{\partial y}\right) \cdot (a_ix + b_iy)^d = 0,$$

the first part of the theorem is clear. Let S_i denote the vector space of binary forms of degree i, where we set $S_i = 0$ for $i < 0$. Then $D(S_d) \subset S_{d-s}$ and it is easy to see that actually $D(S_d) = S_{d-s}$ (for instance, use induction on s). Therefore $\dim \operatorname{Ker} D|_{S_d} = s'$, where $s' = \min\{d+1, s\}$. Since the forms

$$(a_1x + b_1y)^d, \ \ldots, \ (a_{s'}x + b_{s'}y)^d$$

are linearly independent (consider the Vandermond determinant), they span $\operatorname{Ker} D|_{S_d}$ and everything is proved. □

Let S_i' denote the vector space of differential operators of degree i with constant coefficients. Then any $f \in S_d$ induces a linear map

$$A_f : S_s' \to S_{d-s}, \quad D \mapsto D \cdot f.$$

Example 4.18 Let $d = 2k$, $s = k+1$. Then $\dim S_s' = k+2$ and $\dim S_{d-s} = k$. It can be verified that $\dim \operatorname{Ker} A_f = 2$ for generic f, and generic differential operators in $\operatorname{Ker} A_f$ have no multiple roots. Therefore general forms of degree $2k$ have $\mathbb{P}^1$ presentations as a sum of $k+1$ powers of linear forms.

Example 4.19 Let $d = 2k$, $s = k$. Then $\dim S_s' = k+1$ and $\dim S_{d-s} = k+1$. Then $\det A_f = 0$ is an irreducible hypersurface, and its generic points are exactly forms that admit the presentation as a sum of k powers of linear forms. Therefore $\operatorname{Sec}^{k-1} X_{2k,1}$ is a hypersurface of degree $k+1$.

Example 4.20 Let $d = 2k+1$, $s = k+1$. Then $\dim S_s' = k+2$ and $\dim S_{d-s} = k+1$. It can be verified that $\dim \operatorname{Ker} A_f = 1$ for generic f, and $\operatorname{Ker} A_f$ is spanned by a differential operator without multiple roots. Therefore general forms of degree $2k+1$ have a unique presentation as a sum of $k+1$ powers of linear forms.

Example 4.21 Let $d = 2k+1$, $s = k$. Then $\dim S_s' = k+1$ and $\dim S_{d-s} = k+2$. It can be verified that

$$\{f \mid \operatorname{rk} A_f < k+1\}$$

is an irreducible subvariety of codimension 2 and of degree $(k+2)(k+1)/2$; cf. Example 4.7. Its generic points are exactly forms that admit the presentation as a sum of k powers of linear forms. Therefore $\operatorname{Sec}^{k-1} X_{2k+1,1}$ has codimension 2 and degree $(k+2)(k+1)/2$.

4.3 Zak Theorems

4.3.1 Theorem on Tangencies

Definition 4.22 Suppose that $X \subset \mathbb{P}^N$ is a smooth n-dimensional projective variety. The *tangential variety* $\operatorname{Tan} X$ is the union of all (embedded) tangent spaces to X

Both $\operatorname{Tan} X$ and $\operatorname{Sec} X$ are closed irreducible subvarieties of $\mathbb{P}^N$, with $\operatorname{Tan} X \subset \operatorname{Sec} X$, and

$$\dim \operatorname{Tan} X \leq 2n, \quad \dim \operatorname{Sec} X \leq 2n+1.$$

Theorem 4.23 (**[FH]**, **[Z2]**) *Suppose that $X \subset \mathbb{P}^N$ is a smooth projective variety. Then either*

$$\dim \operatorname{Tan} X = 2n, \quad \dim \operatorname{Sec} X = 2n+1$$

or

$$\operatorname{Tan} X = \operatorname{Sec} X.$$

This theorem has the following generalization. Suppose that $Y \subset X \subset \mathbb{P}^N$ are arbitrary irreducible projective varieties. The *relative tangential variety* $\operatorname{Tan}(Y, X)$ is the union of tangent stars $T^\star_y X$ for $y \in Y$, where the *tangent star* $T^\star_x X$ for $x \in X$ is the union of limit positions of secants $\langle x', x'' \rangle$, where $x', x'' \in X$ and $x', x'' \to x$.

Theorem 4.24 (**[Z2]**) *Suppose that $Y \subset X \subset \mathbb{P}^N$ are irreducible projective varieties. Then either*

$$\dim \operatorname{Tan}(Y, X) = \dim X + \dim Y, \quad \dim S(Y, X) = \dim X + \dim Y + 1$$

or

$$\operatorname{Tan}(Y, X) = S(Y, X).$$

The following theorem is called the Zak Theorem on tangencies in the literature.

Theorem 4.25 (**[Z2]**, **[FL]**)

(i) *Suppose that $X \subset \mathbb{P}^N$ is a smooth nondegenerate projective variety with $\dim X = n$. If L is a k-plane in $\mathbb{P}^N$ and $k \geq n$ then*

$$\dim (L \cap X)_{sing} \leq k - n.$$

(ii) *If $X \subset \mathbb{P}^N$ is a non-linear smooth projective variety then* $\dim X^* \geq \dim X$. *If X^* is smooth, then* $\dim X = \dim X^*$.

Proof. (i) Indeed, let $Y = (L \cap X)_{sing}$. Then $x \in Y$ if and only if $\hat{T}_{x,X} \subset L$. Therefore $\operatorname{Tan}(Y, X) \subset L$. In particular, $\dim \operatorname{Tan}(Y, X) \leq k$. On the other hand, since X is non-degenerate, $X \not\subset L$. Therefore $S(Y, X) \not\subset L$. In particular, $\operatorname{Tan}(Y, X) \neq S(Y, X)$. By Theorem 4.24 it follows that $\dim \operatorname{Tan}(Y, X) = \dim Y + n$. It follows that $\dim Y \leq k - n$.

(ii) Let $H \subset (X^*)_{sm}$. Then $(X \cap H)_{sing}$ is the projective subspace of dimension $\operatorname{def} X$. Therefore, by (i) we have

$$\operatorname{def} X \leq \dim H - \dim X = \operatorname{codim} X - 1.$$

It follows that $\dim X \leq \dim X^*$. If X^* is also smooth then, by the Reflexivity Theorem and the same argument as above, $\dim X^* \leq \dim X$, and therefore $\dim X = \dim X^*$. □

Next we prove that if X is smooth then X is a normalization of its Gauss image $\gamma(X)$:

Proof of Theorem 4.2. Indeed, it follows from Theorem 4.25 that γ_n is finite, i.e. any n-dimensional projective subspace L is tangent to X at finitely many points. On the other hand, Theorem 1.18 implies that a generic tangent space L is tangent to X along a projective subspace. □

The following theorem generalizes Theorem 4.25 (i) to the singular case.

Theorem 4.26 ([Z2]) *Suppose that $X \subset \mathbb{P}^N$ is a nondegenerate projective variety with Gauss images $\gamma_k(X)$. If $L \subset \gamma_k(X)$ then*

$$\dim\{x \in X \,|\, \hat{T}_{x,X} \subset L\} \leq k - n + (\dim X_{sing} + 1),$$

where we set $\dim X_{sing} = -1$ if X is smooth.

4.3.2 Theorem on Linear Normality

Theorem 4.27 ([Z4]) *Suppose that $X \subset \mathbb{P}^N$ is a smooth non-degenerate projective variety with $\dim X = n$. If*

$$\operatorname{codim} X < \frac{N+4}{3}$$

then $\operatorname{Sec} X = \mathbb{P}^N$.

Proof. By Theorem 4.23, either $\operatorname{Tan} X = \operatorname{Sec} X$ or $\dim \operatorname{Sec} X = 2n+1$. Suppose that $\operatorname{Sec} X \neq \mathbb{P}^N$ and $\dim \operatorname{Sec} X = 2n+1$. Then $2n+1 < N$, which contradicts $\operatorname{codim} X < \frac{N+4}{3}$.

Suppose now that $\operatorname{Tan} X = \operatorname{Sec} X$ and $\operatorname{Sec} X \neq \mathbb{P}^N$. Let $z \in \operatorname{Tan} X$ be a generic point and

$$Q = \{x \in X \,|\, z \in \hat{T}_{x,X}\}.$$

Then Q is closed and an easy dimension count shows that

$$\dim Q = 2n - \dim \operatorname{Tan} X.$$

For any point $x \in Q$ we have

$$z \in \hat{T}_{x,X} \subset \operatorname{Tan} X$$

and therefore

$$\hat{T}_{x,X} \subset \hat{T}_{z,\operatorname{Tan} X}.$$

It follows that $\operatorname{Tan}(Q,X) \subset \hat{T}_{z,\operatorname{Sec}(X)}$. However, $S(Q,X) \not\subset \hat{T}_{z,\operatorname{Tan} X}$ because $\hat{T}_{z,\operatorname{Tan} X}$ is a proper subspace and X is not degenerate. By Theorem 4.24 we have

$$\dim S(Q,X) = \dim Q + \dim X + 1 = 3n - \dim \operatorname{Sec} X + 1.$$

Since $S(Q,X) \subset \operatorname{Sec} X$, we have $\dim S(Q,X) \le \dim \operatorname{Sec} X$. Then

$$2 \dim \operatorname{Sec} X \ge 3n + 1$$

Hence $2N > 3n + 1$. But this contradicts $\operatorname{codim} X < \frac{N+4}{3}$. □

Theorem 4.27 is called the theorem on linear normality due to the following corollary first conjectured in [Ha2].

Corollary 4.28 *Suppose that $X \subset \mathbb{P}^N$ is a smooth non-degenerate projective variety. If*

$$\operatorname{codim} X < \frac{N+2}{3}$$

then X is linearly normal.

Proof. Recall from Sect. 1.4 that if X is not linearly normal then there exists a subvariety $X' \subset \mathbb{P}^{N+1}$ and a point $p \in \mathbb{P}^{N+1}$ such that X is isomorphic to X' via the linear projection with center p. In particular, $\operatorname{Sec} X' \ne \mathbb{P}^{N+1}$. But this contradicts Theorem 4.27. □

4.3.3 Theorem on Severi Varieties

Zak [Z2] has also classified the varieties in the borderline case

$$\operatorname{codim} X = \frac{N+4}{3}.$$

Theorem 4.29 ([Z4]) *Suppose that $X \subset \mathbb{P}^N$ is a smooth non-degenerate n-dimensional projective variety such that $\operatorname{codim} X = \frac{N+4}{3}$. Then $\operatorname{Sec}(X) = \mathbb{P}^N$ except for the 4 following cases:*

- $n = 2$, $X = \mathbb{P}^2$, $X \subset \mathbb{P}^5$ *is the Veronese embedding.*
- $n = 4$, $X = \mathbb{P}^2 \times \mathbb{P}^2$, $X \subset \mathbb{P}^8$ *is the Segre embedding.*
- $n = 8$, $X = \operatorname{Gr}(2, \mathbb{C}^6)$, $X \subset \mathbb{P}^{14}$ *is the Plücker embedding.*
- $n = 16$, $X \subset \mathbb{P}^{26}$ *is the projectivization of the highest weight vector orbit in the 27-dimensional irreducible representation of E_6.*

The varieties listed in Theorem 4.29 are called Severi varieties after Severi who proved that the unique 2-dimensional Severi variety is the Veronese surface. Scorza and Fujita–Roberts [FR] have shown that the unique 4-dimensional Severi variety is the Segre embedding of $\mathbb{P}^2 \times \mathbb{P}^2$. It is also shown in [FR] that $n \equiv 0 \mod 16$ when $n > 8$. Tango [Tan] proved that if $n > 16$ then either $n = 2^a$ or $n = 3 \cdot 2^b$, where $a \ge 7$ and $b \ge 5$. The 16-dimensional Severi variety was constructed by Lazarsfeld [La]. All Severi varieties (1)–(4) arise from Example 4.8. Another description of Severi varieties can be found in Example 2.17.

4.3.4 Connectedness Theorem of Fulton and Hansen

The rest part of this section is occupied by the proof of Theorem 4.24. We are going to use the Connectedness Theorem of Fulton and Hansen:

Theorem 4.30 ([FH]) *Let $f : X \to \mathbb{P}^N$ be any morphism, where X is irreducible and projective. Then $f^{-1}(L_0)$ is connected for any projective subspace $L_0 \subset \mathbb{P}^N$ such that* $\dim f(X) > \operatorname{codim}_{\mathbb{P}^N} L_0$.

Proof. We may assume that f is finite. Indeed, for a non-finite f, let $f = g \circ h$ be the Stein factorization (see e.g. [Hal]), i.e. $h : X \to X'$ has connected fibers, and $g : X' \to \mathbb{P}^N$ is finite. Then $\dim X' = \dim f(X)$ and if $g^{-1}(L_0)$ is connected then $f^{-1}(L_0)$ is also connected.

Let $k = \dim L_0$. Let $U \subset \mathrm{Gr}(k, \mathbb{P}^N)$ be the open set of subspaces L such that $\operatorname{codim}_{f(X)} f(X) \cap L = \operatorname{codim}_{\mathbb{P}^N} L$, and let $U_0 \subset \mathrm{Gr}(k, \mathbb{P}^N)$ be the set of all subspaces L such that $f^{-1}(L)$ is irreducible. U_0 is non-empty by Bertini's theorem (see [Hal] or [Wei]). It follows from Zariski's principle of degeneration (see [Ful] or [Za]) that $f^{-1}(L)$ is connected, non-empty, of codimension $\dim X + k - N$ for all $L \in U$. For each $L \in U$, there is a positive cycle $f^*(L)$ on X whose support is $f^{-1}(L)$. This determines a morphism from U to the Chow variety Z of cycles on X. Let

$$\Gamma \subset \mathrm{Gr}(k, \mathbb{P}^N) \times Z$$

be the closure of the graph of this morphism, $\pi : \Gamma \to \mathrm{Gr}(k, \mathbb{P}^N)$ the projection. Since Z is projective, π is proper as well as birational. Zariski's principle of degeneration implies that for any $\gamma \in \Gamma$ the corresponding cycle C_γ is connected (i.e. its support $|C_\gamma|$ is connected). Zariski's main theorem implies that $\pi^{-1}(L_0)$ is connected. It follows that the union of $|C_\gamma|$ for $\gamma \in \pi^{-1}(L_0)$ is connected. It remains to prove that

$$f^{-1}(L_0) = \bigcup_{\gamma \in \pi^{-1}(L_0)} |C_\gamma|.$$

The inclusion $\supset$ follows from the continuity of limit cycles. So it remains to prove the inclusion $\subset$. Let $x \in f^{-1}(L_0)$ and let $z = f(x)$. Consider the subvariety G of $\mathrm{Gr}(k, \mathbb{P}^N)$ consisting of all subspaces passing through z (therefore G is isomorphic to $\mathrm{Gr}(k-1, \mathbb{P}^{N-1})$). An easy dimension count shows that G meets U. Let $\Gamma' \subset \Gamma$ be an irreducible subvariety that maps birationally to G. For each $L \subset G \cap U$, the cycle $f^*(L)$ contains x, and therefore for any $\gamma \in \Gamma'$ the cycle C_γ contains x. In particular, there exists the limit cycle C_γ for $\gamma \in \Gamma' \cap \pi^{-1}(L_0)$ that contains x. □

Now we derive several corollaries.

Theorem 4.31 ([FH]) *Let $\Delta \subset \mathbb{P}^N \times \mathbb{P}^N$ be the diagonal. If X is an irreducible variety, $f : X \to \mathbb{P}^N \times \mathbb{P}^N$ a morphism, and* $\dim f(X) > N$ *then $f^{-1}(\Delta)$ is connected.*

Proof. Suppose that f is a finite morphism, a general case is handled using the Stein factorization. Assume that $f^{-1}(\Delta)$ is not connected and choose points x and x' in its different connected components. Choose a hyperplane H in $\mathbb{P}^N$ not containing any of the factors of $f(x)$ or $f(x')$. Let $\mathbb{P}_0$ be the open set of $\mathbb{P}^N \times \mathbb{P}^N$ consisting of those points none of whose factors lie in H. Then $\mathbb{P}_0 = \mathbb{C}^{2N} = \mathbb{P}^{2N} \setminus \{z_0 = 0\}$, where $z_0, \dots, z_{2N}$ are the homogeneous coordinates on $\mathbb{P}^{2N}$. Let $W \subset \mathbb{P}^N \times \mathbb{P}^N \times \mathbb{P}^{2N}$ be the closure of the graph of this birational correspondence. If coordinates are chosen on $\mathbb{P}^N$ such that H is the hyperplane $x_0 = 0$, and the homogeneous coordinates on the k-th copy of $\mathbb{P}^N$ are $x_0^k, \dots, x_N^k$, then a point $(x^1) \times (x^2) \times (z) \in \mathbb{P}^N \times \mathbb{P}^N \times \mathbb{P}^{2N}$ belongs to W if and only if there are constants λ_1, λ_2 such that

$$(z_0, z_1, \dots, z_N) = \lambda_1(x_0^1, \dots, x_N^1),$$
$$(z_0, z_{N+1}, \dots, z_{2N}) = \lambda_2(x_0^2, \dots, x_N^2).$$

Hence W is defined by the equations

$$\begin{cases} x_j^1 z_0 = x_0^1 z_j & 1 \le j \le N \\ x_j^2 z_0 = x_0^2 z_{N+j} & 1 \le j \le N \\ x_i^1 z_j = x_j^1 z_i & 1 \le i < j \le N \\ x_i^2 z_{N+j} = x_j^2 z_{N+i} & 1 \le i < j \le N. \end{cases}$$

Let $\alpha : W \to \mathbb{P}^N \times \mathbb{P}^N$ and $\beta : W \to \mathbb{P}^{2N}$ be the two birational projections. Let L be the N-dimensional linear subspace of $\mathbb{P}^{2N}$ defined by the equations

$$z_j = z_{N+j}, \quad 1 \le j \le N.$$

Then it is easy to check that

$$\alpha^{-1}(\Delta) \supset \beta^{-1}(L) \text{ and } \alpha^{-1}(\Delta \cap \mathbb{P}_0) \subset \beta^{-1}(L). \tag{4.2}$$

Let $\tilde{X}$ be the irreducible component of the fiber product $X \times_{\mathbb{P}^N \times \mathbb{P}^N} W$ that maps onto X. Then there is a diagram

$$\begin{array}{ccccc} \tilde{X} & \xrightarrow{g} & W & \xrightarrow{\beta} & \mathbb{P}^{2N} \\ {\scriptstyle \pi}\downarrow & & \downarrow{\scriptstyle \alpha} & & \\ X & \xrightarrow[f]{} & \mathbb{P}^N \times \mathbb{P}^N & & \end{array}$$

where the square commutes, and π is surjective. Then equations (4.2) imply that

$$f^{-1}(\Delta) \supset \pi(g^{-1}\beta^{-1}(L)), \text{ and } f^{-1}(\Delta \cap \mathbb{P}_0) \subset \pi(g^{-1}\beta^{-1}(L)). \tag{4.3}$$

Now $\dim \beta g(\tilde{X}) = \dim f(X)$ since $f(X)$ meets the open set $\mathbb{P}_0$ which corresponds isomorphically to an open set in $\mathbb{P}^{2N}$. Then by Theorem 4.30 we know that $(\beta g)^{-1}(L)$ is connected. Therefore $\pi((\beta g)^{-1}(L))$ is connected. But by (4.3) $\pi((\beta g)^{-1}(L))$ is then a connected subset of $f^{-1}(\Delta)$ which contains both x and x'. This is a contradiction. □

Corollary 4.32 ([FH])

(i) *Let Y be an irreducible subvariety of the projective space $\mathbb{P}^N$ and let $f : X \to \mathbb{P}^N$ be a morphism from an irreducible variety X to $\mathbb{P}^N$ with* $\dim f(X) > \operatorname{codim}_{\mathbb{P}^N} Y$. *Then $f^{-1}(Y)$ is connected.*
(ii) *If X and Y are irreducible subvarieties of $\mathbb{P}^N$ with* $\dim X + \dim Y > N$ *then $X \cap Y$ is connected.*

Proof. (i) We apply Theorem 4.31 for the map $F : X \times Y \to \mathbb{P}^N \times \mathbb{P}^N$ given by $F = (f, \mathrm{Id})$.
(ii) Follows from (i). □

Corollary 4.33 ([FH])

(i) *Let X be an irreducible variety of dimension n, and $f : X \to \mathbb{P}^N$ an unramified morphism, with $N < 2n$. Then f is a closed embedding.*
(ii) *Let X be an irreducible n-dimensional subvariety of $\mathbb{P}^N$ with $N < 2n$. Then the algebraic fundamental group of X is trivial (i.e. X has no non-trivial étale coverings).*

Proof. (i) If $f : X \to \mathbb{P}^N$ is unramified, consider the product mapping $f \times f : X \times X \to \mathbb{P}^N \times \mathbb{P}^N$. The unramified assumption is equivalent to the assertion that the diagonal Δ_X in $X \times X$ is open as well as closed in $(f \times f)^{-1}(\Delta_{\mathbb{P}^N})$. Since f is automatically finite, $\dim(f \times f)(X \times X) = 2n$, and so $(f \times f)^{-1}(\Delta_{\mathbb{P}^N})$ is connected by Theorem 4.31. Hence

$$\Delta_X = (f \times f)^{-1}(\Delta_{\mathbb{P}^N}),$$

implying that f is one-to-one and is therefore a closed embedding.
(ii) Follows from (i). □

Proof of Theorem 4.24. If $\dim \mathrm{Tan}(Y, X) = \dim Y + \dim X$ then the Theorem is obvious because $S(X, Y)$ is irreducible. So assume that

$$\dim \mathrm{Tan}(Y, X) < \dim Y + \dim X.$$

Assume also that

$$\mathrm{Tan}(Y, X) \neq S(X, Y).$$

Let L be a generic subspace of $\mathbb{P}^N$ such that

$$\dim L + \dim \mathrm{Tan}(Y, X) = N - 1, \quad L \cap \mathrm{Tan}(Y, X) = \emptyset.$$

But L meets $S(X, Y)$ at a point

$$z_0 \in \langle x_0, y_0 \rangle, \quad x_0 \in X, \ y_0 \in Y, \quad x_0 \neq y_0.$$

Let $\pi : \mathbb{P}^N \to \mathbb{P}^{N'}$ be the projection with center L, where $N' = \dim \mathrm{Tan}(Y, X)$. Let $X' = \pi(X)$, $Y' = \pi(Y)$. Since $X \cap L = \emptyset$, $\pi|_X$ is finite and therefore

$\dim X' + \dim Y' = \dim X + \dim Y > N'$. Hence by Theorem 4.31, $Y \times_{X'} X$ is connected. We claim that $Y \times_{X'} X$ is supported on $Y \times Y$. This will be a contradiction because clearly $(y_0, x_0) \in Y \times_{X'} X$. Suppose that $Y \times_{X'} X$ is not supported on $Y \times Y$. Then for any $(y, x) \in (Y \times_{X'} X) \setminus (Y \times Y)$, the secant $\langle x, y \rangle$ intersects L. Therefore, for each point

$$(y, y) \in (Y \times Y) \cap \overline{(Y \times_{X'} X) \setminus (Y \times Y)}$$

(the set of these points is non-empty by connectedness!), the embedded tangent star at y intersects L, which contradicts $L \cap \mathrm{Tan}(Y, X) = \emptyset$. □

4.4 Chow Forms

Let $X \subset \mathbb{P}(V)$ be an irreducible subvariety and $\mathbb{P}^l = \mathbb{P}(U)$. By Theorem 3.5, $(X \times \mathbb{P}^l)^*$ is a hypersurface in $\mathbb{P}(V \otimes U)^\vee$ for $\operatorname{def} X \leq l \leq \dim X$. Since we always have $\operatorname{def} X \leq \dim X$, this gives a system of *associated hypersurfaces.*

We can identify $\mathbb{P}(V \otimes U)^\vee$ with the projectivization of the space of linear maps $\mathrm{Hom}(U, V^\vee)$. We abuse notation and denote a non-zero linear map and its projectivization by the same letter. Let $\mathbb{P}(V \otimes U)^\vee_0$ be an open subset consisting of operators having maximal possible rank $l + 1$. For any $f \in \mathbb{P}(V \otimes U)^\vee_0$ we can associate two projective subspaces

$$U(f) = \mathbb{P}(\mathrm{Im}(f)) \subset \mathbb{P}(V)^\vee$$

and the orthogonal subspace $U(f)^\perp \subset \mathbb{P}(V)$. Recall that $I^0_X \subset \mathbb{P}(V) \times \mathbb{P}(V)^\vee$ is an open part of the conormal variety consisting of pairs (x, H) such that $x \in X_{sm}$ and H is the hyperplane tangent to X at x.

Theorem 4.34 ([WZ]) $(X \times \mathbb{P}^l)^*$ *is the closure of the set of points* $f \in \mathbb{P}(V \otimes U)^\vee_0$ *such that*

$$\left(U(f)^\perp \times U(f)\right) \cap I^0_X \neq \emptyset.$$

In two extreme cases $l = \dim X$ or $l = \operatorname{def} X$ this result can be made slightly more precise. Consider the set $\mathcal{Z}(X)$ of all $(N - n - 1)$-dimensional projective subspaces in $\mathbb{P}^N$ that intersect X. This is a subvariety in the Grassmannian $\mathrm{Gr}(N - n - 1, \mathbb{P}^N)$.

Theorem 4.35 $\mathcal{Z}(X)$ *is an irreducible hypersurface.*

Proof. Let $B(X) \subset X \times \mathcal{Z}(X)$ be an incidence variety consisting of pairs (x, L) such that $x \in X$, $L \in \mathrm{Gr}(N - n - 1, \mathbb{P}^N)$, and $x \in L$. Consider projections $p : B(X) \to X$ and $q : B(X) \to \mathcal{Z}(X)$. Then p is a Grassmannian fibration, and any fiber of p is isomorphic to $\mathrm{Gr}(N - n - 2, \mathbb{P}^{N-1})$. In particular, $B(X)$ (and hence $\mathcal{Z}(X)$) is irreducible. On the other hand, q is birational, because a generic $(N-n-1)$-dimensional projective subspace intersecting X meets X in exactly one point. Now an easy dimension count shows that $\mathcal{Z}(X)$ is a hypersurface. □

Definition 4.36 *$\mathcal{Z}(X)$ is called the Chow form of X.*

Now we can relate Chow forms and dual varieties. We continue to use the notation of Theorem 4.34. The following theorem is called a *Cayley trick.*

Theorem 4.37 ([WZ]) *Let $n = \dim X$ and $n^* = \dim X^*$.*

Let $l = \dim X$, and $p_1 : \mathbb{P}(V \otimes U)_0^\vee \to \mathrm{Gr}(N-n-1, \mathbb{P}^N)$ be the map $f \mapsto U(f)^\perp$. Then

$$(X \times \mathbb{P}^l)^* = \overline{p_1^{-1}(\mathcal{Z}(X))}.$$

Let $l = \operatorname{def} X$, and $p_2 : \mathbb{P}(V \otimes U)_0^\vee \to \mathrm{Gr}(N-n^*-1, \mathbb{P}^N)$ be the map $f \mapsto U(f)$. Then

$$(X \times \mathbb{P}^l)^* = \overline{p_2^{-1}(\mathcal{Z}(X^*))}.$$

We refer the reader to [GKZ2] for further results on Chow forms.

5

Vector Bundles Methods

The dual variety is the image of the conormal variety which is the projectivized conormal bundle if the variety is smooth. In this chapter we exploit this relation between duality and vector bundles. In Sect. 5.1 we prove a theorem of Holme and Ein about the defect of a smooth effective very ample divisor. We deduce this result from a theorem of Munoz about the dimension of the linear span of a tangential variety. We discuss the related notion of projective extendability. In Sect. 5.2 we apply Hartshorne's ample vector bundles to prove a theorem of Ein that a dual variety of a smooth complete intersection is a hypersurface. We introduce resultants and prove a classical theorem that they are well-defined. We also explain the Cayley trick for resultants. In the last Sect. 5.3 we describe the "Cayley method" developed by Gelfand, Kapranov, and Zelevinsky. The idea is to show that the dual variety is represented in the derived category by Koszul complexes of jet bundles. The discriminant is then equal to the "Cayley determinant" of a generically exact complex. As an application we deduce some classical formulas for discriminants and their degrees.

5.1 Dual Varieties of Smooth Divisors

5.1.1 Linear Envelope of a Tangential Variety

Here we study the relation of a dual variety of a projective variety to a dual variety of its hyperplane section. We start with an auxilliary result.

Definition 5.1 Let $X \subset \mathbb{P}^N = \mathbb{P}(V)$ be a smooth projective variety. Let $\gamma_X : X \to \mathrm{Gr}(n, \mathbb{P}^N)$, $n = \dim X$, be the Gauss map, let $\mathcal{S} \subset \mathcal{O} \otimes V$ be the universal bundle on $\mathrm{Gr}(n, \mathbb{P}^N)$, and let $G_X = \gamma^*(\mathcal{S})$. The fiber of G_X over $x \in X$ is equal to the tangent space to $\mathrm{Cone}\, X$ along the generatrix corresponding to x.

Theorem 5.2 ([Mun3]) *Let $X \subset \mathbb{P}^N$ be a smooth projective variety and $Y \subset X$ a connected subvariety of X. Then*

$$\dim\langle \mathrm{Tan}(Y,X)\rangle = N - h^0(N^\vee_{X,\mathbb{P}^N}(1)|_Y).$$

Proof. The obvious exact sequences

$$0 \to \mathcal{O}_{\mathbb{P}^N}(-1) \to \mathcal{O}_{\mathbb{P}^N} \otimes V \to T_{\mathbb{P}^N}(-1) \to 0,$$

$$0 \to \mathcal{O}_X(-1) \to G_X \to T_X(-1) \to 0$$

give rise to the exact sequence

$$0 \to G_X \to \mathcal{O}_X \otimes V \to N_{X,\mathbb{P}^N}(-1) \to 0.$$

Let $\langle \mathrm{Tan}(Y,X)\rangle = \mathbb{P}(W) \subset \mathbb{P}(V)$. Then there is a commutative diagram

$$\begin{array}{ccccccccc}
 & & & & 0 & & 0 & & \\
 & & & & \downarrow & & \downarrow & & \\
0 & \to & G_X|_Y & \to & \mathcal{O}_Y \otimes W & \to & F & \to & 0 \\
 & & \| & & \downarrow & & \downarrow & & \\
0 & \to & G_X|_Y & \to & \mathcal{O}_Y \otimes V & \to & N_{X,\mathbb{P}^N}(-1)|_Y & \to & 0 \\
 & & & & \downarrow & & \downarrow & & \\
 & & & & \mathcal{O}_Y \otimes V/W & = & \mathcal{O}_Y \otimes V/W & & \\
 & & & & \downarrow & & \downarrow & & \\
 & & & & 0 & & 0 & &
\end{array}$$

Any choice of splitting

$$\mathcal{O}_Y \otimes V = (\mathcal{O}_Y \otimes W) \oplus (\mathcal{O}_Y \otimes V/W)$$

induces the splitting of $N_{X,\mathbb{P}^N}(-1)|_Y$ as a direct sum of F and $\mathcal{O}_Y \otimes V/W$. This implies that

$$h^0(N^\vee_{X,\mathbb{P}^N}(1)|_Y) \geq h^0(\mathcal{O}_Y \otimes V/W) = N - \dim\langle \mathrm{Tan}(Y,X)\rangle.$$

Since there is an injective map

$$N^\vee_{X,\mathbb{P}^N}(1)|_Y \to \mathcal{O}_Y \otimes V^\vee,$$

$H^0(N^\vee_{X,\mathbb{P}^N}(1)|_Y)$ can be seen as a vector subspace $U^\vee \subset V^\vee$ giving a split sequence

$$0 \to F_1 \to N_{X,\mathbb{P}^N}(-1)|_Y \to U \otimes \mathcal{O}_Y \to 0.$$

So we can build a diagram as before:

$$\begin{array}{ccccccccc}
 & & & & 0 & & 0 & & \\
 & & & & \downarrow & & \downarrow & & \\
0 & \to & G_X|_Y & \to & \mathcal{O}_Y \otimes \mathrm{Ker}(V \to U) & \to & F_1 & \to & 0 \\
 & & \| & & \downarrow & & \downarrow & & \\
0 & \to & G_X|_Y & \to & \mathcal{O}_Y \otimes V & \to & N_{X,\mathbb{P}^N}(-1)|_Y & \to & 0 \\
 & & & & \downarrow & & \downarrow & & \\
 & & & & \mathcal{O}_Y \otimes U & = & \mathcal{O}_Y \otimes U & & \\
 & & & & \downarrow & & \downarrow & & \\
 & & & & 0 & & 0 & &
\end{array}$$

It shows that

$$\langle \mathrm{Tan}(Y, X) \rangle \subset \mathbb{P}(\mathrm{Ker}(V \to U)).$$

Therefore

$$h^0(N^\vee_{X,\mathbb{P}^N}(1)|_Y) = \dim U \leq N - \dim\langle \mathrm{Tan}(Y, X) \rangle.$$

The theorem is now proved. □

5.1.2 Dual Varieties of Smooth Divisors

Suppose that $X \subset \mathbb{P}^N$ is a smooth projective variety. The following theorem gives the defect of smooth hyperplane or hypersurface sections of X.

Theorem 5.3 ([E2, HK, Hol, Mun3])

(i) *Let $Y = X \cap H$ be a smooth hyperplane section of X. Then*

$$\operatorname{def} Y = \max\{0, \operatorname{def} X - 1\}.$$

Moreover, if X^* is not a hypersurface, then Y^* is a linear projection of X^* with center H.

(ii) *If Y is a smooth divisor corresponding to a section of $\mathcal{O}_X(d)$ for $d \geq 2$ then Y^* is a hypersurface.*

Proof. (i) Let us suppose first that $\operatorname{def} X > 0$. Take a linear projection

$$\Pi : (\mathbb{P}^N)^\vee \setminus H \to (\mathbb{P}^{N-1})^\vee$$

associating to any hyperplane H' of $\mathbb{P}^N$ the intersection $H' \cap H$.

Since $\operatorname{def} X > 0$, the contact locus $L_1 = (X \cap H_1)_{sing}$ of an $H_1 \in X^*$ with X intersects Y. Hence H_1 is tangent to X at a point of Y, and so $H_1 \cap H$ is tangent to Y. This implies that $\Pi(X^*) \subset Y^*$.

On the other hand, a general $H_1 \in Y^*$ is tangent to Y at $y \in Y$. Then the hyperplane $M_1 = \langle \hat{T}_{y,X}, H_1 \rangle$ is tangent to X at x and $\Pi(M_1) = H_1$. Hence $\Pi(X^*) = Y^*$.

Let us show that the map $\Pi|_{X^*}$ is generically one-to-one. Suppose that $H' \in Y^*$ is a general tangent hyperplane, $L = (X \cap H')_{sing}$ is a contact locus, and $H_1, H_2 \in X^*$ are such that $\Pi(H_1) = \Pi(H_2) = H'$. Then each H_i is tangent to X at a point $y_i \in L$ and

$$H_1 = \langle H', \hat{T}_{y_1,X} \rangle, \quad H_2 = \langle H', \hat{T}_{y_2,X} \rangle.$$

But since

$$\dim\langle \mathrm{Tan}(L, Y) \rangle = (N-1) - h^0(N^\vee_{Y,\mathbb{P}^{N-1}}|_L(1)) =$$
$$= \left(N - h^0(N^\vee_{X,\mathbb{P}^N}|_L(1))\right) - 1 = \dim\langle \mathrm{Tan}(L, X) \rangle - 1$$

by Theorem 5.2, we have

$$\langle \mathrm{Tan}(L,X)\rangle = \langle \mathrm{Tan}(L,Y), \hat{T}_{y,X}\rangle$$

for any point $y \in L$. This implies that $H_1 = H_2$. It follows that $\dim X^* = \dim Y^*$, and hence $\operatorname{def} Y = \operatorname{def} X - 1$.

Now we are going to show that a smooth hyperplane section of a zero defect variety is a zero defect variety. Suppose the opposite and take the following divisor in the conormal variety I_X:

$$I_X|_Y = \{(x, H') \in I_X \mid x \in Y\}$$

in the linear system $|\operatorname{pr}_1^*(\mathcal{O}_X(1))|$.

Since H' is tangent to X at y for any $(y, H') \in I_X|_Y$, $H' \cap H$ is tangent to Y at y. But $k = \operatorname{def} Y > 0$ and so it is tangent to Y along a k-dimensional projective subspace L for which

$$\langle \mathrm{Tan}(L,X)\rangle = \langle \mathrm{Tan}(L,Y), \hat{T}_{y,X}\rangle.$$

$I_X|_Y$ is contained in the set of points

$$\{a \in I_X \mid \dim(\operatorname{pr}_2^{-1}\operatorname{pr}_2(a)) > 0\}.$$

Then

$$\dim \operatorname{pr}_2(I_X|_Y) = N - 2 - \dim \operatorname{pr}_2^{-1}(F)$$

with F a generic point in the image, and therefore $\dim \operatorname{pr}_2(I_X|_Y) \le N-2-k$. It follows that we can choose an irreducible curve C in X^* such that

$$I_X|_Y \cap \operatorname{pr}_2^{-1}(C) = \emptyset$$

and $\pi_2^{-1}(C)$ is again a curve because, if $H' \in X^*$ is such that the contact locus of X with H' has a poisitive dimension, then this contact locus must intersect Y and then $H' \in \operatorname{pr}_2(I_X|_Y)$.

This implies that

$$[\operatorname{pr}_1^*(\mathcal{O}_X(1))] \cdot [\operatorname{pr}^{-1}(C)] = 0.$$

By the projection formula we have the contradiction

$$[(\mathcal{O}_X(1))] \cdot [\operatorname{pr}_1 \operatorname{pr}_2^{-1}(C)] = 0,$$

because $\operatorname{pr}_1 \operatorname{pr}_2^{-1}(C)$ is again a curve.

(ii) Here we shall consider Y as a subvariety in $\mathbb{P}^N$, not in H. Then I_Y contains $I_X|_Y$ as a divisor D. Using the exact sequence

$$0 \to N^\vee_{X,\mathbb{P}^N}|_Y \to N^\vee_{Y,\mathbb{P}^N} \to N^\vee_{Y,X} \to 0$$

and $I_Y = \mathbb{P}(N^\vee_{Y,\mathbb{P}^N})$, $I_X|_Y = \mathbb{P}(N^\vee_{X,\mathbb{P}^N}|_Y)$, $N_{Y,X} = \mathcal{O}_Y(-d)$, we see that

$$\mathcal{O}_{I_Y}(D) \simeq \operatorname{pr}_1^* \mathcal{O}_Y(1-d) \otimes \operatorname{pr}_2^* \mathcal{O}_{Y^*}(1). \qquad (*)$$

Consider the morphism g given by the complete linear system $|\operatorname{pr}_2^* \mathcal{O}_{Y^*}(1)|$. Then Y^* is a linear projection of $g(I_Y)$. Thus

$$\dim Y^* = \dim g(I_Y).$$

We claim that $g|_{I_Y \setminus D}$ is one to one and, therefore, $\dim Y^* = \dim g(I_Y) = \dim I_Y = N - 1$ and Y^* is a hypersurface. Indeed, let y_1 and y_2 be two distinct points in $I_Y \setminus D$. We need to show that the linear system $|\operatorname{pr}_2^* \mathcal{O}_{Y^*}(1)|$ separates y_1 and y_2. If $\operatorname{pr}_1(y_1) = \operatorname{pr}_1(y_2)$ then $\operatorname{pr}_2(y_1) \neq \operatorname{pr}_2(y_2)$ because pr_2 maps $\operatorname{pr}_1^{-1}(\operatorname{pr}_1(y_1))$ isomorphically to a projective subspace in Y^*. So we may assume that $\operatorname{pr}_1(y_1) \neq \operatorname{pr}_1(y_2)$. Then we can find a hypersurface S of degree $d-1$ in $\mathbb{P}^N$ which contains $\operatorname{pr}_1(y_1)$ but not $\operatorname{pr}_1(y_2)$. By $(*)$, $D + \operatorname{pr}_1^{-1}(S) \in |\operatorname{pr}_2^* \mathcal{O}_{Y^*}(1)|$. But

$$y_1 \in D + \operatorname{pr}_1^{-1}(S) \quad \text{and} \quad y_2 \notin D + \operatorname{pr}_1^{-1}(S).$$

Thus y_1 and y_2 are separated by $|\operatorname{pr}_2^* \mathcal{O}_{Y^*}(1)|$. □

5.1.3 Projective Extendability

Definition 5.4 A nonsingular variety $X \subset \mathbb{P}^N$ is called *projectively extendable* if there exist a variety $X' \subset \mathbb{P}^{N+1}$ and a hyperplane $L \subset \mathbb{P}^{N+1}$ such that L intersects X' transversally, $L \cap X' = X$ and X' is not a cone. In this case X' is called a *projective extension* of X. $X \subset \mathbb{P}^N$ is called *smoothly extendable* if X' is smooth.

In general X can be extended in many different ways:

Example 5.5 ([Z6]) Let $X = v_4(\mathbb{P}^1) \subset \mathbb{P}^4$ be the rational normal curve of degree four. Then X is a hyperplane section of the Veronese surface $v_2(\mathbb{P}^2)$ which is projectively inextendable (see below). On the other hand, X is a hyperplane section of the rational scroll F_2 of degree four in $\mathbb{P}^5$ and of $\mathbb{P}^1 \times v_2(\mathbb{P}^1)$. Both F_2 and $\mathbb{P}^1 \times v_2(\mathbb{P}^1)$ are hyperplane sections of a scroll $X \subset \mathbb{P}^6$ with fiber $\mathbb{P}^2$ over $\mathbb{P}^1$ which in its turn is a hyperplane section of the Segre variety $\mathbb{P}^1 \times \mathbb{P}^3 \subset \mathbb{P}^7$; generic sections of $\mathbb{P}^1 \times \mathbb{P}^3$ by five-dimensional linear subspaces of $\mathbb{P}^7$ are projectively isomorphic to $\mathbb{P}^1 \times v_2(\mathbb{P}^1)$ while special nonsingular sections are isomorphic to F_2.

It is not true in general that extendability implies smooth extendability:

Example 5.6 ([Z6]) Let $X' = v_2(S_z(v_3(\mathbb{P}^2)))$, where v_m is the Veronese map of order m and $S_z(v_3(\mathbb{P}^2)) \subset \mathbb{P}^{10}$ is the cone with vertex $z \notin v_3(\mathbb{P}^2))$ over the Del Pezzo surface $v_3(\mathbb{P}^2) \subset \mathbb{P}^9$. Then $K_{X'} \simeq \mathcal{O}_{X'}(-1)$ and $X' \subset \mathbb{P}^{38}$ is a Fano threefold of degree 72 with a unique singular point. Let $X \subset \mathbb{P}^{37}$ be a generic hyperplane section of X'. Then $H^1(X, \mathcal{O}_X) = 0$, $K_X = 0$ and X is a nonsingular K3-surface. If X had a nonsingular projective extension $\tilde{X}'$, then $\tilde{X}'$ would be a nonsingular Fano threefold of degree 72 while by Iskovskih's theorem the degree of such a threefold does not exceed 64. Therefore the K3-surface X is smoothly inextendable.

Recall the definition of the vector bundle G_X from the proof of Theorem 5.2. It is clear that $H^0(X, G_X) = 0$ since a nonzero section of G_X would yield a point from $\cap_{x \in X} \hat{T}_x X = \emptyset$. Thus the exact sequence

$$0 \to G_X \to \mathcal{O}_X^{N+1} \to N_{X,\mathbb{P}^N}(-1) \to 0$$

shows that the natural homomorphism $\alpha : V \to H^0(X, N_{X,\mathbb{P}^N}(-1))$, where $V = \mathbb{C}^{N+1}$ and $\mathbb{P}(V) = \mathbb{P}^N$, is a monomorphism and

$$h^0(X, N_{X,\mathbb{P}^N}(-1)) \geq N + 1.$$

Theorem 5.7 **([Z6])** *Suppose that $h^0(X, N_{X,\mathbb{P}^N}(-1)) = N + 1$. Then either X is a quadric or X is inextendable.*

Sketch of the proof. We observe that, if X is a hyperplane section of a variety $X' \subset \mathbb{P}^{N+1}$, then the condition of the theorem is satisfied for almost all hyperplane sections of X'. This means that for a general hyperplane $M \subset \mathbb{P}^{N+1}$ the variety $\mathrm{pr}_2(I_{X'}|_{X' \cap M})$ is contained in a hyperplane in $(\mathbb{P}^{N+1})^\vee$. Since X' is not a cone, $(X')^*$ is non-degenerate, and from the Bertini theorem applied to the preimage of the linear system $|\mathcal{O}_{(\mathbb{P}^{N+1})^\vee}(1)|$ on $I_{X'}$ it follows that X is a hypersurface. On the other hand, it is clear that quadric is the only hypersurface satisfying the condition of the theorem. □

The connection with dual varieties is also clarified by the following theorem.

Theorem 5.8 **([Z6])** *If X^* is normal, then X is inextendable.*

It should be noted that normality of X^* is not necessary for inextendability of X. In fact, let $X = v_m(\mathbb{P}^n)$, where $n \geq 2$, $m \geq 3$. Then X^* is not normal, but X is inextendable. The last assertion (due to Scorza) can be derived from the following theorem.

Theorem 5.9 **([Z6])** *If $H^1(X, T_X(-1)) = 0$, then either X is a twisted cubic curve or a quadric or X is inextendable.*

5.2 Ample Vector Bundles

5.2.1 Definitions

The notion of ampleness was extended to vector bundles by Hartshorne [Ha3]. Let E be a vector bundle on X. We consider the variety $X_E = \mathbb{P}(E^\vee)$, the projectivization of the bundle $E^\vee$. There is a projection $p : X_E \to X$ whose fibers are projectivizations of fibers of $E^\vee$, and a natural projection $\pi : E^\vee \setminus X \to X_E$, where X is embedded into the total space of $E^\vee$ as the zero section. We denote by $\xi(E)$ the tautological line bundle on X_E defined as follows. For open $U \subset X_E$, a section of $\xi(E)$ over U is a regular function on $\pi^{-1}(U)$ which is homogeneous of degree 1 with respect to dilations of $E^\vee$. The restriction of $\xi(E)$ to every fiber $p^{-1}(x) = \mathbb{P}(E_x^\vee)$ is the tautological line bundle $\mathcal{O}(1)$ of the projective space $\mathbb{P}(E_x^\vee)$.

Definition 5.10 E is called *ample* if $\xi(E)$ is an ample line bundle on X_E.

If E is itself a line bundle then $X_E = X$, $\xi(E) = E$, and this definition gives nothing new. Basic properties and criteria for ampleness were obtained in [Ha3]. For example, the Grothendieck definition of ample line bundles also holds for vector bundles: E is ample if and only if, for every coherent sheaf F, there is an integer $n_0 > 0$, such that, for every $n \geq n_0$, the sheaf $F \otimes \mathrm{Sym}^n(E)$ is generated by its global sections as an $\mathcal{O}_X$-module . Using this description it is easy to see that the direct sum of ample vector bundles is ample, etc.

5.2.2 Dual Varieties of Smooth Complete Intersections

Theorem 5.11 ([E1]) *If $X \subset \mathbb{P}^N$ is a nonlinear smooth complete intersection then X^* is a hypersurface.*

Proof. We may assume that X is nondegenerate. Let $\dim X = n$. Then X is scheme-theoretic intersection of $N-n$ hypersurfaces $H_1, \ldots, H_{N-n}$ such that $\deg H_k = d_k \geq 2$. Therefore

$$N_{X,\mathbb{P}^N} = N_{H_1,\mathbb{P}^N} \oplus \ldots \oplus N_{H_{N-n},\mathbb{P}^N} = \mathcal{O}_X(d_1) \oplus \ldots \oplus \mathcal{O}_X(d_{N-n}).$$

It follows that

$$N_{X,\mathbb{P}^N}(-1) = \mathcal{O}_X(d_1 - 1) \oplus \ldots \oplus \mathcal{O}_X(d_{N-n} - 1)$$

is an ample vector bundle. Let

$$I_X = \mathbb{P}(N^\vee_{X,\mathbb{P}^N}(1)) \subset \mathbb{P}^N \times (\mathbb{P}^N)^\vee$$

be the conormal variety and let $\mathrm{pr} : I_X \to (\mathbb{P}^N)^\vee$ be the second projection. Then $\mathrm{pr}^* \mathcal{O}(1)$ is the tautological line bundle L on $\mathbb{P}(N^\vee_{X,\mathbb{P}^N}(1))$. Since $N_{X,\mathbb{P}^N}(-1)$ is ample, it follows by the standard argument that pr is a finite morphism onto its image X^*. Indeed, if pr is not finite then it contracts a curve C. Therefore $L \cdot C = 0$ but then $nL \cdot C = 0$ for any n and nL cannot be very ample.

It follows that

$$\dim X^* = \dim I_X = N - 1$$

and X^* is a hypersurface. □

5.2.3 Resultants

Classically, discriminants were studied together with resultants. Let X be a smooth irreducible projective variety and let E be a vector bundle on X of rank $k = \dim X+1$. Set $V = H^0(X,E)$. We shall assume that E is very ample, i.e. $\xi(E)$ is a very ample line bundle on X_E. In particular, $\xi(E)$ (and hence E) is generated by global sections. Notice that $H^0(X,E) = H^0(X_E, \xi(E))$, and therefore $\xi(E)$ embeds X_E in $\mathbb{P}(V^\vee)$.

Definition 5.12 The *resultant variety* $\nabla_X \subset \mathbb{P}(V)$ is the set of all sections vanishing at some point $x \in X$.

Example 5.13 Suppose that $X = \mathbb{P}^{k-1} = \mathbb{P}(\mathbb{C}^k)$ and

$$E = \mathcal{O}(d_1) \oplus \ldots \oplus \mathcal{O}(d_k).$$

Then $V = S^{d_1}(\mathbb{C}^k)^\vee \oplus \ldots \oplus S^{d_k}(\mathbb{C}^k)^\vee$ and ∇ is the classical resultant variety parametrizing k-tuples of homogeneous forms on $\mathbb{C}^k$ of degrees $d_1, \ldots, d_k$ having a common non-zero root.

The assertion (ii) of the following theorem is sometimes called the Cayley trick since Cayley first noticed that the resultant can be written as a discriminant.

Theorem 5.14 ([GKZ2])

(i) ∇_X *is an irreducible hypersurface.*
(ii) ∇_X *is projectively dual to* X_E.

Proof. $\xi(E)$ embeds X_E in $\mathbb{P}(V^\vee)$ in such a way that all fibers become projective subspaces of dimension $k-1$. Geometrically, ∇_X parametrizes hyperplanes in $\mathbb{P}(V^\vee)$ that contain at least one fiber of X_E. Since the embedded tangent space to any point of X_E contains the fiber through it, it follows that $\nabla \supset (X_E)^*$. On the contrary, suppose that $v \in \nabla$. Then the corresponding section $s \in H^0(X,E)$ vanishes at some point $x \in X$. Therefore s defines a linear map $ds : T_{x,X} \to T_{x,E} = T_{x,X} \oplus E_x$. Let $\mathrm{pr} : T_{x,E} \to E_x$ be the projection. Since $\dim X = \dim E - 1$, $\mathrm{pr} \circ ds(T_{x,X})$ is contained in some hyperplane $H \subset E_x$. An easy local calculation shows that the hyperplane in $\mathbb{P}(V^\vee)$ corresponding to v contains the embedded tangent space to X_E at the point corresponding to H. Therefore $\nabla = (X_E)^*$.

Now let us prove that ∇ is a hypersurface. A dimension count shows that it suffices to prove that a generic section s of $H^0(X,E)$ vanishing at $x \in X$ does not vanish at other points. In other words, a generic hyperplane in $P = \mathbb{P}(V^\vee)$ containing some fiber F of E_X does not contain other fibers. Consider the linear projection $\pi_F : P \setminus F \to P/F$. If F' is any fiber of X_E distinct from F then $\pi_F(F')$ is a projective subspace of dimension $\dim F' = k-1$. Therefore images of fibers of X_E distinct from F form an at most $(k-1)$-dimensional family of projective subspaces of dimension $k-1$. An easy dimension count shows that there exists a hyperplane $H \subset P/F$ that does not contain any projective subspace from this family. The preimage of this hyperplane in P contains only one fiber F. □

5.3 Cayley Method

5.3.1 Jet Bundles and Koszul Complexes

Let X be a smooth irreducible algebraic variety with an algebraic line bundle $\mathcal{L}$. We consider the bundle $J(\mathcal{L})$ of *first jets* of sections of $\mathcal{L}$. By definition, the fiber of $J(\mathcal{L})$ at $x \in X$ is the quotient of the space of all sections of $\mathcal{L}$ near x by the subspace of sections which vanish at x with their first derivatives. In other words,

$$J(\mathcal{L})_x = \mathcal{L}/\mathcal{I}_x^2\mathcal{L},$$

where $\mathcal{I}_x$ is the ideal of functions vanishing at x. Thus $J(\mathcal{L})$ is a vector bundle of rank $\dim X + 1$.

To any section f of $\mathcal{L}$, we associate a section $j(f)$ of $J(\mathcal{L})$, called the first jet of f. Namely, the value of $j(f)$ at x is the class of f modulo $I_x^2\mathcal{L}$. The correspondence $f \mapsto j(f)$ is $\mathbb{C}$-linear but, being a differential operator, it is not $\mathcal{O}_X$-linear. Let us summarize without proofs some properties of jet bundles.

Theorem 5.15 ([GKZ2, KS])

(i) *For any line bundles $\mathcal{L}$ and $\mathcal{M}$ on X, there exists a canonical isomorphism*

$$J(\mathcal{L} \otimes \mathcal{M}) \simeq J(\mathcal{L}) \otimes \mathcal{M}.$$

(ii) *There exists a canonical exact sequence of vector bundles*

$$0 \to T_X^\vee \otimes \mathcal{L} \to J(\mathcal{L}) \to \mathcal{L} \to 0. \tag{5.1}$$

(iii) *Let $X = \mathbb{P}(V)$. Then the $J(\mathcal{O}_X(1))$ is a trivial vector bundle $\mathcal{O}_X \otimes V^\vee$*

The relevance of jets to dual varieties is as follows. Suppose that $X \subset \mathbb{P}(V)$ is an irreducible projective variety. We take $\mathcal{L} = \mathcal{O}_X(1)$. Then any $f \in V^\vee$ is a global section of $\mathcal{L}$. The following result is an immediate consequence of the definitions.

Proposition 5.16 *$f \in V^\vee$ represents a point in the dual variety X^* if and only if $j(f)$ vanishes at some point $x \in X$.*

Let E be an algebraic vector bundle of rank r on an irreducible algebraic variety X. For any global section s of E consider the following *Koszul complexes* of sheaves on X:

$$\mathcal{K}_+(E,s) = \left\{0 \to \mathcal{O}_X \xrightarrow{s} E \xrightarrow{\wedge s} \Lambda^2 E \xrightarrow{\wedge s} \ldots \xrightarrow{\wedge s} \Lambda^r E \to 0\right\},$$

$$\mathcal{K}_-(E,s) = \left\{0 \to \Lambda^r E^\vee \xrightarrow{i_s} \dots \xrightarrow{i_s} \Lambda^2 E^\vee \xrightarrow{i_s} E^\vee \xrightarrow{i_s} \mathcal{O}_X \to 0\right\}.$$

The differential in $\mathcal{K}_+$ is given by the exterior multiplication with s, and the differential in $\mathcal{K}_-$ is given by the contraction with s, that is, by the map $\Lambda^j E^\vee \to \Lambda^{j-1} E^\vee$ dual to the map $\Lambda^{j-1} E \to \Lambda^j E$ given by $\wedge s$. We fix gradings in $\mathcal{K}_+$ and $\mathcal{K}_-$ by assigning the degree j to $\Lambda^j E$ in $\mathcal{K}_+$ and by assigning the degree $-j$ to $\Lambda^j E^\vee$ in $\mathcal{K}_-$. Notice that we have

$$\mathcal{K}_-(E,s) \simeq \mathcal{K}_+(E,s) \otimes \Lambda^r E^\vee[r],$$

where $[r]$ means the shift in the grading by r. The following well-known theorem shows that the cohomology of Koszul complexes 'represents' the vanishing set of s.

Theorem 5.17

(i) *$\mathcal{K}_+(E,s)$ or $\mathcal{K}_-(E,s)$ are exact if and only if the scheme of zeros $Z(s)$ is empty.*

(ii) *Suppose that X is smooth and that s vanishes along a smooth subvariety $Z(s)$ of codimension $r = \operatorname{rank} E$ and, moreover, s is transversal to the zero section. Then $\mathcal{K}_-(E,s)$ has only one non-trivial cohomology sheaf, namely $\mathcal{O}_{Z(s)}$ (regarded as a sheaf on X) in highest degree 0. The complex $\mathcal{K}_+(E,s)$ in this situation has the only non-trivial cohomology sheaf in the highest degree r, and this sheaf is the restriction $\det E|_{Z(s)} = \Lambda^r E|_{Z(s)}$ regarded as a sheaf on X.*

Sketch of the proof. We shall prove only (i), the proof of (ii) follows the same ideas but is more technically involved; see e.g. [Ful, GH1]. If V is a vector space and $v \in V$ is a non-zero vector then the differential in the exterior algebra $\Lambda^\bullet V$ given by $\wedge v$ is exact. Indeed, we can include v in some basis as a first vector, and then both the kernel and the image of the differential are spanned by monomials containing v. Therefore the dual differential i_v in $\Lambda^\bullet V^\vee$ is also exact. Applying this to our situation we see that if s does not vanish anywhere then both Koszul complexes are exact fiberwise and, therefore, are exact as complexes of sheaves. However, if $Z(s)$ is not empty then the cokernel of the last differential in Koszul complexes is not trivial. Indeed, if $x \in Z(s)$ then, trivializing E near x, we represent s as a collection of functions $(f_1, \dots, f_r) \in \mathcal{O}^r$. Then the last differential in both complexes has a form $\mathcal{O}^r \to \mathcal{O}$, $(u_1, \dots, u_r) \mapsto \sum u_i f_i$. If s vanishes at x then all $f_i = 0$ and this map is not surjective. □

Let now $X \subset \mathbb{P}(V)$ be a smooth projective variety. Then we can apply Theorem 5.17 to the jet bundle $E = J(\mathcal{L})$, where $\mathcal{L} = \mathcal{O}_X(1)$. Combining Proposition 5.16 and Theorem 5.17 we get the following theorem.

Theorem 5.18 ([GKZ2]) *For any $f \in V^\vee$ let $j = j(f)$ denote its first jet. Then f represents a point in X^* if and only if any of the following complexes of sheaves on X is not exact:*

$$\mathcal{K}_+(J(\mathcal{L}),j) = \left\{0 \to \mathcal{O}_X \xrightarrow{j} J(\mathcal{L}) \xrightarrow{\wedge j} \Lambda^2 J(\mathcal{L}) \xrightarrow{\wedge j} \dots \xrightarrow{\wedge j} \Lambda^r J(\mathcal{L}) \to 0\right\},$$

$$\mathcal{K}_-(J(\mathcal{L}),j) = \left\{0 \to \Lambda^r J(\mathcal{L})^\vee \xrightarrow{i_j} \dots \xrightarrow{i_j} \Lambda^2 J(\mathcal{L})^\vee \xrightarrow{i_j} J(\mathcal{L})^\vee \xrightarrow{i_j} \mathcal{O}_X \to 0\right\}.$$

5.3.2 Cayley Determinants of Exact Complexes

The determinants of exact complexes (in the implicit form) were first introduced by Cayley in his paper [Ca] on resultants. A systematic early treatment of this subject was undertaken by Fisher [Fi] whose aim was to give a rigorous proof of the Cayley results. In topology, determinants of complexes were introduced in 1935 by Reidermeister and Franz [Fra]. They used the word 'torsion' for the determinant-type invariants constructed. In this section we give only the definitions that are necessary for the formulation of results related to dual varieties. More details could be found in [GKZ2, KnMu, Del, Q, RS].

The base field k can be arbitrary. Suppose that V is a finite-dimensional vector space. Then the top-degree component of the exterior algebra $\Lambda^{\dim V} V$ is called the determinant of V, denoted by $\operatorname{Det} V$. If $V = 0$ then we set $\operatorname{Det} V = k$. It is easy to see that for any exact triple $0 \to U \to V \to W \to 0$ we have a natural isomorphism $\operatorname{Det} V \simeq \operatorname{Det} U \otimes \operatorname{Det} W$.

Suppose now that $V = V_0 \oplus V_1$ is a finite-dimensional supervector space. Then, by definition, $\operatorname{Det} V$ is set to be $\operatorname{Det} V_0 \otimes (\operatorname{Det} V_1)^\vee$. Once again, for any exact triple of supervector spaces $0 \to U \to V \to W \to 0$ we have $\operatorname{Det} V \simeq \operatorname{Det} U \otimes \operatorname{Det} W$. For any supervector space V we denote by $\tilde{V}$ the new supervector space given by $\tilde{V}_0 = V_1$, $\tilde{V}_1 = V_0$. Clearly, we have a natural isomorphism

$$\operatorname{Det} \tilde{V} = (\operatorname{Det} V)^\vee. \tag{5.2}$$

Now let (V, ∂) be a finite-dimensional supervector space with a differential ∂ such that $\partial V_0 \subset V_1$, $\partial V_1 \subset V_0$, $\partial^2 = 0$. Then $\operatorname{Ker} \partial$, $\operatorname{Im} \partial$, and the cohomology space $H(V) = \operatorname{Ker} \partial / \operatorname{Im} \partial$ are again supervector spaces. We claim that there exists a natural isomorphism

$$\operatorname{Det} V \simeq \operatorname{Det} H(V). \tag{5.3}$$

Indeed, from the exact sequence

$$0 \to \operatorname{Ker} \partial \to V \xrightarrow{\partial} \widetilde{\operatorname{Im} \partial} \to 0$$

we see that $\operatorname{Det} V \simeq \operatorname{Det}(\operatorname{Ker} \partial) \otimes \operatorname{Det}(\operatorname{Im} \partial)^\vee$. From the exact sequence

$$0 \to \operatorname{Im} \partial \to \operatorname{Ker} \partial \to H(V) \to 0$$

we deduce that $\operatorname{Det} H(V) \simeq \operatorname{Det}(\operatorname{Ker} \partial) \otimes \operatorname{Det}(\operatorname{Im} \partial)^\vee$. This gives (5.3). In particular, if ∂ is exact (i.e. $H(V) = 0$) then we have a natural isomorphism

$$\operatorname{Det} V \simeq k. \tag{5.4}$$

Let us fix some bases $\{e_1, \ldots, e_{\dim V_0}\}$ in V_0 and $\{e'_1, \ldots, e'_{\dim V_1}\}$ in V_1. Let $\{f'_1, \ldots, f'_{\dim V_1}\}$ be a dual basis in $V_1^\vee$. Then we have a basis vector

$$(e_1 \wedge \ldots \wedge e_{\dim V_0}) \otimes (f'_1 \wedge \ldots \wedge f'_{\dim V_1}) \in \operatorname{Det} V.$$

Therefore, if (V, ∂, e) is a based supervector space with an exact differential, then by (5.4) we get a number $\det(V, \partial, e)$ called the *Cayley determinant* of a based supervector space with an exact differential. If we fix other bases $\{\tilde{e}_1, \ldots, \tilde{e}_{\dim V_0}\}$ in V_0 and $\{\tilde{e}'_1, \ldots, \tilde{e}'_{\dim V_1}\}$ in V_1 then, clearly,

$$\det(V, \partial, \tilde{e}) = \det A_0 (\det A_1)^{-1} \det(V, \partial, e), \tag{5.5}$$

where $(A_0, A_1) \in \mathrm{GL}(V_0) \times \mathrm{GL}(V_1)$ are transition matrices from bases e to bases $\tilde{e}$. One upshot of this is the fact that if bases e and $\tilde{e}$ are equivalent over some subfield $k_0 \subset k$ then the Cayley determinants with respect to these bases are equal up to a non-zero multiple from k_0. The homogeneity condition (5.5) can be reformulated as follows. For any based supervector space (V, ∂, e) with an exact differential and any $(A_0, A_1) \in \mathrm{GL}(V_0) \times \mathrm{GL}(V_1)$, we can consider a new based supervector space $(V, A \cdot \partial, e)$ with the exact differential $A \cdot \partial$ given by

$$A \cdot \partial|_{V_0} = A_1 \cdot \partial \cdot A_0^{-1}, \quad A \cdot \partial|_{V_1} = A_0 \cdot \partial \cdot A_1^{-1}.$$

Then we have

$$\det(V, A \cdot \partial, e) = (\det A_0)^{-1} \det A_1 \det(V, \partial, e).$$

In particular, it is easy to see that the following formula is valid:

$$\det(V, \lambda\partial, e) = \lambda^{\dim(\operatorname{Ker} \partial)_1 - \dim(\operatorname{Ker} \partial)_0} \det(V, \partial, e). \tag{5.6}$$

In the matrix form the Cayley determinant can be calculated as follows. Suppose that (V, ∂, e) is a based supervector space with an exact differential,

$$\dim V_0 = \dim \operatorname{Ker} \partial_0 + \dim \operatorname{Im} \partial_0 = \dim \operatorname{Im} \partial_1 + \dim \operatorname{Ker} \partial_1 = \dim V_1 = n.$$

Therefore both ∂_0 and ∂_1 are represented by $(n \times n)$ matrices D_0 and D_1. For any subsets $I, J \subset B = \{1, \ldots, n\}$ we denote by D_0^{IJ} and D_1^{IJ} the submatrices of D_0 and D_1 formed by columns indexed by I and rows indexed by J. Then it is easy to see that there exist subsets $I_0, I_1 \subset B$ such that the submatrices $D_0^{(B \setminus I_0), I_1}$, $D_1^{(B \setminus I_1), I_0}$ are invertible. In particular,

$$|I_1| = \operatorname{rk} \partial_0, \quad |I_0| = \operatorname{rk} \partial_1 = n - |I_1|.$$

The formula (5.5) implies that

$$\det(V, \partial, e) = \det D_0^{(B \setminus I_0), I_1} (\det D_1^{(B \setminus I_1), I_0})^{-1}. \tag{5.7}$$

This formula gives an explicit matrix description of the Cayley determinant.

Now we consider a finite complex of finite-dimensional vector spaces

$$\dots \stackrel{\partial_{i-1}}{\to} V^i \stackrel{\partial_i}{\to} V^{i+1} \stackrel{\partial_{i+1}}{\to} \dots .$$

Then we can define the finite-dimensional supervector space $V = V_0 \oplus V_1$,

$$V_0 = \bigoplus_{i \equiv 0 \bmod 2} V^i, \quad V_1 = \bigoplus_{i \equiv 1 \bmod 2} V^i,$$

with an induced differential ∂. In particular, all previous considerations are valid. Therefore, if the complex $(V^\bullet, \partial)$ is exact and there are some fixed bases $\{e_1^i, \dots, e_{\dim V^i}^i\}$ in each component V^i, then we have the corresponding Cayley determinant $\det(V^\bullet, \partial, e) \in k^*$. For example, if L and M are based vector spaces and $A : L \to M$ is an invertible operator then the complex $0 \to L \stackrel{A}{\to} M \to 0$ is exact and the corresponding Cayley determinant is equal to $\det A$ (if L is located in the even degree of the complex).

In fact, Cayley in [Ca] has found some matrix representation of his determinant. In order to give his formula we shall write our complex in the explicit coordinate form

$$V^\bullet = \{0 \to k^{B_0} \stackrel{D_0}{\to} k^{B_1} \stackrel{D_1}{\to} \to \dots \stackrel{D_{r-1}}{\to} k^{B_r} \to 0\}, \tag{5.8}$$

where for simplicity we assume that the non-zero terms of the complex are located in degrees between 0 and $r > 0$. In general, if $V^\bullet[m]$ is the same complex as $V^\bullet$ but with a grading shifted by m, then by (5.2) we have $\det(V^\bullet[m], \partial, e) = (\det(V^\bullet[m], \partial, e))^{(-1)^m}$ In the formula (5.8) we may suppose that all B_i are finite sets. Therefore D_i is a matrix with columns indexed by B_i and rows indexed by B_{i+1}. For any subsets $X \subset B_i$ and $Y \subset B_{i+1}$ we denote by $(D_i)_{XY}$ the submatrix in D_i with columns from X and rows from Y.

A collection of subsets $I_i \subset B_i$ is called *admissible* if $I_0 = \emptyset$, $I_r = B_r$, and $|B_i \setminus I_i| = |I_{i+1}|$ for any $i = 0, \dots, r-1$, and finally the submatrix $(D_i)_{(B_i \setminus I_i), I_{i+1}}$ is invertible. The following theorem follows easily from (5.7).

Theorem 5.19 ([GKZ2])

(i) *Admissible collections exist. For any admissible collection we have*

$$|I_i| = \sum_{j=0}^{i-1} (-1)^{i-1-j} |B_j|.$$

(ii) *Let $\{I_i\}$ be an admissible collection. Denote by Δ_i the determinant of the matrix $(D_i)_{(B_i \setminus I_i), I_{i+1}}$. Then*

$$\det(V^\bullet, D, B) = \prod_{i=0}^{r-1} \Delta_i^{(-1)^i}.$$

In particular, this theorem (or formula (5.6)) implies the following corollary

Corollary 5.20 *For any based exact complex* $(V^\bullet, d, e)$ *we have*

$$\det(V^\bullet, \lambda d, e) = \lambda^{\sum_i (-1)^{i+1} \cdot i \cdot \dim V^i} \det(V^\bullet, d, e).$$

5.3.3 Discriminant Complexes

Now let $X \subset \mathbb{P}(V)$ be a smooth projective variety. We denote $\mathcal{L} = \mathcal{O}_X(1)$, so that any $f \in V^\vee$ can be regarded as a section of $\mathcal{L}$. Let $\mathcal{M}$ be another line bundle on X. We define discriminant complexes $C_+^\bullet(X, \mathcal{M})$ and $C_-^\bullet(X, \mathcal{M})$ as complexes of global sections of Koszul complexes $\mathcal{K}_+(J(\mathcal{L}), j(f))$ and $\mathcal{K}_-(J(\mathcal{L}), j(f))$ tensored by $\mathcal{M}$:

$$C_+^i(X, \mathcal{M}) = H^0(X, \Lambda^i J(\mathcal{L}) \otimes \mathcal{M}),$$

$$C_-^i(X, \mathcal{M}) = H^0(X, \Lambda^{-i} J(\mathcal{L})^\vee \otimes \mathcal{M}).$$

Thus the terms of discriminant complexes are fixed and the differentials depend on $f \in V^\vee$. We shall denote this differential by ∂_f. By Theorem 5.18 a vector $f \in V^\vee$ represents a point in X^* if and only if any one of two Koszul complexes is not exact. We want to get the same condition but for discriminant complexes, which are complexes of finite-dimensional vector spaces instead of complexes of sheaves. However, the exactness of a complex of sheaves does not necessarily imply the exactness of the corresponding complex of global sections. The obstruction to this is given by the higher cohomology of the sheaves of the complex.

Definition 5.21 The discriminant complexes are called *stably twisted* if all terms of the corresponding Koszul complexes have no higher cohomology.

For example, suppose that $\mathcal{M}$ is any ample line bundle on X. Ampleness is equivalent to the fact that for any coherent sheaf $\mathcal{F}$ on X, the sheaves $\mathcal{F} \otimes \mathcal{M}^{\otimes n}$ have no higher cohomology for $n \gg 0$. Therefore, for sufficiently big n the discriminant complexes $C_+^*(X, \mathcal{M}^{\otimes n})$ and $C_-^*(X, \mathcal{M}^{\otimes n})$ will be stably twisted. In this situation we have the following Theorem that immediately follows from Theorem 5.18:

Theorem 5.22 ([GKZ2]) *Suppose that the discriminant complex* $C_+^\bullet(X, \mathcal{M})$ *(resp.* $C_-^\bullet(X, \mathcal{M})$*) is stably twisted. Let* $f \in V^\vee$ *be such that its projectivization does not belong to the dual variety* $X^* \subset \mathbb{P}(V^\vee)$*. Then* $(C_+^\bullet(X, \mathcal{M}), \partial_f)$ *(resp.* $(C_-^\bullet(X, \mathcal{M}), \partial_f)$*) is exact.*

Remarkably, it turns out that much more is true. Suppose that the discriminant complex $C_+^\bullet(X, \mathcal{M})$ (or $C_-^\bullet(X, \mathcal{M})$) is stably twisted. Then for a generic $f \in V^\vee$, $C_+^\bullet(X, \mathcal{M})$ (or $C_-^\bullet(X, \mathcal{M})$) is exact by the previous theorem. Let $f_1, \ldots, f_n$ be a basis of $V^\vee$. Let $\mathbb{C}(x_1, \ldots, x_n) = \mathbb{C}(V^\vee)$ be a field of rational functions on $V^\vee$. Consider the complex $C_+^\bullet(X,\mathcal{M}) \otimes \mathbb{C}(V^\vee)$ (or $C_-^\bullet(X,\mathcal{M}) \otimes \mathbb{C}(V^\vee)$) with the differential given by $\partial = \sum x_i \partial_{f_i}$. Then this complex is exact, since its generic specialization is exact. We fix some bases over $\mathbb{C}$ in each component $C_+^i(X,\mathcal{M})$ (or $C_-^{-i}(X,\mathcal{M})$) and consider them as bases over $\mathbb{C}(V^\vee)$ in each $C_+^i(X,\mathcal{M}) \otimes \mathbb{C}(V^\vee)$ (or $C_-^{-i}(X,\mathcal{M}) \otimes \mathbb{C}(V^\vee)$). Then we can calculate the Cayley determinant $\Delta^+_{X,\mathcal{M}}$ (or $\Delta^-_{X,\mathcal{M}}$) of these based exact complexes, which will be the non-zero element of $\mathbb{C}(V^\vee)$. By the homogeneity, Cayley determinants depend only on the discriminant complexes (up to a complex multiple) and do not depend on the choice of bases. The following theorem shows that the Cayley determinant is equal to the discriminant Δ_X:

Theorem 5.23 ([GKZ2]) *If $C_-^\bullet(X,\mathcal{M})$ is stably twisted then*

$$\Delta^-_{X,\mathcal{M}} = \Delta_X$$

up to a non-zero scalar. If $C_+^\bullet(X,\mathcal{M})$ is stably twisted then

$$(\Delta^+_{X,\mathcal{M}})^{(-1)^{\dim X+1}} = \Delta_X$$

up to a non-zero scalar.

The proof of this theorem involves a lot of homological algebra and the theory of Cayley determinants for complexes over arbitrary Noetherian integral domains; see [GKZ2]. Let us consider several examples.

Example 5.24 Let $X = \mathbb{P}^1$ be embedded into $\mathbb{P}(S^d\mathbb{C}^2)$ via the Veronese embedding. The space $V^\vee = (S^d\mathbb{C}^2)^\vee$ is the space of binary forms

$$f(x_0,x_1) = a_0x_0^d + a_1x_0^{d-1}x_1 + \ldots + a_dx_1^d$$

and the discriminant $\Delta_X(f)$ is the classical discriminant of this binary form. It is quite easy to implement Theorem 5.23 in this case. We take a twisting bundle $\mathcal{M} = \mathcal{O}_{\mathbb{P}^1}(2d-3)$. The discriminant complex $C_-^\bullet(X,\mathcal{M})$ has only two non-zero terms and, therefore, the Cayley determinant is reduced to a determinant of a square matrix. More precisely, the Cayley determinant in this case is equal to the determinant of the linear map

$$\partial_f : S^{d-2}\mathbb{C}^2 \oplus S^{d-2}\mathbb{C}^2 \to S^{2d-3}\mathbb{C}^2,$$

given by

$$\partial_f(u,v) = \frac{\partial f}{\partial x_0}u + \frac{\partial f}{\partial x_1}v.$$

The final formula is the classical *Sylvester formula* for the discriminant of a binary form:

$$\Delta_X(f) = \frac{(-1)^{d-1}}{d^{d-2}}D,$$

where D is the determinant of the following matrix

$$\begin{vmatrix} a_1 & 2a_2 & \dots & (d-1)a_{d-1} & da_d & \dots & 0 \\ 0 & a_1 & \dots & (d-2)a_{d-2} & (d-1)a_{d-1} & \dots & 0 \\ \vdots & \vdots & \ddots & \vdots & \vdots & \ddots & \vdots \\ 0 & 0 & \dots & a_1 & 2a_2 & \dots & da_d \\ da_0 & (d-1)a_1 & \dots & 2a_{d-2} & a_{d-1} & \dots & 0 \\ 0 & da_0 & \dots & 3a_{d-3} & 2a_{d-2} & \dots & 0 \\ \vdots & \vdots & \ddots & \vdots & \vdots & \ddots & \vdots \\ 0 & 0 & \dots & da_0 & (d-1)a_1 & \dots & a_{d-1} \end{vmatrix}.$$

Example 5.25 Let $X \subset \mathbb{P}(V)$ be a smooth algebraic curve of degree d and genus g. We take the twisting bundle $\mathcal{M}$ to be a generic line bundle on X of degree $2d+3g-3$. The discriminant complex $C^\bullet_-(X, \mathcal{M})$ has only two non-zero terms and, therefore, the Cayley determinant is given by the determinant of a square matrix. The size of this matrix (and hence the degree of the dual variety X^*) is equal to $\dim H^0(X, \mathcal{M})$. The latter is equal to $2d+2g-2$ by the Riemann-Roch theorem.

One disadvantage of Theorem 5.23 is the restricting condition on the twisting line bundle $\mathcal{M}$. It is not always convenient to make the discriminant complexes $C^\bullet_\pm$ stably twisted. Another possibility is to take into account the higher cohomology as well. The standard tool for this is the spectral sequence. Recall [GH1] that if $\mathcal{F}^\bullet$ is a finite complex of sheaves on a topological space X, then one can define the *hypercohomology groups* $\mathbf{H}^i(X, \mathcal{F}^\bullet)$. Namely, consider the complexes of Abelian groups $C^\bullet_j$ calculating the cohomology of every individual $\mathcal{F}^j$ (for example, the Čech complexes with respect to an appropriate open covering of X). The differentials $\mathcal{F}^i \to \mathcal{F}^{i+1}$ make this collection of complexes into a double complex $C^{\bullet\bullet}$. The hypercohomology groups $\mathbf{H}^i(X, \mathcal{F}^\bullet)$ are the cohomology groups of its total complex. In particular, we have the spectral sequence of the double complex $C^{\bullet\bullet}$

$$E_1^{pq} = H^q(X, \mathcal{F}^p) \Rightarrow \mathbf{H}^{p+q}(X, \mathcal{F}^\bullet).$$

The first differential d_1 is induced by the differential in $\mathcal{F}^\bullet$. In particular, the complex of global sections

$$\dots \to H^0(X, \mathcal{F}^i) \to H^0(X, \mathcal{F}^{i+1}) \to \dots$$

is just the bottom row of the term E_1.

If the complex of sheaves $\mathcal{F}^\bullet$ is exact then all hypercohomology groups $\mathbf{H}^i(X, \mathcal{F}^\bullet)$ vanish. Indeed, the hypercohomology can be calculated using another spectral sequence 'dual' to the first:

$$'E_2^{pq} = H^q(X, \mathcal{H}^p(\mathcal{F}^\bullet)) \Rightarrow \mathbf{H}^{p+q}(X, \mathcal{F}^\bullet),$$

where $\mathcal{H}^p(\mathcal{F}^\bullet)$ is the p-th cohomology sheaf of $\mathcal{F}^\bullet$

$$\mathcal{H}^p(\mathcal{F}^\bullet) = \frac{\mathrm{Ker}\{\mathcal{F}^p \xrightarrow{d} \mathcal{F}^{p+1}\}}{\mathrm{Im}\{\mathcal{F}^{p-1} \xrightarrow{d} \mathcal{F}^p\}}.$$

Now let $X \subset \mathbb{P}(V)$ be a smooth projective variety and $\mathcal{L} = \mathcal{O}_X(1)$. Let $\mathcal{M}$ be any line bundle on X. We define discriminant spectral sequences $C^{pq}_{r,\pm}(X, \mathcal{M}, f)$ to be the spectral sequences of complexes $\mathcal{K}_\pm(J(\mathcal{L}), j(f))$ tensored by $\mathcal{M}$. Then the following theorem is a consequence of the discussion above and Theorem 5.18.

Theorem 5.26 ([GKZ2]) *Suppose that the projectivization of $f \in V^\vee$ does not belong to the dual variety X^*. Then the discriminant spectral sequences $C^{pq}_{r,\pm}(X, \mathcal{M}, f)$ are exact, i.e. they converge to zero.*

Moreover, it is possible to define the Cayley determinants of exact spectral sequences and to prove an analogue of Theorem 5.23; see [GKZ2] for details.

5.3.4 Cayley Method for Resultants

The Cayley method applies equally well to other elimination problems, in particular to the problem of finding the resultant. This was the original context in which Cayley introduced his method.

Let X be an irreducible projective variety of dimension $k-1$ and let $\mathcal{L}_1, \ldots, \mathcal{L}_k$ be invertible very ample sheaves on X. Consider the vector bundle $E = \mathcal{L}_1 \oplus \ldots \oplus \mathcal{L}_k$. For any choice of sections $f_i \in H^0(X, \mathcal{L}_i)$, the collection $(f_1, \ldots, f_k)$ defines a section s of E. The correspondent resultant is denoted by $R_{\mathcal{L}_1,\ldots,\mathcal{L}_k}(f_1, \ldots, f_k)$.

We can form two Koszul complexes $\mathcal{K}_\pm(E, s)$ of sheaves on X:

$$\mathcal{K}_+(E, s) = \left\{0 \to \mathcal{O}_X \xrightarrow{s} E \xrightarrow{\wedge s} \Lambda^2 E \xrightarrow{\wedge s} \ldots \xrightarrow{\wedge s} \Lambda^k E \to 0\right\},$$

$$\mathcal{K}_-(E, s) = \left\{0 \to \Lambda^k E^\vee \xrightarrow{i_s} \ldots \xrightarrow{i_s} \Lambda^2 E^\vee \xrightarrow{i_s} E^\vee \xrightarrow{i_s} \mathcal{O}_X \to 0\right\}.$$

Let $\mathcal{M}$ be another line bundle on X. We construct the twisted Koszul complexes $\mathcal{K}_\pm(E, s) \otimes \mathcal{M}$. The complexes of global sections of these complexes of sheaves will be denoted $C^\bullet_\pm(\mathcal{L}_1, \ldots, \mathcal{L}_k \,|\, \mathcal{M})$ and called the resultant complexes associated with $\mathcal{L}_1, \ldots, \mathcal{L}_k$ and $\mathcal{M}$. We notice that

$$\Lambda^p(\mathcal{L}_1 \oplus \mathcal{L}_k) = \bigoplus_{1 \le i_1 < \ldots < i_p \le k} \mathcal{L}_{i_1} \otimes \ldots \otimes \mathcal{L}_{i_p}.$$

Therefore the resultant complex $C^\bullet_+(\mathcal{L}_1, \ldots, \mathcal{L}_k \,|\, \mathcal{M})$ has the form

$$H^0(X, \mathcal{M}) \to \bigoplus_i H^0(X, \mathcal{L}_i \otimes \mathcal{M}) \to \bigoplus_{i<j} H^0(X, \mathcal{L}_i \otimes \mathcal{L}_j \otimes \mathcal{M}) \to \ldots,$$

and the resultant complex $C^\bullet_-(\mathcal{L}_1, \ldots, \mathcal{L}_k \,|\, \mathcal{M})$ has the form

$$\ldots \to \bigoplus_{i<j} H^0(X, \mathcal{L}_i^\vee \otimes \mathcal{L}_j^\vee \otimes \mathcal{M}) \to \bigoplus_i H^0(X, \mathcal{L}_i^\vee \otimes \mathcal{M}) \to H^0(X, \mathcal{M}).$$

We say that each of the complexes $C^\bullet_\pm(\mathcal{L}_1, \ldots, \mathcal{L}_k \,|\, \mathcal{M})$ is stably twisted if all terms of the above complex $\mathcal{K}_\pm(E, s) \otimes \mathcal{M})$ (of which it is a complex of global sections) have no higher cohomology.

The differential in any of these complexes will be denoted by $\partial_{f_1,\ldots,f_k}$. As in section 5.3.2, we have the following theorem:

Theorem 5.27 ([GKZ2]) *If the resultant complex $C^\bullet_\pm(\mathcal{L}_1, \ldots, \mathcal{L}_k \,|\, \mathcal{M})$ is stably twisted, then the differential $\partial_{f_1,\ldots,f_k}$ in this complex is exact for any $(f_1, \ldots, f_k)$ such that $R_{\mathcal{L}_1,\ldots,\mathcal{L}_k}(f_1, \ldots, f_k) \neq 0$.*

As in section 5.3.2, we define determinants of the resultant complexes

$$R^\pm_{\mathcal{L}_1,\ldots,\mathcal{L}_k \,|\, \mathcal{M}}(f_1, \ldots, f_k) = \det\left(C^\bullet_\pm(\mathcal{L}_1, \ldots, \mathcal{L}_k \,|\, \mathcal{M}), \partial_{f_1,\ldots,f_k}, e\right),$$

where e is some system of bases of the terms of the complex. These are rational functions on the space $V_1 \oplus \ldots \oplus V_k$, where $V_i = H^0(X, \mathcal{L}_i)$.

Theorem 5.28 ([GKZ2]) *If the resultant complex $C^\bullet_+(\mathcal{L}_1, \ldots, \mathcal{L}_k \,|\, \mathcal{M})$ is stably twisted, then*

$$R+_{\mathcal{L}_1,\ldots,\mathcal{L}_k \,|\, \mathcal{M}} = R^{(-1)^k}_{\mathcal{L}_1,\ldots,\mathcal{L}_k},$$

where $R_{\mathcal{L}_1,\ldots,\mathcal{L}_k}$ is the resultant. If the resultant complex $C^\bullet_-(\mathcal{L}_1, \ldots, \mathcal{L}_k \,|\, \mathcal{M})$ is stably twisted, then

$$R-_{\mathcal{L}_1,\ldots,\mathcal{L}_k \,|\, \mathcal{M}} = R_{\mathcal{L}_1,\ldots,\mathcal{L}_k}.$$

Example 5.29 Let $X = \mathbb{P}^1$ and $\mathcal{L}_i = \mathcal{O}(d_i)$, $i = 1, 2$. For any d, we denote by S^d the space of homogeneous polynomials of degree d in two variables x_0, x_1. Thus $V_i = H^0(X, \mathcal{L}_i)$ is identified with S^{d_i}. The resultant of $f_1 \in S^{d_1}$ and $f_2 \in S^{d_2}$ is the classical resultant $R(f_1, f_2)$ of two polynomials. Let us apply Theorem 5.28. We choose the twisting line bundle $\mathcal{M}$ to be $\mathcal{O}(d_1 + d_2 - 1)$. The resultant complex $C^\bullet_-(\mathcal{L}_1, \mathcal{L}_2 \,|\, \mathcal{M})$ is then stably twisted and has the form

$$S^{d_2-1} \oplus S^{d_1-1} \overset{\partial_{f_1,f_2}}{\longrightarrow} S^{d_1+d_2-1},$$

where $\partial_{f_1,f_2}(u, v) = f_1 u + f_2 v$. Then Theorem 5.28 implies that $R(f_1, f_2)$ is equal to the determinant of the matrix ∂_{f_1,f_2} with respect to some fixed bases in S^{d_1-1}, S^{d_2-1}, and $S^{d_1+d_2-1}$. We suppose that

$$f_1 = a_0 x_0^{d_1} + a_1 x_0^{d_1-1} x_1 + \ldots + a_{d_1} x_1^{d_1},$$

$$f_2 = b_0 x_0^{d_2} + b_1 x_0^{d_2-1} x_1 + \ldots + b_{d_2} x_1^{d_2},$$

and choose in each of the spaces S^d the basis of monomials $x_0^i x_1^{d-i}$. Then we get the classical *Sylvester formula:* $R(f_1, f_2)$ is equal to

$$\begin{vmatrix} a_0 & a_1 & a_2 & \dots & a_{d_1-1} & a_{d_1} & 0 & 0 & \dots & 0 \\ 0 & a_0 & a_1 & \dots & a_{d_1-2} & a_{d_1-1} & a_{d_1} & 0 & \dots & 0 \\ \vdots & \ddots & \ddots & \ddots & \ddots & \ddots & \ddots & \ddots & \ddots & \vdots \\ 0 & 0 & 0 & a_0 & a_1 & \dots & a_{d_1-2} & a_{d_1-1} & a_{d_1} & 0 \\ 0 & 0 & 0 & 0 & a_0 & \dots & a_{d_1-3} & a_{d_1-2} & a_{d_1-1} & a_{d_1} \\ b_0 & b_1 & b_2 & \dots & b_{d_2-1} & b_{d_2} & 0 & 0 & \dots & 0 \\ 0 & b_0 & b_1 & \dots & b_{d_2-2} & b_{d_2-1} & b_{d_2} & 0 & \dots & 0 \\ \vdots & \ddots & \ddots & \ddots & \ddots & \ddots & \ddots & \ddots & \ddots & \vdots \\ 0 & 0 & 0 & b_0 & b_1 & \dots & b_{d_2-2} & b_{d_2-1} & b_{d_2} & 0 \\ 0 & 0 & 0 & 0 & b_0 & \dots & b_{d_2-3} & b_{d_2-2} & b_{d_2-1} & b_{d_2} \end{vmatrix}.$$

Example 5.30 ([GKZ2]) Let $X = \mathbb{P}^1$, $\mathcal{L}_1 = \mathcal{L}_2 = \mathcal{O}(d)$. The space $V = H^0(\mathbb{P}^1, \mathcal{O}(d))$ is identified with S^d, the space of binary forms of degree d. Let g, h be two such forms. Then R_{f_1,f_2} is the classical resultant of g and h. Consider the following bihomogeneous polynomial $F(x_0, x_1, y_0, y_1)$ of bidegree $(d-1, d-1)$:

$$F(x_0, x_1, y_0, y_1) = \frac{g(x_0, x_1)h(y_0, y_1) - h(x_0, x_1)g(y_0, y_1)}{x_o y_1 - y_0 x_1}.$$

Then F can be considered as an element of $S^{d-1} \otimes S^{d-1}$. The determinant of this matrix is equal to $R(f_1, f_2)$. This is the *Bézout formula* for the classical resultant.

Example 5.31 ([GKZ2]) Let $X = \mathbb{P}^2$, $\mathcal{L}_1 = \mathcal{L}_2 = \mathcal{O}(d)$. The space $H^0(\mathbb{P}^2, \mathcal{O}(d))$ is equal to $S^d\mathbb{C}^3$, the space of ternary forms of degree d. There is a remarkable formula for the resultant $R_{\mathcal{L}_1,\mathcal{L}_2,\mathcal{L}_3}$ due to Sylvester. Let k be an integer equal to $d-2$ or $d-1$. Let $f_1, f_2, f_3 \in S^d\mathbb{C}^3$ be three ternary forms. We first consider the operator

$$T_{f_1,f_2,f_3} : S^{2d-3-k}\mathbb{C}^3 \oplus S^{2d-3-k}\mathbb{C}^3 \oplus S^{2d-3-k}\mathbb{C}^3 \to S^{3d-3-k}\mathbb{C}^3,$$

$$(u_1, u_2, u_3) \mapsto u_1 f_1 + u_2 f_2 + u_3 f_3.$$

Then, for any three non-negative integers α, β, γ such that $\alpha + \beta + \gamma = k$, let us write

$$f_1 = x_1^{\alpha+1} P^1_{\alpha\beta\gamma} + x_2^{\beta+1} Q^1_{\alpha\beta\gamma} + x_3^{\gamma+1} R^1_{\alpha\beta\gamma},$$
$$f_2 = x_1^{\alpha+1} P^2_{\alpha\beta\gamma} + x_2^{\beta+1} Q^2_{\alpha\beta\gamma} + x_3^{\gamma+1} R^2_{\alpha\beta\gamma},$$
$$f_3 = x_1^{\alpha+1} P^3_{\alpha\beta\gamma} + x_2^{\beta+1} Q^3_{\alpha\beta\gamma} + x_3^{\gamma+1} R^3_{\alpha\beta\gamma},$$

where $P^i_{\alpha\beta\gamma}$, $Q^i_{\alpha\beta\gamma}$, $R^i_{\alpha\beta\gamma}$ are some homogeneous polynomials of degrees respectively $d-\alpha-1$, $d-\beta-1$, $d-\gamma-1$. This representation of the f_i's is clearly possible although not unique. We define the polynomial

$$D_{\alpha\beta\gamma} = \det \begin{vmatrix} P^1_{\alpha\beta\gamma} & Q^1_{\alpha\beta\gamma} & R^1_{\alpha\beta\gamma} \\ P^2_{\alpha\beta\gamma} & Q^2_{\alpha\beta\gamma} & R^2_{\alpha\beta\gamma} \\ P^3_{\alpha\beta\gamma} & Q^3_{\alpha\beta\gamma} & R^3_{\alpha\beta\gamma} \end{vmatrix}$$

of degree $3d-3-k$. Define the linear map

$$D : (S^k\mathbb{C}^3)^\vee \to S^{3d-3-k}\mathbb{C}^3$$

by sending $\delta_{\alpha\beta\gamma} \mapsto D_{\alpha\beta\gamma}$, where $\{\delta_{\alpha\beta\gamma}\}$ is the basis dual to the monomial basis in $S^k\mathbb{C}^3$.

Finally, consider the linear operator

$$T_{f_1,f_2,f_3}+D : S^{2d-3-k}\mathbb{C}^3 \oplus S^{2d-3-k}\mathbb{C}^3 \oplus S^{2d-3-k}\mathbb{C}^3 \oplus (S^k\mathbb{C}^3)^\vee \to S^{3d-3-k}\mathbb{C}^3.$$

An elementary calculation shows that, for $k = d-2$ or $k = d-1$, the dimension of the source and the target of this operator are the same.

Let us show that the determinant of $T_{f_1,f_2,f_3} + D$ is independent of the arbitrary choices made in the definition of D, more precisely, that a different choice of $P^i_{\alpha\beta\gamma}$, $Q^i_{\alpha\beta\gamma}$, $R^i_{\alpha\beta\gamma}$ changes $D_{\alpha\beta\gamma}$ by adding a polynomial from the image of T_{f_1,f_2,f_3}. Indeed, suppose that we choose a different representation for, say f_1, i.e., we replace

$$P^i_{\alpha\beta\gamma} \mapsto P^i_{\alpha\beta\gamma} + P, \quad Q^i_{\alpha\beta\gamma} \mapsto Q^i_{\alpha\beta\gamma} + Q, \quad R^i_{\alpha\beta\gamma} \mapsto R^i_{\alpha\beta\gamma} + Q,$$

where

$$x_1^{\alpha+1}P + x_2^{\beta+1}Q + x_3^{\gamma+1}R = 0.$$

Then the difference between the new and the old value of $D_{\alpha\beta\gamma}$ is equal to

$$\det \begin{vmatrix} P & Q & R \\ P^2_{\alpha\beta\gamma} & Q^2_{\alpha\beta\gamma} & R^2_{\alpha\beta\gamma} \\ P^3_{\alpha\beta\gamma} & Q^3_{\alpha\beta\gamma} & R^3_{\alpha\beta\gamma} \end{vmatrix} \qquad (*)$$

Since the Koszul complex associated to $x_1^{\alpha+1}, x_2^{\beta+1}, x_3^{\gamma+1}$ is exact, the space of triples (P,Q,R) such that $x_1^{\alpha+1}P + x_2^{\beta+1}Q + x_3^{\gamma+1}R = 0$ is linearly generated by triples of three types:

$$(Ux_2^{\beta+1}, -Ux_1^{\alpha+1}, 0), \quad (Vx_3^{\gamma+1}, 0, -Vx_1^{\alpha+1}, 0), \quad (0, Wx_3^{\gamma+1}, -Wx_2^{\beta+1}, 0),$$

where U, V, W are some homogeneous polynomials. For example, if we take (P,Q,R) of the first type then $(*)$ is equal to

$$\det \begin{vmatrix} Ux_2^{\beta+1} & -Ux_1^{\alpha+1} & 0 \\ P^2_{\alpha\beta\gamma} & Q^2_{\alpha\beta\gamma} & R^2_{\alpha\beta\gamma} \\ P^3_{\alpha\beta\gamma} & Q^3_{\alpha\beta\gamma} & R^3_{\alpha\beta\gamma} \end{vmatrix}.$$

The Laplace expansion along the first row shows that the determinant is equal to $U(R^3_{\alpha\beta\gamma}f_2 - R^2_{\alpha\beta\gamma}f_3)$ and therefore belongs to the image of T_{f_1,f_2,f_3}. The other types of (P,Q,R) are considered similarly.

Now the result of Sylvester is as follows:

$$R_{\mathcal{L}_1,\mathcal{L}_2,\mathcal{L}_3}(f_1,f_2,f_3) = \det(T_{f_1,f_2,f_3} + D).$$

The operator T_{f_1,f_2,f_3} is equal to the differential ∂_1 of the E^1 term of the corresponding spectral sequence (with twisting bundle $\mathcal{O}(3d-3-k)$), and the map D is the differential ∂_3 in E^3, whose values are defined only modulo the image of ∂_1 (this is why the definition of D involves arbitrary choices).

6

Degree of the Dual Variety

Here we discuss the degree of dual varieties and resultants. We start in Sect. 6.1 by recalling Chern classes and then proving a formula of Katz, Kleiman, and Holme that expresses the degree of a dual variety in terms of Chern classes of the cotangent bundle. We give many examples and generalizations. We prove a theorem of De Concini and Weyman about the formula for the degree with non-negative coefficients. In Sect. 6.2 we discuss formulas for the codegree and ranks related to the Cayley method, such as a formula due to Lascoux.

6.1 Katz–Kleiman–Holme Formula

6.1.1 Chern Classes

Let $X \subset \mathbb{P}(V)$ be a smooth projective variety and let $\mathcal{L} = \mathcal{O}_X(1)$. Let us recall the definitions of the Grothendieck ring $K(X)$, Chow groups $A^i(X)$, and Chern classes. $K(X)$ is generated by classes $[\mathcal{F}]$ of coherent sheaves. For any short exact sequence

$$0 \to \mathcal{F} \to \mathcal{G} \to \mathcal{H} \to 0$$

we impose the relation

$$[\mathcal{G}] = [\mathcal{F}] + [\mathcal{H}].$$

By the Hilbert syzygy theorem [GH1] any coherent sheaf $\mathcal{F}$ on X has a finite resolution by locally free sheaves. Therefore $K(X)$ is generated by classes of vector bundles and we can extend to $K(X)$ most constructions available for locally free sheaves. For example, the ring structure on $K(X)$ is first defined on vector bundles using tensor multiplication and then extended to the whole of $K(X)$. The function taking a vector bundle to its rank extends to a homomorphism $\mathrm{rk} : K(X) \to \mathbb{Z}$ called the generic rank. The Euler characteristic χ defines another homomorphism from $K(X)$ to $\mathbb{Z}$.

The group $K(X)$ is equipped with the codimension filtration

$$K(X) = K(X)^0 \supset K(X)^1 \supset \dots,$$

where $K(X)^i$ is generated by classes of sheaves with support on a subvariety of codimension at least i. The quotient $K(X)^i/K(X)^{i+1}$ is isomorphic to the Chow group $A^i(X)$ of codimension i cycles on X modulo rational equivalence. More precisely, $A^i(X)$ is an abelian group generated by classes $[Z]$ of irreducible subvarieties $Z \subset X$ of codimension i with relations $\sum_i [Z_i] = 0$ if there exists a subvariety $Y \subset X$ of codimension $i-1$ such that $\sum Z_i$ is the divisor of a rational function on Y. The isomorphism $A^i(X) \to K(X)^i/K(X)^{i+1}$ is established by assigning to each irreducible subvariety $Z \subset X$ of codimension i its structure sheaf $\mathcal{O}_Z$ regarded as a sheaf on X.

Consider the exterior power operations $\Lambda^i : K(X) \to K(X)$. For classes of vector bundles, Λ^i is the usual exterior power and, for an exact sequence of vector bundles $0 \to \mathcal{F} \to \mathcal{G} \to \mathcal{H} \to 0$, we have $[\Lambda^p \mathcal{G}] = \sum_{i+j=p} [\Lambda^i \mathcal{F}] \cdot [\Lambda^j \mathcal{H}]$. For any vector bundle $\mathcal{F}$, consider the formal power series $\lambda_{\mathcal{F}}(t) = \sum_i [\Lambda^i \mathcal{F}] t^i$. Then we define, for any $\alpha = \sum m_i [\mathcal{F}_i] \in K(X)$, the element $\Lambda^p \alpha$ as the p-th coefficient of $\prod \lambda_{\mathcal{F}_i}(t)^{m_i}$.

There is a natural homomorphism $A^r(X) \to H_{2 \dim X - 2r}(X, \mathbb{Z})$, which takes an algebraic cycle into its homology class. The Poincaré duality on X identifies $H_{2 \dim X - 2r}(X, \mathbb{Z})$ with $H^{2r}(X, \mathbb{Z})$. Grothendieck has defined Chern classes $c_r : K(X) \to A^r(X)$ which lift the ordinary Chern classes with values in $H^{2r}(X, \mathbb{Z})$ (see e.g. [Ful]). By definition, if $\alpha \in K(X)$ is an element of generic rank p, then

$$c_r(\alpha) = \Lambda^r(\alpha - (p - r + 1)[\mathcal{O}_X]) \mod K(X)^{r+1}.$$

If E is a spanned vector bundle, i.e. its fibers are generated by restrictions of global sections, then the Chern classes $c_r(E)$ have a fairly transparent geometric interpretation (see e.g. [Ful]). Namely, $c_r(E)$ is represented by the cycle

$$\{x \in X \mid s_1(x), \dots, s_{p-r+1}(x) \text{ are linearly dependent}\},$$

where $s_1, \dots, s_{p-r+1} \in H^0(X, E)$ are generic global sections. For instance, let $r = p = \operatorname{rk} E$ and suppose that E is spanned. Then we see that $c_r(E)$ is represented by the zero locus $Z(s)$ of a generic global section $s \in H^0(X, E)$. For any 0-dimensional cycle Z on X we denote its degree by $\int_X Z$, i.e. $\int_X$ is the notation for the natural map $A^n(X) \to H_0(X, \mathbb{Z})$.

6.1.2 Top Chern Class of the Jet Bundle

Let $\dim X = n$.

Theorem 6.1 ([BFS1])

(i) *X^* is a hypersurface if and only if $c_n(J(\mathcal{L})) \neq 0$. Moreover,*

$$\deg \Delta_X = \int_X c_n(J(\mathcal{L})).$$

(ii) *More generally,*

$$\deg X^* = \int_X c_{n-\operatorname{def} X}(J(\mathcal{L})) \cdot H^{\operatorname{def} X}$$

and

$$\operatorname{def} X = \min\{k \mid c_{n-k}(J(\mathcal{L})) \neq 0\}.$$

Proof. Suppose that $\operatorname{def} X = k$. Notice that jets $j(f)$, $f \in V^\vee$, span $J(\mathcal{L})$. We choose a generic basis

$$f_1, \ldots, f_v \in V^\vee.$$

$c_{n-k+1}(J(\mathcal{L}))$ is represented by the class

$$\{x \in X \mid j(f_1)(x), \ldots, j(f_{k+1})(x) \text{ are linearly dependent}\}.$$

Since $\operatorname{codim}_{\mathbb{P}(V^\vee)} X^* = k+1$, a generic k-dimensional projective subspace in $\mathbb{P}(V^\vee)$ does not meet X^*. Therefore any non-trivial linear combination of $f_1, \ldots, f_{k+1}$ does not belong to $\operatorname{Cone}(X^*)$. But this is equivalent to

$$\dim \operatorname{Span}\{j(f_1)(x), \ldots, j(f_{k+1})(x)\} = k+1$$

for any $x \in X$. This implies that

$$c_r(J(\mathcal{L})) = 0 \quad \text{for} \quad r \geq n-k+1.$$

Now, the Chern class $c_{n-k}(J(\mathcal{L}))$ is represented by the class

$$\{x \in X \mid j(f_1)(x), \ldots, j(f_{k+2})(x) \text{ are linearly dependent}\}.$$

This class is a union of singular loci of hyperplane sections $H \cap X$, where $H \in X^* \cap L$ for a generic $(k+1)$-dimensional subspace $L \subset \mathbb{P}(V^\vee)$. Notice that the number of these hyperplane sections is equal to the degree of the dual variety $\deg X^*$. Moreover, for dimension reasons we may suppose that $X^* \cap L \subset (X^*)_{sm}$. By Theorem 1.18 these singular loci are projective subspaces of dimension k. Therefore $c_{n-k}(J(\mathcal{L}))$ is represented by $\deg X^*$ k-dimensional projective subspaces. In particular, $c_{n-k}(J(\mathcal{L})) \neq 0$ and $\deg X^* = \int_X c_{n-k}(J(\mathcal{L})) \cdot H^k$. □

Using the exact sequence (5.1) it is possible to substitute Chern classes of the jet bundle by Chern classes of the cotangent bundle $T^\vee_X$. The resulting formula was found by N. Katz and S. Kleiman [Ka, Kl1] in the case where X^* is a hypersurface and by A. Holme [Ho1, Ho2] in the general case. Consider the Chern polynomial of X with respect to the given projective embedding

$$c_X(q) = \sum_{i=0}^{n} q^{i+1} \int_X c_{n-i}(T^\vee_X) \cdot H^i.$$

Theorem 6.2

(i)

$$\deg \Delta_X = c'_X(1) = \sum_{i=0}^{n} (i+1) \int_X c_{n-i}(T_X^\vee) \cdot H^i.$$

(ii) *The codimension of* X^* *equals the order of the zero of the polynomial* $c_X(q) - c_X(1)$ *at* $q = 1$. *If this order is* μ *then* $\deg X^* = c_X^{(\mu)}(1)/\mu!$.

The formula in (i) can be rewritten in numerous ways.

Example 6.3 ([Kl1])

$$\deg \Delta_X = \int_X \frac{1}{c(N^\vee_{X,\mathbb{P}^N}(1))}.$$

If X is a hypersurface of degree d in $\mathbb{P}^N$, then $N^\vee_{X,\mathbb{P}^N}(1) = \mathcal{O}_X(1-d)$ and

$$\deg \Delta_X = \int_X \frac{1}{1-(d-1)H} = \int_X (d-1)^n H^n = d(d-1)^n.$$

In particular, X^* is a hypersurface (if X is not a hyperplane). In the same way one can show that X^* is a hypersurface if X is a non-degenerate smooth complete intersection (c.f. Theorem 5.11).

Example 6.4 Suppose that

$$X = \mathbb{P}^n \subset \mathbb{P}(S^d\mathbb{C}^{n+1})$$

is the Veronese embedding. Then elements of X^* parametrize hypersurfaces with singularities of degree d in $\mathbb{P}^n$, and therefore Δ_X is the classical discriminant of homogeneous forms of degree d in $n+1$ variables. Let us find its degree. The Chow ring of X is equal to $\mathbb{Z}[t]/t^{n+1}$, the element t being the class of the hyperplane. The first Chern class of the line bundle corresponding to the Veronese embedding is equal to dt. The total Chern class

$$\sum_i c_i(T^\vee_{\mathbb{P}^n}) = (1-t)^{n+1},$$

and therefore

$$c_i(T^\vee_{\mathbb{P}^n}) = (-1)^i \binom{n+1}{i} t^i.$$

Clearly $\int_X t^n = 1$. Hence we have

$$c_X(q) = \sum_{i=0}^{n} q^{i+1}(-1)^{n-i}\binom{n+1}{i+1} d^i = \frac{(qd-1)^{n+1} - (-1)^{n+1}}{d}.$$

Finally, we get the *Boole formula*

$$\deg \Delta_X = c'_X(1) = (n+1)(d-1)^n.$$

Example 6.5 Let X be a n-dimensional smooth projective variety in $\mathbb{P}^N$ and let $\mathcal{L} = \mathcal{O}_X(1)$. Then the s-th *rank* of X with respect to the projective embedding is defined as

$$\delta_s = \sum_{i=s}^{n} \binom{i+1}{s+1} e_{n-i},$$

where e_j denotes the degree of $c_j(T^\vee X)$ with respect to the given projective embedding. Then Theorem 6.2 can be reformulated as follows:

$$\operatorname{def} X = \min\{r \mid \delta_r \neq 0\} \quad \text{and} \quad \deg X^* = \delta_{\operatorname{def} X}.$$

6.1.3 Formulas with Positive Coefficients

Theorem 6.6 *If $\mathcal{L}$ and $\mathcal{M}$ are very ample line bundles on X then*

$$\operatorname{def}(X, \mathcal{L} \otimes \mathcal{M}) = 0.$$

Proof. Indeed, we have

$$\deg \Delta_{(X,\mathcal{L}\otimes\mathcal{M})} = \int_X c_n(J(\mathcal{L} \otimes \mathcal{M}))$$

by Theorem 6.1. Therefore

$$\deg \Delta_{(X,\mathcal{L}\otimes\mathcal{M})} = \int_X c_n(J(\mathcal{L}) \otimes \mathcal{M}) = \sum_{i=0}^{n} (i+1) \int_X c_1(\mathcal{M})^i \cdot c_{n-i}(J(\mathcal{L})).$$

Since $J(\mathcal{L})$ is spanned, all summands are non-negative. Since $\int_X c_1(\mathcal{M})^n > 0$ (being equal to the degree of X in the embedding determined by $\mathcal{M}$), the total sum is positive. □

Example 6.7 *If X is embedded in $\mathbb{P}(V)$ and then re-embedded in $\mathbb{P}(S^d V)$ via the Veronese embedding then the dual variety X^* of this re-embedding is a hypersurface.*

Suppose now that $\mathcal{L}_1, \ldots, \mathcal{L}_r$ are line bundles on X such that the corresponding linar systems $|\mathcal{L}_i|$ have no base points. Suppose further that any line bundle $\mathcal{L}$ of the form $\mathcal{L} = \mathcal{L}_1^{\otimes n_1} \otimes \ldots \otimes \mathcal{L}_r^{\otimes n_r}$ is very ample for positive n_i. Then $\deg \Delta_{(X,\mathcal{L})}$ is a function in $n_1, \ldots, n_r$. Let us introduce new non-negative integers $m_i = n_i - 1$. Then $\deg \Delta_{(X,\mathcal{L})} = f(m_1, \ldots, m_r)$.

Theorem 6.8 ([CW])

(i) *f is a non-trivial polynomial with non-negative coefficients.*
(ii) *If each $n_i \geq 2$ then $(X, \mathcal{L})^*$ is a hypersurface.*

Proof. Let $n = \dim X$. Recall that by (6.2) we have

$$\deg \Delta_{(X,\mathcal{L})} = \sum_{i=0}^{n} (i+1) \int_X c_{n-i}(T_X^\vee) \cdot \mathcal{L}^i = P'(\mathcal{L}),$$

where

$$P(t) = \sum_{i=0}^{n} c_{n-i}(T_X^\vee) t^{i+1}.$$

This clearly implies that f is a polynomial in n_i and, hence, in m_i. This polynomial is non-trivial by Theorem 6.6. It remains to check positivity. Using the Taylor expansion, it suffices to check the following claim. For each $j = 1, \ldots, n$ and for each collection of nonnegative integers $u_1, \ldots, u_r$ such that $u_1 + \ldots + u_r = j - 1$, we have

$$\int_X P^{(j)}(\mathcal{L}) \cdot \mathcal{L}_1^{u_1} \ldots \mathcal{L}_r^{u_r} \geq 0.$$

The conditions imposed on $\mathcal{L}_i$ and the Bertini theorem together imply that any intersection of the form $\mathcal{L}_1^{u_1} \ldots \mathcal{L}_r^{u_r}$ (as an element of a Chow ring) is represented by a smooth subvariety of codimension $j-1$. Therefore it suffices to check that for any such subvariety $Y \subset X$ of codimension $j-1$ we have

$$\int_X P^{(j)}(\mathcal{L}) \cdot [Y] \geq 0.$$

Using the exact sequence 5.1 it is easy to see that

$$P^{(j)}(\mathcal{L}) = c_{n+1-j}(J(\mathcal{L})) \cdot j!$$

Therefore it remains to show that

$$\int_X c_{n+1-j}(J(\mathcal{L})) \cdot [Y] \geq 0.$$

But this is clear because $J(\mathcal{L})$ is spanned. □

6.1.4 Degree of the Resultant

Let X be a smooth irreducible projective variety and let E be a vector bundle on X of rank $k = \dim X + 1$. Recall that the resultant variety $\nabla_X \subset \mathbb{P}(V)$ is the set of all sections vanishing at some point $x \in X$. ∇_X is a hypersurface by Theorem 5.14.

Theorem 6.9 ([GKZ2]) $\deg \nabla_X = \int_X c_{\dim X}(E)$.

Proof. Take two generic sections s_1 and s_2 of V and find the number of values $\lambda \in \mathbb{C}$ such that $s_1 + \lambda s_2$ vanishes at some $x \in X$. In other words, we need to find the number of points (with multiplicities) of a zero-dimensional cycle $Z \subset X$ consisting of all points $x \in X$ where s_1 and s_2 are linearly dependent. But this is exactly the geometric definition of $c_{k-1}(E)$; see [Ful]. □

6.2 Formulas Related to the Cayley Method

6.2.1 Degree of the Discrminant

Theorem 5.23 identifies Δ_X with the determinant of any of the discriminant complexes $C_{\pm}^{\bullet}(X, \mathcal{M})$, provided the complex is stably twisted. By Corollary 5.20 this gives rise to a formula for $\deg \Delta_X$:

Theorem 6.10 ([GKZ2]) *If $C_{-}^{\bullet}(X, \mathcal{M})$ is stably twisted then*

$$\deg \Delta_X = \sum_{i=-\dim X-1}^{0} (-1)^i \cdot i \cdot \dim C_{-}^{i}(X, \mathcal{M}).$$

If $C_{+}^{\bullet}(X, \mathcal{M})$ is stably twisted then

$$\deg \Delta_X = (-1)^{\dim X+1} \sum_{i=0}^{\dim X+1} (-1)^i \cdot i \cdot \dim C_{+}^{i}(X, \mathcal{M}).$$

Thus $\deg \Delta_X$ can be expressed through the dimensions of vector spaces

$$H^0\left(X, \Lambda^{-i} J(\mathcal{L})^{\vee} \otimes \mathcal{M}\right) \text{ or } H^0\left(X, \Lambda^{i} J(\mathcal{L}) \otimes \mathcal{M}\right).$$

It is possible to rewrite these dimensions using simpler quantities associated with X, at least in the case $\mathcal{M} = \mathcal{O}_X(l)$, $l \gg 0$. For any coherent sheaf $\mathcal{F}$ on X, we write

$$\mathcal{F}(l) = \mathcal{F} \otimes \mathcal{O}_X(l),\ h^i(\mathcal{F}) = \dim H^i(X, \mathcal{F}),\ \chi(\mathcal{F}) = \sum (-1)^i h^i(\mathcal{F}).$$

The number $\chi(\mathcal{F})$ is called the *Euler characteristic of* $\mathcal{F}$. It follows from the Riemann-Roch-Hirzebruch theorem [Hir] that $\chi(\mathcal{F}(l))$ is a polynomial in l. This polynomial is called the *Hilbert polynomial* of $\mathcal{F}$ and is denoted by $h_{\mathcal{F}}(l)$. For $l \gg 0$, the higher cohomology of $\mathcal{F}(l)$ vanish and we have

$$h_{\mathcal{F}}(l) = \chi(\mathcal{F}(l) = h^0(\mathcal{F}(l)).$$

Simple calculations show that Theorem 6.10 can be reformulated as follows. Recall that T_X is the tangent bundle of X and $\Omega_X^i = \Lambda^i T_X^*$ is the bundle of differential i-forms on X.

Theorem 6.11 ([GKZ2]) *For any $l \in \mathbb{Z}$ we have*

$$\deg \Delta_X = \sum_{i=-\dim X-1}^{0} (-1)^i \cdot i \cdot \left(h_{\Lambda^{-i}T_X}(i+l) + h_{\Lambda^{-i-1}T_X}(i+l)\right),$$

$$\deg \Delta_X = (-1)^{\dim X+1} \sum_{i=0}^{\dim X+1} (-1)^i \cdot i \cdot \left(h_{\Omega_X^i}(i+l) + h_{\Omega_X^{i-1}}(i+l)\right).$$

This theorem was generalized in [GK] to handle the case of positive def X. For any $l \in \mathbb{Z}$ consider the polynomial $f_l(q)$ in a formal variable q

$$f_l(q) = \sum_{i=0}^{\dim X+1} (-1)^i \left(h_{\Lambda^i T_X}(i+l) + h_{\Lambda^{i-1} T_X}(i+l)\right) q^i,$$

where all polynomials that do not make sense are equal to 0. Clearly $f_l(q)$ is also a polynomial in l.

Theorem 6.12 ([GK, GKZ2]) *Let $l \in \mathbb{Z}_+$ be any non-negative integer. The codimension of X^* equals the order of the zero of $f_l(q)$ at $q = 1$. Let this order be μ and let*

$$f_l(q) = a_\mu(l)(1-q)^\mu + O((1-q)^{\mu+1}).$$

Then

$$a_\mu(l) = \binom{\mu+l-1}{\mu-1} \cdot \deg X^*.$$

6.2.2 Lascoux Formula

If the coherent sheaf $\mathcal{F}$ is supported on the subvariety of codimension n, i.e. on the set of points $\{x_i\}$, then $\mathcal{F}_{x_i}$ is a vector space of dimension n_i and the corresponding element of $A^n(X)$ is just $\sum n_i x_i$. The corresponding element of $H_0(X, \mathbb{Z}) = \mathbb{Z}$ is just $\sum n_i$, and therefore the homomorphism

$$K(X)^n \to A^n(X) \to H_0(X, \mathbb{Z})$$

is obtained by taking the Euler characteristic. It follows that if $\alpha \in K(X)$ is an element of generic rank p then

$$\int_X c_n(\alpha) = \chi\left(\Lambda^n(\alpha - (p-n+1)[\mathcal{O}_X]\right).$$

Theorem 6.13 ([Las]) *For any $p \in \mathbb{Z}$,*

$$\deg \Delta_X = \chi\left(\Lambda^n([T_X^\vee] - 2[\mathcal{L}^\vee] + [\mathcal{O}_X])[\mathcal{L}]^p\right).$$

Proof. Indeed, by Theorem 6.1, we have

$$\deg \Delta_X = \int_X c_n(J(\mathcal{L})).$$

By the calculations above, we have

$$\int_X c_n(J(\mathcal{L})) = \chi\left(\Lambda^n([J(\mathcal{L})] - 2[\mathcal{O}_X]\right).$$

From the exact sequence

$$0 \to T_X^\vee \otimes \mathcal{L} \to J(\mathcal{L}) \to \mathcal{L} \to 0$$

we get

$$[J(\mathcal{L})] = [\mathcal{L}] + [\mathcal{L}][T_X^\vee],$$

and therefore

$$\deg \Delta_X = \chi\left(\Lambda^n([\mathcal{L}] + [\mathcal{L}][T_X^\vee] - 2[\mathcal{O}_X]\right).$$

It remains to tensor the right hand side with $[\mathcal{L}^\vee]^n \cdot [\mathcal{L}]^p$. This is possible, because if $\mathcal{F}$ is supported on a finite set. Then, obviously, $\chi([\mathcal{F}] \cdot M) = \chi([\mathcal{F}])$ for any line bundle M. □

7

Varieties with Positive Defect

In this chapter we study varieties with positive defect. In Sect. 7.1 we focus on beautiful theorems of Ein about the normal bundle of a generic contact locus. Since this locus is a projective subspace, it is possible to use the machinery of vector bundles on projective spaces. We prove Ein's theorem that this normal bundle is symmetric and uniform, which explains among other things a parity theorem of Landman. We introduce the Beilinson spectral sequence and use it to calculate the normal bundle to a generic contact locus in small dimensions. Finally, we study dual varieties of scrolls and prove a theorem of Ein that a variety of defect at least 2 is a scroll if and only if the normal bundle to a generic contact locus splits. In Sect. 7.2 we follow [IL] and discuss linear systems of quadrics of constant rank and how they are related to dual varieties via the second fundamental form. In Sect. 7.3 we prove a theorem of Beltrametti, Fania, and Sommese that relates the defect of a projective variety and its Mori-theoretic characteristic called the nef value. We give a brief survey of necessary results from Mori theory. We finish by giving a classification of smooth varieties of positive defect up to dimension 10 obtained by many authors and initiated by Ein. Finally, in Sect. 7.4 we use this connection with Mori theory to classify all flag varieties with positive defect. This approach was developed by Snow in contrast with the original proof of Knop and Menzel that used the Katz dimension formula.

7.1 Normal Bundle of the Contact Locus

7.1.1 Ein Theorems

Suppose that $X \subset \mathbb{P}^N$ is a non-degenerate smooth projective variety, $\dim X = n$. For any hyperplane $H \subset \mathbb{P}^N$ the contact locus $(X \cap H)_{sing}$ is a subvariety of X consisting of all points $x \in X$ such that the embedded tangent space $\hat{T}_{x,X}$ is contained in H. One can use the Jacobian ideal of $X \cap H$ to define a scheme structure on the contact locus. This scheme could however be not reduced. Clearly the contact locus is non-empty if and only if H belongs to the dual variety X^*. By Theorem 1.18, if $\operatorname{def} X = d$ then for any $H \in (X^*)_{sm}$ the contact locus $(H \cap X)_{sing}$ is a projective subspace of dimension d, and a union of these projective subspaces is dense in X. In this section we study further properties of the contact locus.

Theorem 7.1 ([E1]) *Suppose that q is a generic point of X, that H is generic tangent hyperplane of X at q, and that $L = \mathbb{P}^d$ is the contact locus of H with X and $d > 0$. Then there exists a symmetric isomorphism $N_{L,X} \cong N^\vee_{L,X}(1)$.*

Proof. Let s_h be the section of $\mathcal{O}_X(1)$ defining $H \cap X$ and let $I_L \subset \mathcal{O}_X$ be the ideal sheaf of L in X. We choose a local coordinate system $\{x_1, \ldots, x_n\}$ of X near p. We may assume that I_L is generated by $x_1, \ldots, x_{n-d}$. Since $L \subset H \cap X$, we can write the power series of s_h in the following form:

$$s_h = x_1 f_1 + \ldots + x_{n-d} f_{n-d} + \sum_{i,j=1}^{n-d} x_i x_j g_{ij},$$

where $f_1, \ldots, f_{n-d}$ are power series in variables $x_{n-d+1}, \ldots, x_n$. But $(H \cap X)_{sing} = L$, and therefore $(\partial s_h / \partial x_i)|_L = 0$ for $i = 1, \ldots, n-d$. Hence

$$f_1 = \ldots = f_{n-d} = 0 \quad \text{and} \quad s_h = \sum_{i,j=1}^{n-d} x_i x_j g_{ij}.$$

Thus $s_h \in I_L^2(1)$. Therefore s_h gives a section of

$$I_L^2/I_L^3(1) = S^2(N^\vee_{L,X})(1) \subset \operatorname{Hom}(N_{L,X}, N^\vee_{L,X}(1))$$

and induces a map from $N_{L,X}$ to $N_{L,X}(1)$. It remains to show that this map is an isomorphism. We can write $g_{ij} = a_{ij} + h_{ij}$, where the a_{ij} are constants and the h_{ij} are power series without the constant term. Now

$$s_h = \sum_{i,j=1}^{n-d} x_i x_j (a_{ij} + h_{ij}).$$

It follows from Theorem 3.1, using Sard's Lemma for algebraic varieties, that the tangent cone of the hyperplane section $H \cap X$ at p is a quadric hypersurface of rank $n-d$ in $\hat{T}_{p,X}$. Therefore $\sum_{i,j=1}^{n-d} x_i x_j a_{ij}$ is a quadratic form of rank $n-d$ and this is also the equation for the quadric hypersurface in $\mathbb{P}(N_{L,X}|_p)$ induced by s_h. □

Recall that by a theorem of Grothendieck any vector bundle on $\mathbb{P}^1$ has a form $\oplus_i \mathcal{O}(a_i)$ for some integers a_i (see e.g. [Ha1]).

Definition 7.2 A vector bundle E on a projective space $\mathbb{P}^N$ is called *uniform* if the restriction $E|_T$ is a fixed vector bundle $\oplus_i \mathcal{O}(a_i)$ for any line $T \subset \mathbb{P}^N$.

Theorem 7.3 ([E1]) *Suppose that q is a generic point of X, that H is generic tangent hyperplane of X at q, and that $L = \mathbb{P}^d$ is the contact locus of H with X and $d > 0$. Then $N_{L,X}$ is a uniform vector bundle and*

$$N_{L,X}|_T \cong \mathcal{O}_T^{\oplus(n-d)/2} \oplus \mathcal{O}_T(1)^{\oplus(n-d)/2}$$

for any line T in L.

Proof. Consider the exact sequence

$$0 \to N_{L,X}|_T \to N_{L,\mathbb{P}^N}|_T \to N_{X,\mathbb{P}^N}|_T \to 0,$$

where $N_{L,\mathbb{P}^N}|_T = (N-d)\mathcal{O}_T(1)$. Suppose that

$$N_{L,X}|_T = \bigoplus_{i=1}^{n-d} \mathcal{O}_T(a_i).$$

Then exactness implies that all $a_i \leq 1$. Using the isomorphism between $N_{L,X}$ and $N_{L,X}^{\vee}(1)$ we observe that each $a_i \geq 0$. Therefore

$$N_{L,X}|_T \cong \mathcal{O}_T^{\oplus(n-d)/2} \oplus \mathcal{O}_T(1)^{\oplus(n-d)/2}$$

and $N_L X$ is a uniform vector bundle. This implies that

$$N_{T,X} \cong \mathcal{O}_T^{\oplus(n-d)/2} \oplus \mathcal{O}_T(1)^{\oplus(n+d-2)/2}$$

and the theorem is proved. □

As a formal consequence of Theorem 7.3 we get the following parity theorem that was first proved by A. Landman, using the Picard–Lefschetz theory (unpublished):

Theorem 7.4 ([E1]) *If* $\operatorname{def} X > 0$ *then* $\dim X \equiv \operatorname{def} X \mod 2$.

Example 7.5 ([GH]) Suppose that $X \subset \mathbb{P}^N$ is a smooth non-linear surface. Then X^* is a hypersurface. Indeed, if $\operatorname{def} X > 0$ then $\operatorname{def} X$ is even by Theorem 7.4. But X obviously can not contain an even-dimensional projective subspace $\mathbb{P}^{\operatorname{def} X}$.

Example 7.6 Suppose that $X \subset \mathbb{P}^N$ is a smooth projective nonlinear variety with $\dim X \geq 2$. Then $\operatorname{def} X \leq \dim X - 2$. Indeed, it is clear that $\operatorname{def} X \leq \dim X - 1$. Therefore, by Theorem 7.4 we have $\operatorname{def} X \leq \dim X - 2$. Moreover, it can be shown [E1] that $\operatorname{def} X = \dim X - 2$ if and only if X is a projective bundle over a curve C, $X = \mathbb{P}_C(F)$, where F is a rank n vector bundle on C and all fibers are embedded into $\mathbb{P}^N$ linearly (cf. Theorem 5.14 and remarks after it).

Example 7.7 Suppose that $X \subset \mathbb{P}^N$ is a smooth projective nonlinear variety, $n = \dim X \geq 3$, $N = 2n - 1$, and $\dim X = \dim X^*$. Since in this case $\operatorname{def} X = n - 2$, by Example 7.6 X is a projective bundle over a curve C, $X = \mathbb{P}_C(F)$, where F is a rank n vector bundle on C and all fibers are embedded into $\mathbb{P}^N$ linearly. Moreover, by a theorem of S. Kleiman [Kl5] X is the Segre embedding of $\mathbb{P}^1 \times \mathbb{P}^{n-1}$ in this case.

Theorem 7.8 ([Mun3]) *Let $X \subset \mathbb{P}^N$ be a variety with positive defect d, H a generic point of X^* and L the contact locus of X with H. Then*

$$N_{L,X}^\vee = [\mathcal{O}_L \oplus \mathcal{O}_L(-1)]^{\oplus h^0(N_{L,X}^\vee)} \oplus F,$$

where F is a vector bundle such that $h^0(F) = 0$ and $F \simeq F^\vee(-1)$ by a symmetric isomorphism.

Proof. If $h^0(N_{L,X}^\vee) = 0$ the proposition is vacuous. If not, take a global section of $N_{L,X}^\vee$. With this chosen section we can construct a morphism $N_{L,X} \to \mathcal{O}_L \to 0$. This gives the exact sequence

$$0 \to F_1 \to N_{L,X} \to \mathcal{O}_L \to 0. \tag{$*$}$$

From the exact sequence

$$0 \to N_{L,X} \to \mathcal{O}_L(1)^{N-d} \to N_{X,\mathbb{P}^N}|_L \to 0$$

it follows that $N_{L,X}^\vee(1) = N_{L,X}$ is globally generated.

This shows that $(*)$ is split (cf. the proof of Theorem 5.2) and so $N_{L,X} = F_1 \oplus \mathcal{O}_L$. Since $N_{L,X} \simeq N_{L,X}^\vee(-1)$ we have

$$N_{L,X} = \mathcal{O}_L \oplus F_2 \oplus \mathcal{O}_L(1),$$

where F_2 is a vector bundle such that $\operatorname{rk}(F_2) = \operatorname{rk}(N_{L,X}) - 2$. We can iterate this process and obtain the claim. □

Example 7.9 ([Mun3]) If there exists a linear space M of dimension m such that $L \subset M \subset X$ then

$$N_L^\vee X = \mathcal{O}_L^{\oplus m-k} \oplus F \oplus \mathcal{O}_L(-1)^{\oplus m-k)},$$

where F is a vector bundle such that $F \simeq F^\vee(-1)$ by a symmetric isomorphism. Indeed, from the exact sequence

$$0 \to N_{L,M}(-1) \to N_{L,X}(-1) \to N_{M,X}(-1)|_L \to 0$$

we see that $H^0(N_{L,M}(-1))$ is a vector subspace of $H^0(N_{L,X}^\vee)$ and we finish by Theorem 7.8.

It was first observed by Griffiths and Harris [GH] that any smooth projective variety with positive defect has negative Kodaira dimension, i.e.

$$H^0(X, K_X^{\otimes m}) = 0$$

for $m > 0$, where K_X is the canonical line bundle.

Theorem 7.10 ([E1]) *Suppose that q is a generic point of X, that H is generic tangent hyperplane of X at q, and that $L = \mathbb{P}^d$ is the contact locus of H with X and $d > 0$. Then*

$$K_X|_L = \mathcal{O}_L\left(\frac{-n-d-2}{2}\right) \tag{7.1}$$

and the Kodaira dimension of X is negative.

Proof. Let $r = (n-d)/2$. By Theorem 7.3, if $T = \mathbb{P}^1$ is a line in $L = \mathbb{P}^d$, then $N_{L,X}|_T \cong \mathcal{O}_T^{\oplus r} \oplus \mathcal{O}_T(1)^{\oplus r}$. It follows that $\Lambda^{n-d} N_{L,X}|_T = \mathcal{O}_T(r)$, and hence $\Lambda^{n-d} N_{L,X} = \mathcal{O}_L(r)$. Therefore (7.1) follows by the adjunction formula. Since there is such a plane L through a generic point $p \in X$, the linear system $|K_X^m|$ is empty for $m \geq 0$. □

Example 7.11 In the notation of Theorem 7.10, if $K_X = \mathcal{O}_X(a)$, then $a = (-n-d-2)/2$. For example, if $n > (N/2)+1$ then $\operatorname{Pic} X = \mathbb{Z}$ is generated by $\mathcal{O}_X(1)$ by the Barth Theorem [Ba] and therefore $K_X = \mathcal{O}_X((-n-d-2)/2)$.

Example 7.12 ([E1]) Suppose that X is a nonlinear smooth projective variety in $\mathbb{P}^N$ with $\operatorname{codim} X = 2$. Then X^* is a hypersurface, unless X is the Segre embedding of $\mathbb{P}^1 \times \mathbb{P}^2$ in $\mathbb{P}^5$. Indeed, assume first that $\dim X \geq 4$ and $\operatorname{def} X = d > 0$. Then by Example 7.11, the canonical class $K_X = \mathcal{O}_X((-n-d-2)/2)$. However, it follows from results of [BC] that in this case X is automatically a complete intersection. This contradicts Theorem 5.11. Now, if $\dim X = 1$ or $\dim X = 2$ then X^* is a hypersurface by Example 1.19 and Example 7.5. The case $\dim X = 3$ is ruled out by Example 7.7.

The following generalization of Theorem 7.10 holds.

Corollary 7.13 ([BFS1]) *Suppose that X is a smooth n-dimensional projective variety in $\mathbb{P}^N$ with $\operatorname{def} X = d > 0$. Let $H \in X^*$. Then $(H \cap X)_{sing}$ is the union of projective subspaces $\mathbb{P}^d$ with $K_X|_{\mathbb{P}^d} = \mathcal{O}_{\mathbb{P}^d}\left(\frac{-n-d-2}{2}\right)$.*

Proof. Let $I_X \subset X \times (\mathbb{P}^N)^\vee$ be the conormal variety with projections $\pi : I_X \to X^*$ and $p : I_X \to X$. The general fibers F of π are isomorphic to the contact loci of X with generic tangent hyperplanes, and therefore they are isomorphic to the projective subspaces $\mathbb{P}^d$, and by Theorem 7.10 we have $p^*(K_X + ((n+d)/2+1)L)_F \simeq \mathcal{O}_F$. It follows by continuity that every fiber of π is a union of projective subspaces $\mathbb{P}^d$ with

$$p^*(K_X + ((n+d)/2+1)L)_{\mathbb{P}^d} \simeq \mathcal{O}_{\mathbb{P}^d}.$$

The theorem is now proved. □

7.1.2 Monotonicity Theorem

The following simple but useful result first appeared in [LS] (where it was attributed to the referee).

Theorem 7.14 ([LS]) *Let $X \subset \mathbb{P}^N$ be a smooth projective n-dimensional variety. Suppose that through its generic point there passes a smooth subvariety Y of dimension h and defect θ. Then* $\operatorname{def} X \geq \theta - n + h$. *In other words,*

$$\dim X + \operatorname{def} X \geq \dim Y + \operatorname{def} Y.$$

Proof. Suppose that $x \in X$ is a generic point and H is a generic hyperplane tangent to X at x. Let Y be an h-dimensional submanifold passing through x such that $\operatorname{def} Y = \theta$. Since the defect is the dimension of a generic contact locus, which is a projective subspace, there exists a θ-dimensional projective subspace $Z \subset Y$ containing x such that H is tangent to Y along Z (though H is not a generic tangent hyperplane to Y). Let $f = 0$ be a local equation for $H \cap X$ at x. In a neighborhood U of x, the differential df annihilates the tangent spaces $T_{x,X}$ and $T_{z,Y}$ for every $z \in Z \cap U$. Hence df defines on $Z \cap U$ a section of the conormal bundle $N^\vee_{Y,X}|_Z$ vanishing at x. Since $N^\vee_{Y,X}$ is a $(n-h)$-dimensional vector bundle, it follows that this section in fact vanishes on a subvariety $Z' \subset Z \cap U$ of codimension less then or equal to $n-h$. Therefore H is tangent to X along Z'. But $(X \cap H)_{sing}$ is $(\operatorname{def} X)$-dimensional, and therefore $\operatorname{def} X \geq \theta - (n-h)$. □

7.1.3 Beilinson Spectral Sequence

Theorem 7.15 ([OSS]) *Let E be a vector bundle over $\mathbb{P}^N$. There is a spectral sequence E_r^{pq} with $E_1^{pq} = H^q(\mathbb{P}^N, E(p)) \otimes \Omega^{-p}_{\mathbb{P}^N}(-p)$ which converges to*

$$E^i = \begin{cases} E & \text{for } i = 0 \\ 0 & \text{otherwise,} \end{cases}$$

i.e. $E^{pq}_\infty = 0$ for $p + q \neq 0$ and $E^{0,0}_\infty \oplus E^{-1,1}_\infty \oplus \ldots \oplus E^{-N,N}_\infty$ is the associated graded sheaf of some filtration of E.

Proof. Consider projections $p_1, p_2 : \mathbb{P}^N \times \mathbb{P}^N \to \mathbb{P}^N$ and abbreviate $A \boxtimes B = p_1^* A \otimes p_2^* B$ for any bundles A, B over $\mathbb{P}^N$. Consider the bundle

$$I = \mathcal{O}_{\mathbb{P}^N}(1) \boxtimes T_{\mathbb{P}^N}(-1) = \operatorname{Hom}(p_1^*\mathcal{O}_{\mathbb{P}^N}(-1), p_2^* T_{\mathbb{P}^N}(-1))$$

over $\mathbb{P}^N \times \mathbb{P}^N$. Using the obvious exact sequence

$$0 \to \mathcal{O}_{\mathbb{P}^N}(-1) \to \mathcal{O}^{N+1}_{\mathbb{P}^N} \to T_{\mathbb{P}^N}(-1) \to 0$$

and Künneth formula, we get

$$H^0(\mathbb{P}^N \times \mathbb{P}^N, I) = H^0(\mathbb{P}^N, \mathcal{O}_{\mathbb{P}^N}(1)) \otimes H^0(\mathbb{P}^N, T_{\mathbb{P}^N}(-1))$$

$$= (\mathbb{C}^{N+1})^\vee \otimes \mathbb{C}^{N+1} = \mathrm{Hom}(\mathbb{C}^{N+1}, \mathbb{C}^{N+1}).$$

Therefore I has the canonical section s that corresponds to the identity in $\mathrm{Hom}(\mathbb{C}^{N+1}, \mathbb{C}^{N+1})$. More explicitly, let $v, w \in \mathbb{C}^{N+1} \setminus \{0\}$ and let $x, y \in \mathbb{P}^N$ be corresponding lines. Then

$$s(x,y) \in \mathrm{Hom}(\mathcal{O}_{\mathbb{P}^N}(-1)(x), T_{\mathbb{P}^N}(-1)(y)) = \mathrm{Hom}(\mathbb{C}v, \mathbb{C}^{N+1}/\mathbb{C}w)$$

is defined as

$$s(x,y)(\lambda v) = \lambda v \bmod \mathbb{C}w.$$

Clearly, $s(x,y)$ vanishes if and only if $x = y$, i.e. the zero scheme of s is the diagonal $\Delta \subset \mathbb{P}^N \times \mathbb{P}^N$. Consider the corresponding exact Koszul complex

$$\mathcal{K}_-(I,s) = \left\{0 \to \Lambda^N I^\vee \to \ldots \to \Lambda^2 I^\vee \to I^\vee \to \mathcal{O}_{\mathbb{P}^N \times \mathbb{P}^N} \to \mathcal{O}_\Delta \to 0\right\},$$

which can be rewritten as

$$0 \to \mathcal{O}(-N) \boxtimes \Omega^N(N) \to \ldots \to \mathcal{O}(-1) \boxtimes \Omega^1(1) \to \mathcal{O} \boxtimes \mathcal{O} \to \mathcal{O}_\Delta \to 0.$$

By tensoring this exact sequence with $p_1^* E$, we get the complex $C^\bullet$ of sheaves

$$0 \to E(-N) \boxtimes \Omega^N(N) \to \ldots \to E(-1) \boxtimes \Omega^1(1) \to E \boxtimes \mathcal{O} \to 0$$

with cohomology

$$H^q(C^\bullet) = \begin{cases} p_1^* E|_\Delta & \text{for } q = 0 \\ 0 & \text{otherwise.} \end{cases}$$

Consider the complexes of Abelian groups $L^{\bullet,j}$ calculating the cohomology of every individual C^j (for example, the Čech complexes with respect to an appropriate open covering of $\mathbb{P}^N \times \mathbb{P}^N$). The differential in $C^\bullet$ makes this collection of complexes into a double complex $L^{\bullet\bullet}$. We define the i-th hyperdirect image $\mathbb{R}^i(p_2)_*(C^\bullet)$ as the i-th cohomology sheaf of the total chain complex associated to $(p_2)_*(L^{\bullet\bullet})$.

There are two spectral sequences $'E$ and $''E$ associated to the double complex $(p_2)_*(L^{\bullet\bullet})$

$$'E_1^{pq} = R^q(p_2)_*(C^\bullet), \quad ''E_2^{pq} = R^p(p_2)_*(H^q(C^\bullet)).$$

Since

$$''E_2^{pq} = R^p(p_2)_*(H^q(C^\bullet)) = \begin{cases} E & \text{for } p = q = 0 \\ 0 & \text{otherwise,} \end{cases}$$

we have

$$\mathbb{R}^i(p_2)_*(C^\bullet) = \begin{cases} E & \text{for } i = 0 \\ 0 & \text{otherwise.} \end{cases}$$

However,

$$'E_1^{pq} = R^q(p_2)_*(C^p) = R^q(p_2)_*(p_1^* E(p) \otimes p_2^* \Omega^{-p}(-p)) =$$
$$H^q(\mathbb{P}^N, E(p)) \otimes \Omega^{-p}(-p).$$

It follows that $'E$ is the spectral sequence we were looking for. □

Definition 7.16 A *monad* over $\mathbb{P}^k$ is a complex

$$0 \to A \xrightarrow{a} B \xrightarrow{b} C \to 0$$

of vector bundles which is exact at A and C, such that $\mathrm{Im}(a)$ is a subbundle of B. The vector bundle $E = \mathrm{Ker}(b)/\mathrm{Im}(a)$ is called the cohomology of the monad.

7.1.4 Planes in the Contact Locus

Now we return to our situation: $X \subset \mathbb{P}^N$ is a non-degenerate smooth projective variety of positive defect $d = \operatorname{def} X$ and $\dim X = n$. Let $L = (H \cap X)_{sing}$ be a generic contact locus. We would like to use the Beilinson spectral sequence for the study of the bundle $N_{L,X}$. Let us describe the restriction of $N^\vee_{L,X}$ to a plane in L.

Theorem 7.17 **([Mun3])** *Let $X \subset \mathbb{P}^N$ be a variety of dimension n with positive defect $d \geq 2$. Let $H \subset X^*$ be a generic tangent hyperplane and L the contact locus of X with H. Let $R \subset L$ be a plane.*

(i) *$N^\vee_{L,X}|_R$ has the following structure:*

$$N^\vee_{L,X}|_R = \mathcal{M} \oplus [\mathcal{O}_R(-1) \oplus \mathcal{O}_R]^{\oplus h^0(N^\vee_{L,X}|_R)},$$

where $\mathcal{M}$ is the cohomology of a monad

$$0 \to \mathcal{O}_R(-1)^{\oplus h^1(N^\vee_{L,X}|_R)} \xrightarrow{f} T^\vee_R(1)^{\oplus h^1(N^\vee_{L,X}|_R(-1))} \xrightarrow{f'} \mathcal{O}_R^{\oplus h^1(N^\vee_{L,X}|_R)} \to 0.$$

(ii) *$h^1(N^\vee_{L,X}(-1)|_R)$ is an even number*

Proof. (i) The symmetric isomorphism between $N^\vee_{L,X}$ and $N_{L,X}(1)$ produces a symmetric isomorphism between the corresponding restrictions to R:

$$N^\vee_{L,X}|_R \simeq N_{L,X}(-1)|_R.$$

Since $N_{L,X}|_R$ is globally generated then, as in Theorem 7.8, we have the splitting

$$N^\vee_{L,X}|_R = \mathcal{M} \oplus [\mathcal{O}_R(-1) \oplus \mathcal{O}_R]^{\oplus h^0(N^\vee_{L,X}|_R)},$$

Using the Serre duality, we see that the first term

$$E_1^{pq} = H^q(\mathcal{M}(p)) \otimes \Omega^{-p}(-p)$$

of the Beilinson spectral sequence of the vector bundle $\mathcal{M}$ is:

$$\begin{array}{ccccc} 0 & \to & 0 & \to & 0 \\ \mathcal{O}_R(-1)^{\oplus h^1(N^\vee_{L,X}|_R)} & \xrightarrow{f} & T^\vee_R(1)^{\oplus h^1(N^\vee_{L,X}|_R(-1))} & \xrightarrow{g} & \mathcal{O}_R^{\oplus h^1(N^\vee_{L,X}|_R)} \\ 0 & \to & 0 & \to & 0. \end{array}$$

This implies that $E_2^{pq} = E_\infty^{pq}$ and hence, by the convergence of the Beilinson spectral sequence to $\mathcal{M}$, the latter is the cohomology of the monad corresponding to the second row in E_1^{pq}.

(ii) Since $\mathcal{M} \simeq \mathcal{M}^\vee(-1)$ and

$$\mathrm{Hom}\left(T_R^\vee(1)^{\oplus h^1(N_{L,X}^\vee|_R(-1))}, \mathcal{O}_R(-1)^{\oplus h^1(N_{L,X}^\vee|_R)}\right) = 0,$$

$$\mathrm{Hom}\left(\mathcal{O}_R^{\oplus h^1(N_{L,X}^\vee|_R)}, T_R^\vee(1)^{\oplus h^1(N_{L,X}^\vee|_R(-1))}\right) = 0,$$

by [BH] the isomorphism between $\mathcal{M}$ and $\mathcal{M}^*(-1)$ extends to a morphism between the monads

$$\begin{array}{ccccc} \mathcal{O}_R(-1)^{\oplus h^1(N_{L,X}^*|_R)} & \xrightarrow{f} & T_R^*(1)^{\oplus h^1(N_{L,X}^*|_R(-1))} & \xrightarrow{g} & \mathcal{O}_R^{\oplus h^1(N_{L,X}^*|_R)} \\ \downarrow & & \downarrow H & & \downarrow \\ \mathcal{O}_R(-1)^{\oplus h^1(N_{L,X}^*|_R)} & \xrightarrow{f'} & T_R(-2)^{\oplus h^1(N_{L,X}^*|_R(-1))} & \xrightarrow{g'} & \mathcal{O}_R^{\oplus h^1(N_{L,X}^*|_R)}(-1). \end{array}$$

By [BH], H is a symmetric isomorphism. It remains to notice that a symmetric isomorphism between $T_R^\vee(1)^k$ and $T_R(-2)^k$ exists if and only if k is even. Indeed, this follows from two facts. Firstly, an isomorphism between $T_R^\vee(1)$ and $T_R(-2)$ is unique and skew-symmetric (it is given by a usual product of 1-forms). Secondly, if a block matrix

$$M = \begin{pmatrix} M_{11} & \dots & M_{1k} \\ \vdots & \ddots & \vdots \\ M_{k1} & \dots & M_{kk} \end{pmatrix}$$

is symmetric, k is odd, and each M_{ij} is a skew-symmetric (2×2)-matrix, then it is easy to check that $\det M = 0$. □

Theorem 7.18 ([Mun3]) *Let $X \subset \mathbb{P}^N$ be an n-dimensional variety with positive defect d. Let $H \subset X^*$ be a generic tangent hyperplane and L the contact locus of X with H. Then $h^0(N_{L,X}) \geq n$. If $h^0(N_{L,X}) = n$ then* $\mathrm{Tan}(L, X)$ *is a linear subspace.*

Proof. Take the exact sequence

$$0 \to N_{X,\mathbb{P}^N}^\vee(1)|_L \to \mathcal{O}_L^{\oplus(N-d)} \to N_{L,X} \to 0$$

and then

$$h^0(N_{L,X}) \geq N - d - h^0(N_{X,\mathbb{P}^N}^\vee(1)|_L).$$

Let $s = h^0(N_{X,\mathbb{P}^N}^\vee(1)|_L)$. By Theorem 5.2, we have $\dim\langle\mathrm{Tan}(L,X)\rangle = N - s$ and this implies our theorem, because $\dim Tan(L,X) = n+d$ by Theorem 4.24 (otherwise

$$X \subset \mathrm{Sec}(L, X) = \mathrm{Tan}(L, X) \subset H$$

which contradicts the hypothesis of X being not degenerate). □

These results can be used to get a list of all possible candidates for the restriction E of $N^\vee_{L,X}$ to a generic plane R in L. This classification was obtained in [Mun3].

- If $\operatorname{rk} E = 2$ then
$$E = \mathcal{O}_R \oplus \mathcal{O}_R(-1).$$
- If $\operatorname{rk} E = 4$ then
$$E = \begin{cases} \mathcal{O}_R^{\oplus 2} \oplus \mathcal{O}_R(-1)^{\oplus 2} \\ T_R^\vee(1)^{\oplus 2}. \end{cases}$$
- If $\operatorname{rk} E = 6$ then
$$E = \begin{cases} \mathcal{O}_R^{\oplus 3} \oplus \mathcal{O}_R(-1)^{\oplus 3} \\ \mathcal{O}_R \oplus T_R^\vee(1)^{\oplus 2} \oplus \mathcal{O}_R(-1) \\ E_3, \end{cases}$$
where E_3 appears in an exact sequence of the form
$$0 \to E_3 \to \mathcal{O}_R^{\oplus 8} \to \mathcal{O}_R(1) \oplus \mathcal{O}_R(2) \to 0.$$
- If $\operatorname{rk} E = 8$ then
$$E = \begin{cases} \mathcal{O}_R^{\oplus 4} \oplus \mathcal{O}_R(-1)^{\oplus 4} \\ \mathcal{O}_R^{\oplus 2} \oplus T_R^\vee(1)^{\oplus 2} \oplus \mathcal{O}_R(-1)^{\oplus 2} \\ \mathcal{O}_R \oplus E_3 \oplus \mathcal{O}_R(-1) \\ E_4 \\ T_R^\vee(1)^{\oplus 4}, \end{cases}$$
where E_4 appears in an exact sequence of the form
$$0 \to E_3 \to \mathcal{O}_R^{\oplus 10} \to \mathcal{O}_R(2) \oplus \mathcal{O}_R(2) \to 0.$$
- If $\operatorname{rk} E = 10$ then
$$E = \begin{cases} \mathcal{O}_R^{\oplus 5} \oplus \mathcal{O}_R(-1)^{\oplus 5} \\ \mathcal{O}_R^{\oplus 3} \oplus T_R^\vee(1)^{\oplus 2} \oplus \mathcal{O}_R(-1)^{\oplus 3} \\ \mathcal{O}_R^{\oplus 2} \oplus E_3 \oplus \mathcal{O}_R(-1)^{\oplus 2} \\ \mathcal{O}_R \oplus T_R^\vee(1)^{\oplus 4} \oplus \mathcal{O}_R(-1) \\ \mathcal{O}_R \oplus E_4 \oplus \mathcal{O}_R(-1) \\ E_5, E_6, E_7, \end{cases}$$
where E_5, E_6, and E_7 appear in the following exact sequences
$$0 \to E_5 \to \mathcal{O}_R^{\oplus 14} \to \mathcal{O}_R(2) \oplus \mathcal{O}_R(1)^{\oplus 3} \to 0,$$
$$0 \to E_6 \to \mathcal{O}_R^{\oplus 13} \to \mathcal{O}_R(3) \oplus \mathcal{O}_R(1)^{\oplus 2} \to 0,$$
$$0 \to E_7 \to \mathcal{O}_R^{\oplus 12} \to T_R^\vee(4) \to 0.$$

Example 7.19 Suppose that $X = \mathrm{Gr}(2, \mathbb{C}^{2r+1})$ in the Plücker embedding. Then $\dim X = 2(2r-1)$, $\operatorname{def} X = 2$, and $N^\vee_{L,X} = T^\vee_{\mathbb{P}^2}(1)^{\oplus 2r-2}$.

Example 7.20 Consider the spinor variety $X = \mathbb{S}_5$ in its minimal linearly normal embedding. Then $\dim X = 10$, $\operatorname{def} X = 4$, and $N^\vee_{L,X} = \Omega^2_{\mathbb{P}^4}(2)$. Let X' be a generic hyperplane section of $\mathbb{S}_5$. Then $\dim X' = 9$, $\operatorname{def} X' = 3$, and $N^\vee_{L,X'} = \Omega^2_{\mathbb{P}^3}(2) \oplus \Omega^1_{\mathbb{P}^3}(1)$. Let X'' be a generic section of $\mathbb{S}_5$ by two generic hyperplanes. Then $\dim X'' = 8$, $\operatorname{def} X'' = 2$, and $N^\vee_{L,X''} = \mathcal{O}_{\mathbb{P}_2} \oplus T^\vee_{\mathbb{P}^2}(1)^{\oplus 2} \oplus \mathcal{O}_{\mathbb{P}_2}(-1)$.

7.1.5 Scrolls

The following theorem is well-known:

Theorem 7.21 *Let $X \subset \mathbb{P}^N$ be an n-dimensional scroll, i.e. a projective bundle $\mathbb{P}_Y(E)$ over a smooth variety Y such that all fibers are embedded linearly, and let* $\dim Y = m$. *Suppose that $n \geq 2m$. Then* $\operatorname{def} X = n - 2m$ *and $N^\vee_{L,X}$ splits as a sum of line bundles.*

Proof. Take a general $x \in X$ and a general $H \in X^*$ such that $\hat{T}_{x,X} \subset H$. We denote by p the map $p : X \to Y$. The fiber $p^{-1}p(x) = \mathbb{P}^{n-m}$ must be contained in $\hat{T}_{x,X}$. We can take m analytic arcs $y_i(t)$ such that $y_i(0) = p(x)$ and such that their tangent vectors span the linear space of directions of $T_{y,Y}$. These produce m 1-parametric families F^i_t of linear spaces of dimension $n-m$. For any point $y \in p^{-1}p(x)$, any choice of analytic arcs $\alpha_i(t) \in F^i_t$ with $\alpha_i(0) = y$ verifies that $\alpha'_i(0)$ is a set of linearly independent vectors.

The hyperplane H cuts the general member of each one of these families in a linear space of codimension lower than or equal to 1: $F^i_t \cap H = G^i_t$. Then the limit position on $p^{-1}p(x)$ of these G^i_t are m linear spaces f^i of codimension lower than or equal to one in $p^{-1}p(x)$.

Since $\cap f^i$ must be contained in the contact locus of X with H, we have shown that $\operatorname{def} X \geq n - 2m$. If $n = 2m$, see Theorem 5.14. If $n > 2m$, then the general hyperplane section of X is a scroll $X' = \mathbb{P}_Y(E')$, where $\dim E' = \dim E - 1$. We proceed by induction using Theorem 5.3. We see that $\operatorname{def} X = n - 2m$ and $L \subset p^{-1}p(x)$. Now $N_{L,X}$ splits by Example 7.9. □

Theorem 7.22 ([E2, Mun3]) *Let $X \subset \mathbb{P}^N$ be an n-dimensional variety with defect* $\operatorname{def} X > 1$. *If $N_{L,X}$ splits as a sum of line bundles then X is a scroll.*

Proof. Let $m = \frac{n-k}{2}$. Then $N_{L,X} = \mathcal{O}_L^{\oplus m} \oplus \mathcal{O}_L(1)^{\oplus m}$. Take a general $x \in L$, and let $\mathcal{I}_x$ be the ideal sheaf of x in L. Since

$$h^0(N_{L,X} \otimes \mathcal{I}_x) = km, \quad h^1(N_{L,X} \otimes \mathcal{I}_x) = 0,$$

the Hilbert scheme of k-planes in X through x has a single irreducible component $\mathcal{H}$ through the point corresponding to L, and $\dim \mathcal{H} = km$. Take the universal family

$$I = \{(y,S) \in X \times \mathcal{H} \,|\, y \in S\} \xrightarrow{p} X.$$

Then

$$\dim p(I) = km + k - \dim p^{-1}(y)$$

with y a general point of S and S a general point of $\mathcal{H}$. Since a general deformation of a split bundle on $\mathbb{P}^k$ is also a split bundle, we observe that $N_{S,X} = \mathcal{O}_S^{\oplus m} \oplus \mathcal{O}_S(1)^{\oplus m}$.

On the other hand, take $\mathcal{I}_{\langle x,y\rangle}$, the ideal sheaf of the line $\langle x,y\rangle$ in S. Then

$$h^0(N_{S,X} \otimes \mathcal{I}_{\langle x,y\rangle}) = k-1, \quad h^1(N_{S,X} \otimes \mathcal{I}_{\langle x,y\rangle}) = 0.$$

Then $p(I)$ has dimension $k+m$ and contains a family of dimension km of k-planes through x. This family has a maximal possible dimension, which means (see e.g. [Rog]) that $D = p(I)$ is a linear space. Hence $N_{D,X}|_S = \mathcal{O}_L^{\oplus m}$ for any k-plane S in our family, i.e. $N_{D,X}$ is a uniform vector bundle of type $(0,0,\ldots,0)$. Therefore, $N_{D,X} = \mathcal{O}_L^{\oplus m}$, e.g. by [E3]. Hence the claim follows from Theorem 7.23 below. □

Theorem 7.23 ([E2]) *Let X be a projective n-fold in $\mathbb{P}^N$. Assume that there is an m-plane D_0 in X such that $N_{D_0,X} = \mathcal{O}_{D_0}^{\oplus n-m}$ and $n \le 2m-1$. Then $X = \mathbb{P}_Y(F)$ is a scroll on an $(n-m)$-fold Y.*

Theorem 7.22 and Corollary 7.30 imply the following theorem.

Theorem 7.24 ([E2]) *If X is a smooth n-dimensional projective variety in $\mathbb{P}^N$ and $\operatorname{def} X = k \ge n/2$, then X is a projective bundle $X \simeq \mathbb{P}_Y(F)$, where F is a vector bundle of rank $(n+k+2)/2$ on a smooth $(n-k)/2$-dimensional variety Y and the fibers are embedded linearly.*

7.2 Linear Systems of Quadrics of Constant Rank

Definition 7.25 Let $V = \mathbb{C}^m$ and $W = \mathbb{C}^n$. We take a linear subspace

$$A \subset \operatorname{Hom}(V,W) \quad (\text{resp. } A \subset S^2V^\vee, \text{ resp. } A \subset \Lambda^2 V^\vee).$$

A is said to be of *constant rank* r if, for all non-zero $f \in A$, $\operatorname{rank} f = r$. We define numbers $l(r,m,n)$ (resp. $c(r,m)$, resp. $\lambda(r,m)$) as the maximal possible dimension of a linear subspace in $\operatorname{Hom}(V,W)$ (resp. $S^2V^\vee$, resp. $\Lambda^2V^\vee$) of constant rank r.

Given a linear subspace $A \subset \operatorname{Hom}(V,W)$ we can also consider A as a linear subspace A^c of $S^2(V^\vee \oplus W)$ or a linear subspace A^λ of $\Lambda^2(V^\vee \oplus W)$. In matrix form, for any matrix $a \in A$ we associate a symmetric matrix $a^c \in A^c$ or skew-symmetric matrix $a^\lambda \in A^\lambda$ of the form

$$a^c = \begin{pmatrix} 0 & a \\ a^t & 0 \end{pmatrix}, \quad a^\lambda = \begin{pmatrix} 0 & a \\ -a^t & 0 \end{pmatrix}.$$

If A has constant rank r then A^c and A^λ have constant rank $2r$. A^c and A^λ are called *doublings* of A.

Proposition 7.26 ([IL]) (i) *If* $0 < r \leq m \leq n$ *then* $l(r,m,n) \geq n - r + 1$.
(ii) *If* $r \geq 2$ *is even then* $c(r,m) \geq m - r + 1$ *and* $\lambda(r,m) \geq m - r + 1$.

Proof. Let $X_{r-1} \subset \mathbb{P}(\mathrm{Hom}(\mathbb{C}^r, \mathbb{C}^n)$ be the space of matrices of rank less then r. Then $\operatorname{codim} X_{r-1} = n - r + 1$. Thus we can find an $(n-r+1)$-dimensional subspace $A \subset \mathrm{Hom}(\mathbb{C}^r, \mathbb{C}^n)$ of constant rank r and (i) follows. The doublings of A produce $(n-r+1)$-dimensional subspaces $A^c \subset S^2\mathbb{C}^{r+n}$ and $A^\lambda \subset \Lambda^2\mathbb{C}^{r+n}$ of constant rank $2r$. This implies (ii). □

Any linear map $\psi : A \to \mathrm{Hom}(V, W)$ can be viewed as an element of

$$A^\vee \otimes \mathrm{Hom}(V,W) = H^0(\mathbb{P}(A), \mathcal{O}_{\mathbb{P}(A)}(1)) \otimes \mathrm{Hom}(V,W).$$

Thus to give a linear map $A \to \mathrm{Hom}(V,W)$ is the same as to give a vector bundle map

$$\overline{\psi} : V \otimes \mathcal{O}_{\mathbb{P}(A)} \to W \otimes \mathcal{O}_{\mathbb{P}(A)}(1)$$

on $\mathbb{P}(A)$. Let K, N, and E be the kernel, the cokernel, and the image of $\overline{\psi}$. Then it is easy to see that ψ is an embedding of constant rank r iff K, N, and E are vector bundles of ranks $\dim V - r$, $\dim W - r$, and r iff any of K, N, and E is a vector bundle of rank as above. It follows that embeddings $A \subset \mathrm{Hom}(V,W)$ of constant rank are in one-one correspondence with pairs of exact sequences of vector bundles

$$0 \to K \to \mathcal{O}_{\mathbb{P}(A)} \otimes V \to E \to 0$$

and

$$0 \to E \to \mathcal{O}_{\mathbb{P}(A)}(1) \otimes W \to N \to 0.$$

Similarly, all embeddings $A \subset S^2V$ (resp. $A \subset \Lambda^2 V$) of constant rank are in one-one correspondence with exact sequences of vector bundles

$$0 \to K \to \mathcal{O}_{\mathbb{P}(A)} \otimes V^\vee \to E \to 0,$$

where E is a rank r vector bundle such that $E \simeq E^*(1)$ and this isomorphism is symmetric (resp. skew-symmetric).

Theorem 7.27 ([IL]) *Let* $A \subset S^2V$ *or* $A \subset \Lambda^2V$ *be a subspace of constant rank* r *with* $\dim A \geq 2$. *Then* E *is a uniform vector bundle of splitting type* $\mathcal{O}_{\mathbb{P}^1}^{r/2} \oplus \mathcal{O}_{\mathbb{P}^1}^{r/2}(1)$. *In particular,* r *is even.*

Proof. The proof is by the same argument as in the proof of Theorem 7.3. The parity of r was known classically as a consequence of the Kronecker–Weierstrass theory giving a normal form for pencils of symmetric matrices of bounded rank; see [Gan, HP, Me]. □

Theorem 7.28 ([IL]) *A smooth n-dimensional projective variety $X \subset \mathbb{P}^N$ with $d = \operatorname{def} X > 0$ determines a $(d+1)$-dimensional linear subspace $A \subset S^2\mathbb{C}^{N-1-d}$ with the associated short exact sequence*

$$0 \to F^\vee(1) \to \mathcal{O}_L \otimes \mathbb{C}^{N-1-d} \to N_{L,X} \to 0,$$

where $H \in X^*$ is a generic tangent hyperplane and $L = (X \cap H)_{sing}$ is the contact locus.

Proof. Recall that L is a projective subspace of dimension d. We have the following exact sequence of vector bundles

$$0 \to N_{L,X} \to N_{L,H} \to F \to 0$$

which holds because H is tangent to X along L. Dualising and twisting by $\mathcal{O}_L(1)$ we get

$$0 \to F^\vee(1) \to N^\vee_{L,H}(1) \to N^\vee_{L,X}(1) \to 0.$$

Recall that $N_{L,X} \simeq N^\vee_{L,X}(1)$ by Theorem 7.1. It remains to notice that $N_{L,H} \simeq \mathcal{O}_L(1) \otimes \mathbb{C}^{N-1-d}$. □

Theorem 7.29 ([E1, IL]) *Suppose that $A \subset S^2V$ is an $(l+1)$-dimensional subspace of constant rank r. Let*

$$0 \to K \to \mathcal{O}_{\mathbb{P}(A)} \otimes V^\vee \to E \to 0$$

be the corresponding exact sequence of vector bundles. If $r \le l$ then E is isomorphic to $\mathcal{O}^{r/2}_{\mathbb{P}(A)} \oplus \mathcal{O}^{r/2}_{\mathbb{P}(A)}(1)$ unless $r = l = 2$ and $E = T_{\mathbb{P}^2}(-1)$.

Proof. Indeed, E is uniform by Theorem 7.27. Therefore, if $r \le l$, then E either splits into a direct sum of line bundles or E is isomorphic to $T_{\mathbb{P}^l}(a)$ or $T^\vee_{\mathbb{P}^l}(a)$ by [Sa]. If E splits then $E \simeq \mathcal{O}^{r/2}_{\mathbb{P}(A)} \oplus \mathcal{O}^{r/2}_{\mathbb{P}(A)}(1)$ by Theorem 7.27. Suppose now that E is isomorphic to $T_{\mathbb{P}^l}(a)$ or $T^\vee_{\mathbb{P}^l}(a)$. Since $E \simeq E^\vee(1)$, the first Chern class $c_1(E)$ is equal to $(l/2)H$, where H is the class of the hyperplane section. Computing the first Chern classes of $T_{\mathbb{P}^l}(a)$ and $T^\vee_{\mathbb{P}^l}(a)$ we conclude that E is isomorphic either to $T_{\mathbb{P}^2}(-1)$ or to $T^\vee_{\mathbb{P}^2}(2)$. These bundles are in fact isomorphic. □

Corollary 7.30 ([E1, IL]) *Suppose that $X \subset \mathbb{P}^N$ is a smooth projective variety. Let L be a generic contact locus. If $\operatorname{def} X \ge n/2$, then*

$$N_{L,X} = \mathcal{O}_L^{(\dim X - \operatorname{def} X)/2} \oplus \mathcal{O}_L^{(\dim X - \operatorname{def} X)/2}(1).$$

Proof. Let $\dim X = n$ and $\operatorname{def} X = k$. By Theorems 7.29 and 7.28, we see that $N_{L,X}$ is isomorphic to either $\mathcal{O}_L^{(n-k)/2} \oplus \mathcal{O}_L^{(n-k)/2}(1)$ or $T_{\mathbb{P}^2}(-1)$. However, in the last case $n = 4$, $k = 2$, and therefore X is a projective bundle $\mathbb{P}_C(F)$ over a curve C by Example 7.6. Then L is embedded as a 2-plane in a fibre f of $\mathbb{P}_C(F)$. Consider the exact sequence

$$0 \to N_{L,f} \to N_{L,X} \to N_{f,X}|_L \to 0.$$

Since $N_{L,f} = \mathcal{O}_L(1)$ and $N_{f,X} = \mathcal{O}_f$, this exact sequence reduces to

$$0 \to \mathcal{O}_L(1) \to N_{L,X} \to \mathcal{O}_L \to 0.$$

It follows that $N_{L,X} = \mathcal{O}_L \oplus \mathcal{O}_L(1)$. □

The following theorem was proved in [W]. Weaker results (but similar arguments) first appeared in [Sy].

Theorem 7.31 ([W]) *Suppose that* $2 \leq r \leq m \leq n$. *Then*

(i) $l(r,m,n) \leq m+n-2r+1$.
(ii) $l(r,m,n) = n-r+1$ *if* $n-r+1$ *does not divide* $(m-1)!/(r-1)!$.
(iii) $l(r,r+1,2r-1) = r+1$.

Sketch of the proof. Let $A \subset \mathrm{Hom}(V,W)$ be an l-dimensional linear subspace of constant rank r, where $\dim V = m$, $\dim W = n$, and $m \geq n$. Let

$$0 \to K \to \mathcal{O}_{\mathbb{P}(A)} \otimes V \to E \to 0 \text{ and } 0 \to E \to \mathcal{O}_{\mathbb{P}(A)}(1) \otimes W \to N \to 0$$

be the corresponding exact sequences. Then $c(K)c(E) = 1$ and $c(E)c(N) = (1+H)^n$, where H is the class of the hyperplane section. Thus, $c(K)(1+H)^n = c(N)$. If $n-r+1 \leq i \leq l-1$ then $c_i(N) = 0$ and, looking at the coefficient of H^i, we get

$$\sum_{j=0}^{m-r} \binom{n}{i-j} k_j = 0,$$

where $c_j(K) = k_j H^j$ and $\binom{n}{i} = 0$ if $i < 0$ or $i > n$. The coefficient matrix of this system of linear equations is

$$\binom{n}{i-j}_{0 \leq j \leq m-r,\ n-r+1 \leq i \leq l-1}.$$

If $l = m+n-2r+2$ then this is a square invertible matrix with determinant

$$\prod_{j=0}^{m-r} j!$$

Thus $k_0 = 0$ which is a contradiction to $k_0 = 1$. This proves (i).

Proofs of (ii) and (iii) can be found in [W]. By Proposition 7.26 (i) we have $l(r,m,n) \geq n-r+1$, and so only the opposite inequality should be proved. □

Another possible way of proving Theorem 7.31 (i) is to use the following theorem of R. Lazarsfeld (its application is immediate):

Theorem 7.32 ([La1]) *Let X be a projective variety with $\dim X = l$. Let E and F be vector bundles of ranks m and n, respectively. Suppose that $E^\vee \otimes F$ is ample and there is a constant rank r vector bundle map $E \to F$. Then $l \le m + n - 2r$.*

The (partial) symmetric analogue of this theorem was found in [IL]. Theorems 7.32 and 7.33 (and their proofs) are analogous to results related to the problem of the non-emptiness and the connectedness of degeneracy loci, see [La1, FL1, HT, HT1, Tu, IL].

Theorem 7.33 ([IL]) *Let X be a smooth simply connected m-dimensional variety. Let E be a rank n vector bundle on X, and L a line bundle on X. Suppose that $S^2E^\vee \otimes L$ is an ample vector bundle and that there is a constant even rank r symmetric bundle map $E \to E^\vee \otimes L$. Then $m \le n - r$.*

Proof. Consider the projective bundle $\pi : \mathbb{P}(E) \to X$. The symmetric bundle map $E \to E^\vee \otimes L$ gives a section

$$s \in H^0(X, S^2E^* \otimes L)$$

which pulls back to

$$\pi^* s \in H^0(\mathbb{P}(E), \pi^*(S^2E^\vee \otimes L)).$$

There is a natural map $\pi^*\pi_*\mathcal{O}_{\mathbb{P}(E)}(2) \to \mathcal{O}_{\mathbb{P}(E)}(2)$. Since $\pi_*\mathcal{O}_{\mathbb{P}(E)}(2) \simeq S^2E^\vee$ this gives after tensoring with π^*L a map

$$u : \pi^*(S^2E^\vee \otimes L) \to \mathcal{O}_{\mathbb{P}(E)}(2) \otimes \pi^*L.$$

Let

$$t = u \circ \pi^* s \in H^0(\mathbb{P}(E), \mathcal{O}_{\mathbb{P}(E)}(2) \otimes \pi^*L).$$

Let $Y \subset \mathbb{P}(E)$ be the zero locus of t.

Let $x \in X$. The section t at $[v] \in \mathbb{P}(E(x))$ is the linear map

$$\lambda \to s(x)(v, v)\lambda$$

and so vanishes if and only if $s(x)(v,v) = 0$. But by the hypothesis, this defines a rank r quadric hypersurface in $\mathbb{P}(E(x))$. Therefore the fiber of $\pi_Y : Y \to X$ over each $x \in X$ is a rank r quadric hypersurface in $\mathbb{P}(E(x))$.

We claim that $\mathcal{O}_{\mathbb{P}(E)}(2) \otimes \pi^*(L)$ is ample on $\mathbb{P}(E)$. Indeed, since $S^2E^\vee \otimes L$ is ample on X, $\mathcal{O}_{\mathbb{P}(S^2E^\vee \otimes L)}(1)$ is ample on $\mathbb{P}(S^2E^\vee \otimes L)$. Clearly, $\mathbb{P}(S^2E^\vee \otimes L) \simeq \mathbb{P}(S^2E^\vee)$ and $\mathcal{O}_{\mathbb{P}(S^2E^\vee \otimes L)}(1) \simeq \mathcal{O}_{\mathbb{P}(S^2E^\vee)}(1) \otimes \sigma^*L$, where $\sigma : \mathbb{P}(S^2E^\vee) \to X$ is the natural projection. Therefore $\mathcal{O}_{\mathbb{P}(S^2E^\vee)}(1) \otimes \sigma^*L$ is ample on $\mathbb{P}(S^2E^\vee)$. The second Veronese map gives an inclusion $i : \mathbb{P}(E) \subset \mathbb{P}(S^2E^\vee)$ such that $\pi = i \circ \sigma$. Then $i^*\mathcal{O}_{\mathbb{P}(S^2E^\vee)}(1) = \mathcal{O}_{\mathbb{P}(E)}(2)$. Hence $i^*(\mathcal{O}_{\mathbb{P}(S^2E^\vee)}(1) \otimes \sigma^*L) \simeq \mathcal{O}_{\mathbb{P}(E)}(2) \otimes \pi^*L$ is ample on $\mathbb{P}(E)$ as required. In particular, $\mathbb{P}(E) \setminus Y$ is an affine variety and its homology vanishes above its middle dimension:

$$H_i(\mathbb{P}(E) \setminus Y, \mathbb{C}) = 0 \text{ for } i \geq m + n.$$

Let K denote the kernel and F denote the image of the map $E \to E^\vee \otimes L$. Then since the map has constant rank r, K and F are vector bundles on X of ranks $n - r$ and r respectively. Since the map is symmetric, there is a symmetric isomorphism $F \simeq F^\vee \otimes L$. We have a natural map

$$p : \mathbb{P}(E) \setminus \mathbb{P}(K) \to \mathbb{P}(F)$$

given on the fibers by the linear projection centered at $\mathbb{P}(K(x)) \subset \mathbb{P}(E(x))$. Then p is a $\mathbb{C}^{n-r}$-fiber bundle.

The isomorphism $F \simeq F^\vee \otimes L$ determines a hypersurface $Z \subset \mathbb{P}(F)$ such that, if $\rho : \mathbb{P}(F) \to X$ is the natural projection, then the fiber of $\rho|_Z : Z \to X$ over each $x \in X$ is a smooth quadric in $\mathbb{P}(F(x))$. Now $\mathbb{P}(K) \subset Y$ and so p restricts to a $\mathbb{C}^{n-r}$-fiber bundle $\mathbb{P}(E) \setminus Y \to \mathbb{P}(F) \setminus Z$. Thus

$$H_i(\mathbb{P}(E) \setminus Y, \mathbb{C}) \simeq H_i(\mathbb{P}(F) \setminus Z, \mathbb{C})$$

and, by Lefschetz duality, this is isomorphic to $H^{2(m+r-1)-i}(\mathbb{P}(F), Z, \mathbb{C})$. Hence

$$H^i(\mathbb{P}(F), Z, \mathbb{C}) = 0 \text{ for } i \leq 2r + m - n - 2$$

and then using the long exact sequence of the pair $(\mathbb{P}(F), Z)$ we conclude that

$$H^i(Z, \mathbb{C}) \simeq H^i(\mathbb{P}(F), \mathbb{C}) \text{ for } i \leq 2r + m - n - 3.$$

Both bundles $\rho : \mathbb{P}(F) \to X$ and $\rho|_Z : Z \to X$ have smooth fibers and are defined over a smooth simply-connected variety. Therefore, by Deligne' theorem [GH1], the Leray spectral sequence for ρ and $\rho|_Z$ degenerates at the E_2 term. It follows that

$$H^\bullet(\mathbb{P}(F), \mathbb{C}) = H^\bullet(X, \mathbb{C}) \otimes H^\bullet(\mathbb{P}^{r-1}, \mathbb{C})$$

and

$$H^\bullet(Z, \mathbb{C}) = H^\bullet(X, \mathbb{C}) \otimes H^\bullet(Q, \mathbb{C}),$$

where $Q \subset \mathbb{P}^{r-1}$ is a smooth quadric. Let $b_j(B) = \dim H^j(B, \mathbb{C})$ be the Betti numbers. Then it is well-known that

$$b_i(\mathbb{P}^{r-1}) = \begin{cases} 1 & \text{for even } i,\ 0 \leq i \leq 2(r-1) \\ 0 & \text{otherwise.} \end{cases}$$

Since r is even, we have

$$b_i(Q) = \begin{cases} 1 & \text{for even } i,\ 0 \leq i \leq 2(r-2) \\ 2 & i = r-2 \\ 0 & \text{otherwise.} \end{cases}$$

It follows that $b_{r-2}(\mathbb{P}(F)) \neq b_{r-2}(Z)$. Since $b_i(\mathbb{P}(F)) = b_i(Z)$ for $i \leq 2r + m - n - 3$, we finally obtain $r - 2 > 2r + m - n - 3$. Equivalently, $m \leq n - r$. □

As an easy application of Theorem 7.33 we have the following theorem.

Theorem 7.34 (**[IL]**) *Let V be an m-dimensional vector space and let $r > 0$ be an even number. Then*

$$\max\{\dim A \mid A \subset \mathbb{P}(S^2 V^\vee) \text{ is of constant rank } r\} = m - r.$$

Example 7.35 Theorems 7.28 and 7.34 combined furnish a new proof of the formula $\dim X^* \geq \dim X$ (Theorem 4.25 (ii)).

7.3 Defect and Nef Value

7.3.1 Some Results from Mori Theory

Divisors

Let X be a smooth projective variety. If H is a very ample Cartier divisor on X (a hyperplane section in some projective embedding) then a pair (X, H) will be called a *polarized variety*. A divisor D on X is called *semi-ample* if the linear system $|mD|$ is base-point-free for some $m \gg 0$. Let K_X be a canonical divisor.

We denote by $\mathrm{Div}(X)$ a free abelian group of Cartier divisors on X. We write $D_1 \sim D_2$ if D_1 and D_2 are linearly equivalent (differ by a principal divisor). $\mathrm{Pic}(X)$ is the group of algebraic line bundles on X. It is naturally isomorphic to a quotient group of $\mathrm{Div}(X)$ modulo principal divisors. For any $D \in \mathrm{Div}(X)$ the corresponding line bundle is denoted by $\mathcal{O}(D)$.

1-cycles

A 1-*cycle* on X is a finite sum

$$Z = \sum n_i C_i$$

of irreducible curves $C_i \subset X$ with integer multiplicities n_i. Z is called *effective* if all $n_i \geq 0$.

The intersection product

If D is a Cartier divisor on X and $C \subset X$ is an irreducible curve then an *intersection product* $D \cdot C \in \mathbb{Z}$ is defined as $\deg g^* \mathcal{O}(D)$, where $g : \bar{C} \to C$ is a normalisation. This product extends by linearity to a $\mathbb{Z}$-valued pairing of free abelian groups $\mathrm{Div}(X)$ and a group of 1-cycles. A 1-cycle C (resp. a Cartier divisor D) is called *numerically trivial* if $D \cdot C = 0$ for any Cartier divisor D (resp. for any 1-cycle C). Two 1-cycles are *numerically equivalent* (written $Z_1 \equiv Z_2$) if $Z_1 - Z_2$ is numerically trivial. Dually, two Cartier divisors are numerically equivalent (written $D_1 \equiv D_2$) if $D_1 - D_2$ is numerically trivial. The group $N^1_{\mathbb{Z}}(X)$ of Cartier divisors modulo numerical equivalence is a free abelian group of finite rank $\rho(X)$ called a *Picard number* of X. The group $N_1^{\mathbb{Z}}(X)$ of 1-cycles modulo numerical equivalence is also a free abelian group of rank $\rho(X)$.

The cone of effective 1-cycles

We consider two real vector spaces of dimension $\rho(X)$:

$$N_1(X) = N_1^{\mathbb{Z}}(X) \otimes \mathbb{R}, \quad N^1(X) = N_{\mathbb{Z}}^1(X) \otimes \mathbb{R}.$$

These vector spaces are naturally dual to each other by means of the intersection product. The following is the main combinatorial object of the Mori theory:

Definition 7.36

- The *cone of effective 1-cycles* $NE_1(X) \subset N_1(X)$ is a cone generated by classes of effective 1-cycles.
- $\overline{NE}_1(X)$ is its closure in $N_1(X)$.
- The positive part $NE_+(X)$ of $\overline{NE}_1(X)$ is defined as

$$NE_+(X) = \{Z \in \overline{NE}_1(X) \,|\, K_X \cdot Z \geq 0\}.$$

Nef divisors

For $D \in N^1(X)$, we say that D is *nef* (or *numerically effective*) if $D \cdot Z \geq 0$ for any $Z \in NE_1(X)$ (or equivalently for any $Z \in \overline{NE}_1(X)$). This gives a closed cone $NE^1(X)$ in $N^1(X)$ dual to $\overline{NE}_1(X)$. For example, if D is an ample divisor then $C \cdot D > 0$ for any curve $C \subset X$ and therefore D is nef. The following is a useful *Kleiman's criterion*:

Theorem 7.37 ([Kl6]) *D is ample if and only if $D \cdot Z > 0$ for any $Z \in \overline{NE}_1(X)$, $Z \neq 0$.*

In other words, ample divisors correspond to the interior of $NE^1(X)$. All introduced cones are non-degenerate (are not contained in any hyperplane).

$\mathbb{Q}$-divisors

We define *$\mathbb{Q}$-Cartier divisors* as elements of a $\mathbb{Q}$-vector space $\mathrm{Div}(X) \otimes \mathbb{Q}$. The $\mathbb{Q}$-Cartier divisor is called nef if the corresponding element in $N^1(X)$ is nef. The $\mathbb{Q}$-Cartier divisor is called ample if some multiple of it is ample in a usual sense. In the sequel we shall call $\mathbb{Q}$-Cartier divisors simply divisors. This will not lead to any confusion.

Extremal rays

An *extremal ray* is a half line R in $\overline{NE}_1(X)$ such that

- $K_X \cdot R < 0$ ($K_X \cdot Z < 0$ for any non-zero $Z \in R$);
- R is a face of $\overline{NE}_1(X)$, i.e. if $Z_1 + Z_2 \in R$ for any $Z_1, Z_2 \in \overline{NE}_1(X)$ then Z_1 and Z_2 are in R.

An *extremal rational curve* C is a rational curve on X such that $\mathbb{R}_+[C]$ is an extremal ray and $-K_X \cdot C \leq \dim X + 1$. The following is a fundamental result of Mori Theory, usually referred to as the *Mori Cone Theorem:*

Theorem 7.38 ([Mor]) *$\overline{NE}_1(X)$ is the smallest convex cone containing $NE_+(X)$ and all extremal rays. For any open convex cone U containing $NE_+(X) \setminus \{0\}$ there are finitely many extremal rays that are not in $U \cup \{0\}$. Every extremal ray is spanned by an extremal rational curve.*

Another fundamental result in Mori theory is the Kawamata–Shokurov *Contraction Theorem:*

Theorem 7.39 ([Kaw]) *Let D be a nef divisor on X. If, for some $a \geq 1$, the divisor $aD - K_X$ is nef and big (i.e. $(aD - K_X)^n > 0$), then D is semi-ample.*

One of the most useful forms of the Contraction Theorem is as follows.

Theorem 7.40 *Let R be an extremal ray in $\overline{NE}_1(X)$. Then there exists a normal projective variety Y and a surjective morphism $\mathrm{contr}_R : X \to Y$ with connected fibers such that contr_R contracts a curve C to a point if and only if $[C] \in R$.*

Proof. Indeed, let $D \in NE^1(X)$ be such that $D \cdot R = 0$ and $D \cdot Z > 0$ for any $Z \in \overline{NE}_1(X) \setminus R$. It follows from the Mori Cone Theorem that we may assume that D is a divisor. (In fact, usually this theorem is proved as part of the proof of the Mori Cone Theorem. Then the fact that D can be chosen to be a divisor is deduced from the Rationality Theorem; see below.) The Kleiman Criterion implies that for $a \gg 0$ the divisor $aD - K_X$ is ample. Therefore D is semi-ample by the Contraction Theorem. □

The nef value and the nef morphism

Suppose that (X, H) is a smooth polarized projective variety. Assume that K_X is not nef.

Definition 7.41

$$\tau = \min\{t \in \mathbb{R} \mid K_X + tH \text{ is nef}\}$$

is called a *nef value* of (X, H).

If X is fixed then τ will be called a nef value of H, or a nef value of L, where L is a line bundle corresponding to H. Clearly $0 < t < \infty$. Notice also that τ is a nef value of (X, H) if and only if $K_X + \tau H$ is nef, but not ample. The following theorem is known as the Kawamata *Rationality Theorem:*

Theorem 7.42 ([Kaw]) *τ is a rational number.*

The next theorem is a particular case of the Contraction Theorem

Theorem 7.43 *Suppose that τ is a nef value of (X, H). The linear system $|m(K_X + \tau H)|$ is base-point-free for $m \gg 0$. It defines a regular morphism $\Phi : X \to Y$ onto a normal variety Y with connected fibers called a nef value morphism.*

Length of extremal rays

If R is an extremal ray, then its *length* $l(R)$ is

$$l(R) = \min\{-K_X \cdot C \mid C \text{ is a rational curve with } [C] \in R\}.$$

The motivation for this definition is to give an estimate of the dimension of the locus of extremal rational curves in an extremal ray. The idea of the proof of the following theorem is, basically, due to Mori [Mor1]. We will use the same argument several times in this section but the details will be left to the reader.

Theorem 7.44 ([Wis]) *Let C be a rational curve with its class in an extremal ray R, such that $-K_X \cdot C = l(R)$. If Q is a smooth point on C then the locus of points of curves that belong to R and pass through Q has dimension at least $l(R) - 1$.*

Proof. By the local deformation theory the space of deformations of the morphism $f : \mathbb{P}^1 \to X$, $f(\mathbb{P}^1) = C$, has dimension at least

$$h^0(C, f^*T_X) - h^1(C, f^*T_X) = -C \cdot K_X + (1 - g(C)) \dim X = \dim X - C \cdot K_X.$$

Since we want to fix the point Q, we have $\dim X$ more restrictions. Therefore the space of deformations of the morphism $f : \mathbb{P}^1 \to X$ fixing the point Q has dimension at least

$$-K_X \cdot C = l(R).$$

The group of automorphisms of $\mathbb{P}^1$ fixing a point $x \in \mathbb{P}^1$ has dimension 2, and therefore there exists a family of rational curves on X passing through Q of dimension at least $l(R) - 2$. All these curves are deformations of C, and therefore their numerical classes belong to R. To finish the proof it remains to verify that, for a generic point P in the locus of curves from our family, there exists only finitely many curves from the family passing through P. Suppose, on the contrary, that there exists a family of rational curves passing through P and Q, parametrized by some affine curve D with smooth compactification $\overline{D}$. Passing to infinite points $\overline{D} \setminus D$, we get limit positions of these curves that are irreducible and reduced, since otherwise we obtain a contradiction with minimality of $-K_X \cdot C$ for rational curves from the ray R.

Therefore there exists a ruled surface S and a morphism $F : S \to X$ such that $C_1 = F^{-1}(P)$ and $C_2 = F^{-1}(Q)$ are one-dimensional sections of this ruled surface over its base. Moreover, $\dim F(S) = 2$, otherwise all curves in our pencil represent the same 1-cycle. It follows that

$$C_1^2 < 0,\ C_2^2 < 0, \text{ and } C_1 \cdot C_2 = 0, \tag{7.2}$$

the last equality holding because C_1 and C_2 do not intersect. But this is impossible: a Picard number of a ruled surface is 2 and therefore (7.2) contradicts the Hodge index theorem [Ha1]. □

Fano varieties

A smooth variety X is called a *Fano variety* if $-K_X$ is ample. Its *index* is defined as follows:

$$r(X) = \max\{m > 0 \,|\, mH \sim -K_X \text{ for some } H \in \operatorname{Pic}(X)\}.$$

It is well-known that $r(X) \leq \dim X + 1$. Indeed, suppose that $K_X = mH$, where $-H$ is ample and let $\dim X = n$. The Poincare polynomial $\chi(vH)$ has at most n zeros, and therefore $\chi(vH) \neq 0$ for some $1 \leq v \leq n+1$. But $\chi(vH) = \pm h^n(X, vH)$ by the Kodaira vanishing theorem and $h^n(X, vH) = h^0(X, K_X - vH)$ by the Serre duality. Therefore $m \leq n+1$. The following theorem was conjectured by Mukai.

Theorem 7.45 ([Wis1]) *Let X be a n-dimensional Fano variety of index $r(X)$. If $r(X) > \frac{1}{2}\dim X + 1$ then $\operatorname{Pic}(X) = \mathbb{Z}$.*

7.3.2 The Nef Value and the Defect

Suppose that (X, H) is a smooth polarized variety, i.e. $|H|$ gives an embedding $X \subset \mathbb{P}^N$. Recall that $X^* \subset (\mathbb{P}^N)^\vee$ is the dual variety and

$$\operatorname{def}(X, H) = \operatorname{codim} X^* - 1$$

is the defect of (X, H).

Theorem 7.46 *If $\operatorname{def}(X, H) > 0$ then K_X is not nef.*

Proof. By Theorem 1.18 X is covered by lines. But it is well-known that if X is covered by rational curves then K_X is not nef. In our case it also follows directly from Theorem 7.10. □

It follows that, if $\operatorname{def}(X, H)$ is positive, we can apply the technique of the previous section.

Theorem 7.47 ([E1]) *Suppose that X is a nonlinear smooth projective variety in $\mathbb{P}^N$, $\operatorname{def} X = d > 0$, q is a generic point of X and H is generic tangent hyperplane of X at q. Let $L = \mathbb{P}^d$ be the contact locus of H with X. Then there exists an irreducible family of lines in X of dimension $(3n + d - 4)/2$. If p is a generic point in X, then there exists a family of lines in X through p of dimension $(n + d - 2)/2$.*

Proof. The first statement is a consequence of the standard deformation theory. Indeed, let $p \in L$ and let T_0 be a line in L through p. Since

$$N_{T_0,X} \cong \mathcal{O}_{T_0}^{\oplus(n-d)/2} \oplus \mathcal{O}_{T_0}(1)^{\oplus(n+d-2)/2},$$

we have

$$\dim H^0(T_0, N_{T_0,X}) = \frac{n-d}{2} + 2\frac{n+d-2}{2} = \frac{3n+d-4}{2}$$

and

$$H^1(T_0, N_{T_0,X}) = 0.$$

Therefore the Hilbert scheme of lines in X is smooth at the point T_0 and there is a unique irreducible component of the Hilbert scheme containing the point T_0. This component has dimension $(3n+d-4)/2$. The proof of the second statement is similar. □

It turns out that the nef value and the defect of (X, H) are related:

Theorem 7.48 ([BSW, BFS1]) *Assume that* $\operatorname{def}(X) > 0$. *Then*

(i) *The nef value* τ *of* X *is equal to*

$$\tau = \frac{\dim X + \operatorname{def}(X,H)}{2} + 1.$$

(ii) *If* $\operatorname{Sing} X \cap H$, $H \in X^*$, *is a generic contact locus (hence a projective subspace) and* $l \in \operatorname{Sing} X \cap H$ *is any line, then* $R = \mathbb{R}_+[l]$ *is an extremal ray in* $\overline{NE}_1(X)$ *and* contr_R *is the nef value morphism* Φ. *In particular,* Φ *contracts* $\operatorname{Sing} X \cap H$ *to a point.*

(iii) *If* F *is a generic fiber of* Φ *then* $\operatorname{Pic} F = \mathbb{Z}$.

(iv)

$$\dim \Phi(X) \leq \frac{\dim X - \operatorname{def}(X,H)}{2}.$$

Notice that since

$$\dim X \equiv \operatorname{def}(X,H) \mod 2$$

by Theorem 7.4, the nef value in this case is an integer. Theorem 7.48 can be used to show that the defect of a polarized variety is equal to 0. One has only to calculate the nef value directly and compare the result with the formula in (i).

Proof. We denote $n = \dim X$ and $k = \operatorname{def}(X)$.

Step 1

Let $\mathbb{P}^k = \operatorname{Sing} X \cap H$, $H \in X^*$, be a generic contact locus and let $l \in \operatorname{Sing} X \cap H$ be any line. Since $(-K_X) \cdot l = \dfrac{n+k}{2} + 1$ by Theorem 7.10, it follows that

$$\tau \geq \frac{n+k}{2} + 1. \tag{7.3}$$

Step 2

Let Φ be the nef value morphism. Since Φ is not ample, it follows from the Kleiman Criterion and the Mori Cone Theorem that there exists an extremal rational curve C contracted by Φ, $(K_X+\tau H)\cdot C=0$. We claim that $C\cdot H=1$, i.e. C is a projective line. Indeed, if $C\cdot H\geq 2$ then

$$\tau\leq\frac{1}{2}\tau H\cdot C=\frac{1}{2}(-K_X)\cdot C\leq\frac{1}{2}(n+1).$$

This contradicts (7.3). Therefore

$$\tau=(-K_X)\cdot C=l(R),$$

where $R=\mathbb{R}_+[C]$ is an extremal ray. Applying Theorem 7.44, wee see that if F is a positive-dimensional fiber of Φ then

$$\dim F\geq\tau-1. \tag{7.4}$$

Step 3

Suppose that $\tau>\frac{n+k}{2}+1$. Let F be a positive dimensional fiber of Φ, and let $x\in F$. By Theorem 7.13 there exists a line $l\subset X$ passing through x and such that $(-K_X)\cdot l=\frac{n+k}{2}+1$. Therefore Φ does not contract l, and hence $l\not\subset F$. Let $y\in l$ and $y\notin F$. Arguing as in the proof of Theorem 7.44, we see that there exists an $\frac{n+k}{2}$-dimensional subvariety $Y\subset X$ covered by rational curves passing through y that are deformations of l. Applying a similar argument with ruled surfaces once again, it is easy to see that $\dim Y\cap F=0$. However, using (7.4) we get that

$$\dim Y\cap F\geq\dim Y+\dim F-\dim X\geq\frac{n+k}{2}+\frac{n+k}{2}-n>0.$$

This is a contradiction, and (i) is proved.

Step 4

Clearly, Φ contracts any generic contact locus $\operatorname{Sing} X\cap H$, $H\in X^*$. Moreover, arguing as above (or using Theorem 7.47), we see that any fiber of Φ is at least $\frac{n+k}{2}$-dimensional, and therefore $\dim Y\leq n-\frac{n+k}{2}=\frac{n-k}{2}$. This proves (iv). Let F be a generic fiber. Then $K_F=K_X|_F=-\tau H|_F$. Therefore F is a Fano variety of index at least

$$\tau=\frac{n+k}{2}+1>\frac{\dim F}{2}+1.$$

Therefore $\operatorname{Pic} F=\mathbb{Z}$ by Theorem 7.45. This proves (iii).

Step 5

It remains to prove that Φ is a contraction of an extremal ray $\mathbb{R}_+[l]$. Suppose, on the contrary, that there exists an extremal ray R such that $l \notin R$ and Φ contracts any curve from R. Let C be an extremal rational curve from R. Arguing as in Step 2, we see that $l(R) = (-K_X) \cdot C = \tau$. Consider the morphism $\Psi = \mathrm{contr}_R$. By Theorem 7.44, there exists a fiber F of Ψ such that $\dim F \geq \tau - 1$. Let $x \in F$. By Theorem 7.13 there exists a line $l \subset X$ passing through x and such that $(-K_X) \cdot l = \tau$. This line is either a line in a generic contact locus or its limit position, and therefore Ψ does not contract it. Hence $l \not\subset F$. Let $y \in l$, $y \notin F$. Arguing as in the proof of Theorem 7.44, we see that there exists a $(\tau - 1)$-dimensional subvariety $Y \subset X$ covered by rational curves passing through y that are deformations of l. Applying a similar argument once again, it is easy to see that $\dim Y \cap F = 0$. But

$$\dim Y \cap F \geq \dim Y + \dim F - \dim X \geq 2\tau - 2 - n = k > 0.$$

This is the required contradiction. □

Let F be a generic fiber of the nef value morphism. It is a smooth variety with canonical polarization (F, H_F), where H_F is a hyperplane section of F in the embedding $F \subset X \subset \mathbb{P}^N$.

Theorem 7.49 ([BFS1]) *The following conditions are equivalent:*

(i) $\mathrm{def}(X, H) > 0$;
(ii) $\mathrm{def}(F, H_F) > \dim Y$;
(iii) $\mathrm{def}(X, H) = \mathrm{def}(F, H_F) - \dim Y > 0$.

Proof. It follows from Theorem 7.14 that

$$\dim X + \mathrm{def}(X, H) \geq \dim F + \mathrm{def}(F, H_F).$$

Therefore (i) follows from (ii). Obviously (ii) follows from (iii). So it suffices to prove that (iii) follows from (i).

Since $\mathrm{def}(X, H) > 0$ we have

$$\tau(X, H) = \frac{\dim X + \mathrm{def}(X, H)}{2} + 1$$

by Theorem 7.48. Since F is a generic fiber of Φ we deduce that

$$\mathcal{O}(K_F + \tau H_F) = \mathcal{O}(K_X + \tau H)|_F$$

is a trivial bundle. Therefore K_F is not nef and $\tau(F, H_F) = \tau(X, H)$. If $\mathrm{def}(F, H_F) > 0$ then again by Theorem 7.48

$$\mathrm{def}(F, H_F) = \frac{\dim F + \mathrm{def}(F, H_F)}{2} + 1$$

and

$$0 < \operatorname{def}(X, H) = -\dim X + \dim F + \operatorname{def}(F, H_F) = \operatorname{def}(F, H_F) - \dim Y.$$

It remains to show that if $\operatorname{def}(X, H) > 0$ then $\operatorname{def}(F, H_F) > 0$ as well. Let $L = \mathcal{O}(H)$ and $L_F = \mathcal{O}(H_F)$. Recall that for any line bundle L on a smooth variety X we denote by $J_1(X, L)$ the vector bundle of 1-jets of L. By Theorem 6.1 we need to show that $c_{\dim F}(J_1(F, L_F)) = 0$. From the exact sequence

$$0 \to T_F^\vee(X) \otimes L_F \to J_1(X, L)|_F \to J_1(F, L_F) \to 0$$

we see that the total Chern class

$$c(J_1(X, L)|_F) = c(T_F^\vee(X) \otimes L_F) \cdot c(J_1(F, L_F)).$$

Since F is a generic fiber of Φ, $T_F^\vee(X) \otimes L_F = L_F + \ldots + L_F$ ($\dim Y$ copies). Therefore $c(T_F^\vee(X) \otimes L_F) = (1 + H_F)^{\dim Y}$. In particular,

$$c_{\dim F}(J_1(X, L)|_F) = \\ c_{\dim F}(J_1(F, L_F)) + \sum_{i=1}^{\dim Y} \binom{\dim Y}{i} c_{\dim F - i}(J_1(F, L_F)) \cdot H_F^i.$$

All summands on the right are nonnegative since H_F is very ample and $J_1(F, L_F)$ is spanned. Therefore in order to prove that $c_{\dim F}(J_1(F, L_F)) = 0$ it remains to prove that $c_{\dim F}(J_1(X, L)|_F) = 0$.

It will be easier to prove the more general statement that $c_k(J_1(X, L)|_F) = 0$ for $k \geq \dim F - \operatorname{def}(X, H) + 1$. By the classical representation of a Chern class of a spanned vector bundle [Fu1] we see that $c_k(J_1(X, L)) = 0$ is represented by a set $C_k \subset X$ of all points where $\dim X - k + 2$ generic sections of $H^0(X, J_1(X, L))$ are not linearly independent. We may suppose that these sections have a form

$$j_1(s_1), \ldots, j_1(s_{\dim X - k + 2}),$$

where $s_1, \ldots, s_{\dim X - k + 2}$ are generic sections of $H^0(X, L)$ and j_1 is a natural map sending a germ of a section to its 1-jet. It suffices to prove that for $k \geq \dim F - \operatorname{def}(X, H) + 1$ we have $C_k \cap F = \emptyset$. Since $C_k \supset C_{k+1}$, we only need to prove that $C \cap F = \emptyset$, where $C = C_{\dim F - \operatorname{def}(X,H)+1}$. Suppose that this intersection is non-empty. Therefore the restriction map $\Phi_C : C \to Y$ is surjective. Since $\dim C = \dim Y + \operatorname{def}(X, H) - 1$, it follows that a generic fiber of Φ_C has dimension $\operatorname{def}(X, H) - 1$.

Suppose that $x \in C \cap F$. There exist λ_i for $1 \leq i \leq \dim Y + \operatorname{def}(X, H) - 1$, not all zero, such that $\sum_i \lambda_i j_1(s_i(x)) = 0$. Therefore $s = \sum_i \lambda_i s_i$ is a nontrivial global section of L vanishing at x with its 1-jet. Hence the divisor H_s is singular at x. By Theorem 7.13 there exists a linear subspace $\mathbb{P}^{\operatorname{def}(X,H)} \subset X$ containing x, contained in $\operatorname{Sing} H_s$, and such that $\mathcal{O}(K_X + \tau(X, H)H)|_{\mathbb{P}^{\operatorname{def}(X,H)}}$ is a trivial bundle. Since this is a trivial bundle, $\mathbb{P}^{\operatorname{def}(X,H)}$ is contracted by Φ to a point and, therefore, $\mathbb{P}^{\operatorname{def}(X,H)} \subset F$. On the other hand, since $\mathbb{P}^{\operatorname{def}(X,H)}$

is contained in Sing H_s it is also contained in C. Therefore, a generic fiber of Φ_C has dimension greater or equal to def(X, H). This is the required contradiction. □

7.3.3 Varieties with Small Dual Varieties

Here we describe without proofs smooth polarized varieties (X, H) with positive defect for $\dim X \leq 10$. In this section we use the following terminology.

Definition 7.50 A smooth polarized variety (X, H) is called a $\mathbb{P}^d$*-bundle* over a normal variety Y if there exists a surjective morphism $p : X \to Y$ such that all the fibers F of p are projective subspaces $\mathbb{P}^d$ of $\mathbb{P}^N$, where $X \subset \mathbb{P}^N$ is an embedding corresponding to H.

(X, H) is called a *scroll* (resp. a *quadric, Del Pezzo, or Mukai fibration*) over a normal variety Y if there exists a surjective morphism $p : X \to Y$ with connected fibers such that

$$K_X + (\dim X - \dim Y + W)H \sim p^*H',$$

where $W = 1$ (resp. $W = 0$, $W = -1$, or $W = -2$) for some ample Cartier divisor H' on Y.

To justify this definition we note that, by the Kobayashi–Ochiai characterisation of projective spaces and quadrics [KO], if (X, H) is a scroll over Y then a generic fiber F of p is a projective subspace $\mathbb{P}^d$ of $\mathbb{P}^N$, where $X \subset \mathbb{P}^N$ is an embedding corresponding to H. If (X, H) is a quadric fibration then a generic fiber F is isomorphic to a quadric hypersurface Q in some projective space, and the line bundle corresponding to this embedding is $\mathcal{O}(H)|_F$. In fact, quadric fibrations will not appear in the classification because def$(X, H) = 0$ for quadric fibrations by [BFS1].

Now suppose that def$(X, H) > 0$.

$\dim X \leq 2$

Curves and surfaces with positive defect do not exist.

$\dim X = 3$

(X, H) is a $\mathbb{P}^2$-bundle over a smooth curve. In this case def$(X, H) = 1$.

$\dim X = 4$

(X, H) is a $\mathbb{P}^3$-bundle over a smooth curve. In this case def$(X, H) = 2$.

dim $X = 5$

- X is a hyperplane section of the Plücker embedding of the Grassmannian $\mathrm{Gr}(2,5)$ in $\mathbb{P}^9$ and $\mathrm{def}(X,H) = 1$.
- (X,H) is a $\mathbb{P}^4$-bundle over a smooth curve and $\mathrm{def}(X,H) = 3$.
- (X,H) is a $\mathbb{P}^3$-bundle over a smooth surface and $\mathrm{def}(X,H) = 1$.

dim $X = 6$

- (X,H) is the Plücker embedding of the Grassmannian $\mathrm{Gr}(2,5)$ in $\mathbb{P}^9$ and $\mathrm{def}(X,H) = 2$.
- (X,H) is a $\mathbb{P}^5$-bundle over a smooth curve and $\mathrm{def}(X,H) = 4$.
- (X,H) is a $\mathbb{P}^4$-bundle over a smooth surface and $\mathrm{def}(X,H) = 2$.

dim $X = 7$

- (X,H) is a $\mathbb{P}^6$-bundle over a smooth curve and $\mathrm{def}(X,H) = 5$.
- (X,H) is a $\mathbb{P}^5$-bundle over a smooth surface and $\mathrm{def}(X,H) = 3$.
- X is a 7-dimensional Fano variety with $\mathrm{Pic}\, X = \mathbb{Z}$ and $K_X \sim -5H$. In this case $\mathrm{def}(X,H) = 1$. An example of such variety is the section of 10-dimensional spinor variety S in $\mathbb{P}^{15}$ by three generic hyperplanes [Mu, LS].
- (X,H) is a Del Pezzo fibration over a smooth curve such that the general fiber F is isomorphic to $\mathrm{Gr}(2,5)$ and the embedding corresponding to $\mathcal{O}(H)|_F$ is the Plücker embedding in $\mathbb{P}^9$. In this case $\mathrm{def}(X,H) = 1$.
- (X,H) is a scroll over a 3-dimensional variety and $\mathrm{def}(X,H) = 1$.

dim $X = 8$

- (X,H) is a $\mathbb{P}^7$-bundle over a smooth curve and $\mathrm{def}(X,H) = 6$.
- (X,H) is a $\mathbb{P}^6$-bundle over a smooth surface and $\mathrm{def}(X,H) = 4$.
- X is is a 8-dimensional Fano variety with $\mathrm{Pic}\, X = \mathbb{Z}$ and $K_X \sim -6H$. In this case $\mathrm{def}(X,H) = 2$.
- (X,H) is a scroll over a 3-dimensional variety and $\mathrm{def}(X,H) = 2$.

dim $X = 9$

- (X,H) is a $\mathbb{P}^8$-bundle over a smooth curve and $\mathrm{def}(X,H) = 7$.
- (X,H) is a $\mathbb{P}^7$-bundle over a smooth surface and $\mathrm{def}(X,H) = 5$.
- X is a 9-dimensional Fano variety with $\mathrm{Pic}\, X = \mathbb{Z}$ and $K_X \sim -7H$. In this case $\mathrm{def}(X,H) = 3$.
- (X,H) is a scroll over a 3-dimensional variety and $\mathrm{def}(X,H) = 3$.
- X is a Fano 9-dimensional variety with $K_X \sim -6H$ and $\mathrm{Pic}\, X = \mathbb{Z}$. In this case $\mathrm{def}(X,H) = 1$.
- (X,H) is a Mukai fibration over a smooth curve with $\mathrm{Pic}\, F = \mathbb{Z}$ for a generic fiber F and $\mathrm{def}(X,H) = 1$.
- (X,H) is a scroll over a 4-dimensional variety and $\mathrm{def}(X,H) = 1$.

dim $X = 10$

- (X, H) is a $\mathbb{P}^9$-bundle over a smooth curve and $\operatorname{def}(X, H) = 8$.
- (X, H) is a $\mathbb{P}^8$-bundle over a smooth surface and $\operatorname{def}(X, H) = 6$.
- X is a 10-dimensional Fano variety with $\operatorname{Pic} X = \mathbb{Z}$ and $K_X \sim -8H$. In this case $\operatorname{def}(X, H) = 4$.
- (X, H) is a scroll over a 3-dimensional variety and $\operatorname{def}(X, H) = 4$.
- X is a Fano 10-dimensional variety with $K_X \sim -7H$ and $\operatorname{Pic} X = \mathbb{Z}$. In this case $\operatorname{def}(X, H) = 2$.
- (X, H) is a Mukai fibration over a smooth curve with $\operatorname{Pic} F = \mathbb{Z}$ for a generic fiber F and $\operatorname{def}(X, H) = 2$.
- (X, H) is a scroll over a 4-dimensional variety and $\operatorname{def}(X, H) = 2$.

7.4 Flag Varieties with Positive Defect

7.4.1 Nef Cone of a Flag Variety

Let G be a connected simply-connected semisimple complex algebraic group with a Borel subgroup B and a maximal torus $T \subset B$. Let $\mathcal{P}$ be the group of characters of T (the weight lattice). Let $\Delta \subset \mathcal{P}$ be roots of G relative to T. To every $\alpha \in \Delta$ we assign the 1-dimensional unipotent subgroup $U_\alpha = \exp(\mathfrak{g}_\alpha) \subset G$. We define the *negative* roots Δ^- as those roots α such that $U_\alpha \subset B$. Let $\Pi \subset \Delta^+$ be simple roots, $\Pi = \{\alpha_1, \dots, \alpha_n\}$, where $n = \operatorname{rank} G = \dim T$. The *root lattice* $\mathcal{Q} \subset \mathcal{P}$ is a sublattice spanned by Δ. For any simple root $\alpha_i \in \Pi$ we assign a smooth curve $C_{\alpha_i} = \overline{\pi(U_{\alpha_i})} \subset G/B$, where $\pi : G \to G/B$ is a quotient map. We assign a 1-cycle $C_\alpha = \sum n_i C_{\alpha_i}$ in G/B to any $\alpha = \sum n_i \alpha_i \in \mathcal{Q}$ by linearity. C_α is effective iff α belongs to the semigroup $\mathcal{Q}^+$ generated by simple roots.

$\mathcal{P}$ is generated as a $\mathbb{Z}$-module by fundamental weights $\omega_1, \dots, \omega_n$ dual to the simple roots under the Killing form. We denote the dominant weights by $\mathcal{P}^+$ and the strictly dominant weights by $\mathcal{P}^{++}$.

The character group of B is identified with a character group of T. Therefore, for any $\lambda \in \mathcal{P}$, we can assign a 1-dimensional B-module $\mathbb{C}_\lambda$, where B acts on $\mathbb{C}_\lambda$ by a character λ. Now we can define the fiber bundle $G \times_B \mathbb{C}_\lambda$ as the quotient of $G \times E$ by the diagonal action B:

$$b \cdot (g, z) = (gb^{-1}, b \cdot z).$$

The projection onto the first factor induces a map $G \times_B \mathbb{C}_\lambda \to G/B$, which realizes $G \times_B \mathbb{C}_\lambda$ as an equivariant line bundle $\mathcal{L}_\lambda$ on G/B with fiber $\mathbb{C}_\lambda$.

The following theorem is well-known:

Theorem 7.51

- *The correspondence $\alpha \to C_\alpha$ is an isomorphism of $\mathcal{Q}$ and $N_1^{\mathbb{Z}}(G/B)$, the group of 1-cycles modulo numerical equivalence.*

- *The correspondence $\lambda \to L_\lambda$ is an isomorphism of $\mathcal{P}$ and* $\mathrm{Pic}(G/B)$. *The latter group is, in turn, isomorphic to $N^1_{\mathbb{Z}}(G/B)$, the group of Cartier divisors modulo numerical equivalence.*
- *The pairing $\mathcal{P} \times \mathcal{Q} \to \mathbb{Z}$ given by duality corresponds to the intersection product on $N^1_{\mathbb{Z}}(G/B) \times N^{\mathbb{Z}}_1(G/B)$.*
- *L_λ is nef iff $\lambda \in \mathcal{P}^+$. L_λ is ample iff $\lambda \in \mathcal{P}^{++}$. L_λ is ample if and only if L_λ is very ample.*

A similar description is available for flag varieties G/P, where $P \subset G$ is an arbitrary parabolic subgroup. Let $\Pi_P \subset \Pi$ be some subset of simple roots. Let $\Delta^+_P \subset \Delta^+$ be positive roots that are linear combinations of the roots in Π_P. Then we can assume that P is generated by B and the U_α for $\alpha \in \Delta^+_P$. We denote $\Pi \setminus \Pi_P$ by $\Pi_{G/P}$ and $\Delta^+ \setminus \Delta^+_P$ by $\Delta^+_{G/P}$. P is maximal if and only if $\Pi_{G/P}$ is a single simple root. We denote by $\mathcal{Q}_{G/P}$ the sublattice of $\mathcal{Q}$ generated by $\Pi_{G/P}$, and by $\mathcal{Q}^+_{G/P}$ the corresponding subsemigroup. For any $\alpha_i \in \Pi_{G/P}$ we assign a smooth curve

$$C_{\alpha_i} = \overline{\pi(U_{\alpha_i})} \subset G/P,$$

where $\pi : G \to G/P$ is a quotient map. This extends to a map $\mathcal{Q}_{G/P} \to N^{\mathbb{Z}}_1(G/P)$, $\alpha \mapsto C_\alpha$ by linearity. C_α is effective iff $\alpha \in \mathcal{Q}^+_{G/P}$.

The fundamental weights $\omega_{i_1}, \ldots \omega_{i_k}$ dual to the simple roots in $\Pi_{G/P}$ generate the sublattice $\mathcal{P}_{G/P}$ of $\mathcal{P}$. Let $\mathcal{P}^+_{G/P} = \mathcal{P}^+ \cap \mathcal{P}_{G/P}$. The subset $\mathcal{P}^{++}_{G/P} \subset \mathcal{P}^+_{G/P}$ consists of weights $\lambda = \sum n_k \omega_{i_k}$ such that all $n_k > 0$. Any weight $\lambda \in \mathcal{P}_{G/P}$ gives rise to a character of P and therefore to a line bundle L_λ on G/P.

Theorem 7.52

- *The correspondence $\alpha \to C_\alpha$ is an isomorphism of $\mathcal{Q}_{G/P}$ and $N^{\mathbb{Z}}_1(G/P)$.*
- *The correspondence $\lambda \to L_\alpha$ is an isomorphism of $\mathcal{P}_{G/P}$ and* $\mathrm{Pic}(G/P)$. *The latter group is, in turn, isomorphic to $N^1_{\mathbb{Z}}(G/P)$. In particular,* $\mathrm{Pic}(G/P) = \mathbb{Z}$ *if and only if P is maximal.*
- *The pairing $\mathcal{P}_{G/P} \times \mathcal{Q}_{G/P} \to \mathbb{Z}$ given by duality corresponds to the intersection product on $N^1_{\mathbb{Z}}(G/P) \times N^{\mathbb{Z}}_1(G/P)$.*
- *L_λ is nef if and only if $\lambda \in \mathcal{P}^+_{G/P}$. L_λ is ample if and only if L_λ is very ample if and only if $\lambda \in \mathcal{P}^{++}_{G/P}$.*
- *If $\lambda \in \mathcal{P}^+_{G/P}$ then the linear system corresponding to L_λ is base-point-free. The corresponding map given by sections is a quotient $G/P \to G/Q$, where Q is a parabolic subgroup such that Π_Q is a union of Π_P and all simple roots in $\Pi_{G/P}$ orthogonal to λ.*

7.4.2 Nef Values of Flag Varieties

Theorem 7.53 ([Sn])

(i) *The anticanonical divisor* $-K_{G/P}$ *corresponds to the weight*

$$\rho_{G/P} = \sum_{\alpha \in \Delta^+_{G/P}} \alpha, \quad \rho_{G/P} \in \mathcal{P}^{++}_{G/P}.$$

$K_{G/P}$ *is not nef.*

(ii) *The nef value of an ample line bundle* L_λ *given by* $\lambda \in \mathcal{P}^{++}_{G/P}$ *is equal to*

$$\tau = \max_{\alpha \in \Pi_{G/P}} \frac{(\rho_{G/P}, \alpha)}{(\lambda, \alpha)}.$$

(iii) *The nef value morphism* Φ *is a quotient map* $\Phi : G/P \to G/Q$*, where* Q *is the parabolic subgroup defined by* $\Pi_Q = \Pi_P \cup \Pi'$*, where* $\Pi' \subset \Pi_{G/P}$ *are simple roots for which the above maximum occurs.*

Proof. The tangent bundle $T_{G/P}$ is an equivariant vector bundle whose fiber over the identity coset is isomorphic to the quotient of Lie algebras $\mathfrak{g}/\mathfrak{p}$. P acts on it via the adjoint representation on $\mathfrak{g}$ projected to the quotient. T-weights of this representation are given by $\Delta^+_{G/P}$. Therefore the anticanonical bundle $\Lambda^{\dim G/P} T_{G/P}$ corresponds to the weight $\rho_{G/P}$. Notice that for any $\alpha \in \Pi_{G/P}$

$$(\rho_{G/P}, \alpha) = \left(\sum_{\beta \in \Delta +} \beta, \alpha \right) - \left(\sum_{\beta \in \Delta^+_P} \beta, \alpha \right).$$

The first product is equal to 2 (it is well-known that $\sum_{\beta \in \Delta_+} = 2 \sum_{i=1}^n \omega_i$) and the second product is non-positive, because any root from Δ^+_P is a linear combination of simple roots from Π_P with non-negative coefficients and the scalar product of two distinct simple roots is non-positive. Therefore $\rho_{G/P} \in \mathcal{P}^{++}_{G/P}$ and $K_{G/P}$ is not nef.

Let

$$\rho_{G/P} = \sum_{\alpha_i \in \Pi_{G/P}} n_i \omega_i, \quad \lambda = \sum_{\alpha_i \in \Pi_{G/P}} m_i \omega_i.$$

Then, clearly, the nef value τ of L is given by

$$\tau = \max_{\alpha_i \in \Pi_{G/P}} \frac{n_i}{m_i} = \max_{\alpha \in \Pi_{G/P}} \frac{(\rho_{G/P}, \alpha)}{(\lambda, \alpha)}.$$

Part (iii) follows from Theorem 7.52, part (5). □

Using this theorem, it is straightforward to calculate nef values in particular cases. Suppose that G is a simple group and P is a maximal parabolic subgroup such that $\Pi_{G/P} = \{\alpha_i\}$ is a single simple root. Suppose further that L is a generator of $\operatorname{Pic} G/P$, that is, L corresponds to the fundamental weight ω_i. In the Table 7.4.2 we give $\dim G/P$ and the nef value τ of L for each G and i.

G	i	$\dim G/P$	τ
A_l	$1,\dots,l$	$i(l+1-i)$	$l+1$
B_l	$1,\dots,l-1$	$i(4l+1-3i)/2$	$2l-i$
	l	$l(l+1)/2$	$2l$
C_l	i	$i(4l+1-3i)/2$	$2l-i+1$
D_l	$1,\dots,l-2$	$i(4l-1-3i)/2$	$2l-i-1$
	$l-1,l$	$l(l-1)/2$	$2l-2$
E_6	$1,5$	16	12
	$2,4$	25	9
	3	29	17
	6	21	11
E_7	1	27	18
	2	42	13
	3	50	10
	4	53	8
	5	47	11
	6	33	17
	7	42	14
E_8	1	57	29
	2	83	19
	3	97	14
	4	104	11
	5	106	9
	6	98	13
	7	78	23
	8	92	17
F_4	1	15	8
	2	20	5
	3	20	7
	4	15	11
G_2	1	5	5
	2	5	3

Table 7.1.

7.4.3 Flag Varieties of Positive Defect

Now we use these calculations and Theorems 7.48, 7.49 to classify all polarized flag varieties with positive defect.

Suppose first that G is a simple group and $P \subset G$ is a maximal parabolic subgroup corresponding to a simple root α_i. Then all very ample linear bundles on G/P have the form $L^{\otimes k}$ for $k > 0$, where L corresponds to the fundamental weight ω_i. If $k > 1$ then by Theorem 6.6 we have $\operatorname{def}(G/P, L^{\otimes k}) = 0$. Therefore it suffices to find $\operatorname{def}(G/P, L)$. By Theorem 7.48 if $\operatorname{def}(G/P, L) > 0$ then

$$\operatorname{def}(G/P, L) = 2\tau(G/P, L) - 2 - \dim G/P.$$

Therefore we may look through the table of the previous section and pick all entries such that the RHS d_P of this equation is positive. The following table contains all these entries.

No.	G	i	$\tau(G/P, L)$	$\dim G/P$	d_P
1	A_l	$1, l$	$l+1$	l	l
2	A_l, l even	$2, l-1$	$l+1$	$2l-2$	2
3	C_l	1	$2l$	$2l-1$	$2l-1$
4	B_2	2	4	3	3
5	B_4	4	8	10	4
6	D_5	4, 5	8	10	4
7	A_l, l odd	$2, l-1$	$l+1$	$2l-2$	2
8	A_5	3	6	9	1
9	B_l	1	$2l-1$	$2l-1$	$2l-3$
10	B_3	3	6	6	4
11	D_4	3, 4	6	6	4
12	B_5	5	10	15	3
13	D_6	5, 6	10	15	3
14	B_6	6	12	21	1
15	D_7	6, 7	12	21	1
16	C_l	2	$2l-1$	$4l-5$	1
17	D_l	1	$2l-2$	$2l-2$	$2l-4$
18	E_6	1, 5	12	16	6
19	E_7	1	18	27	7
20	F_4	4	11	15	5
21	G_2	1	5	5	3

Table 7.2.

In fact, only the first 6 rows of this table correspond to polarized flag varieties with positive defect, and No. 4 is a particular case of No. 3. This can be verified case-by-case (see [Sn]). The following theorem is an immediate consequence.

Theorem 7.54 ([KM, Sn]) *Suppose that P_i is a maximal parabolic subgroup of a simple algebraic group corresponding to the simple root α_i. Let L be a very ample line bundle on G/P_i corresponding to the fundamental weight ω_i. Then*

(i) $\operatorname{def}(G/P_i, L^{\otimes k}) = 0$ *for* $k > 1$.

(ii) $\mathrm{def}(G/P_i, L) > 0$ *if and only if* G/P_i *is one of the following:*

- A_l/P_1, A_l/P_l, $\mathrm{def}(G/P) = \dim(G/P) = l$.
- A_l/P_2, A_l/P_{l-1}, $l \geq 4$ *even*, $\mathrm{def}(G/P) = 2$, $\dim(G/P) = 2l - 2$.
- C_l/P_1, $\mathrm{def}(G/P) = \dim(G/P) = 2l - 1$.
- B_4/P_4, D_5/P_4, D_5/B_5, $\mathrm{def}(G/P) = 4$, $\dim(G/P) = 10$.

Notice that we consider isomorphic polarized varieties with different group actions as distinct varieties. The next step is to consider flag varieties corresponding to arbitrary parabolic subgroups in a simple group.

Theorem 7.55 ([KM, Sn]) *Suppose that* $(G/P, L)$ *is a polarized flag variety of a simple algebraic group* G*, where* $P \subset G$ *is a non-maximal parabolic subgroup. Then* $\mathrm{def}(G/P, L) = 0$.

Proof. Suppose that $\mathrm{def}(G/P, L) > 0$. By Theorem 7.53 the nef value morphism is a homogeneous fiber bundle $\Phi : G/P \to G/Q$, where $Q \subset G$ is another parabolic subgroup. The fiber F of Φ is isomorphic to Q/P.

By Theorem 7.48 we have $\mathrm{Pic}\, F = \mathbb{Z}$, and therefore F is isomorphic to a quotient of a simple group by a maximal parabolic subgroup and Q is a proper subgroup of G. By Theorem 7.49 $\mathrm{def}(F, L) > \dim G/Q > 0$. Therefore the polarized flag variety (F, L) is one of the varieties from Theorem 7.54. Let us consider three cases.

If F is equal to A_l/P_2 or A_l/P_{l-1} with $l \geq 4$ even, then $\mathrm{def}(F, L) = 2$. Therefore $\dim G/Q = 1$. It follows that $G = \mathrm{SL}_2$, Q is a Borel subgroup, and there are no possibilities for P.

If F is equal to B_4/P_4, D_5/P_4, or D_5/B_5 then $\mathrm{def}(F, L) = 4$. Therefore $\dim G/Q$ is equal to 1, 2, or 3. It follows that G is equal to SL_2, SL_3, SL_4, or Sp_4. Again none of these groups contains B_4 or D_5 as a simple component of a Levi subgroup.

Finally, let F be A_l/P_1, A_l/P_l, or $C_{\frac{l+1}{2}}/P_1$, and so F is isomorphic to $\mathbb{P}^l$. We want to show that $\dim G/Q > k$, which will be a contradiction. The set of simple roots Π_Q contains a subset Ψ with Dynkin diagram of type A_l or $C_{\frac{l+1}{2}}$. Moreover, there exists a simple root $\alpha \in \Pi$ such that $\alpha \notin \Pi_Q$ and the Dynkin diagram of $\Psi' = \Psi \cup \{\alpha\}$ is connected. Then Ψ' is a set of simple roots of a simple subgroup $G' \subset G$ and $\dim G/Q \geq \dim G'/Q'$, where $Q' = G' \cap Q$ is a maximal parabolic subgroup defined by Ψ. It is easy to check that in all arising cases $\dim G'/Q' > k$. □

Now we can handle the case of an arbitrary semisimple group and its arbitrary parabolic subgroup.

Theorem 7.56 *[KM, Sn] Suppose that* $(G/P, L)$ *is a polarized flag variety of a complex semisimple algebraic group* G*. Then* $\mathrm{def}(G/P, L) > 0$ *if and only if either* $(G/P, L)$ *is one of polarized varieties described in Theorem 7.54 or the following conditions hold:*

- $G = G_1 \times G_2$, $P = P_1 \times P_2$, *where* $P_1 \subset G_1$, $P_2 \subset G_2$.

- $L = \mathrm{pr}_1^* L_1 \otimes \mathrm{pr}_2^* L_2$, *where* $\mathrm{pr}_i : G/P \to G_i/P_i$ *are projections and* L_i *are very ample line bundles on* G_i/P_i.
- $(G_1/P_1, L_1)$ *is one of polarized varieties described in Theorem 7.54.*
- $\mathrm{def}(G_1/P_1, L_1) > \dim G_2/P_2$.

In this case $\mathrm{def}(G/P, L) = \mathrm{def}(G_1/P_1, L_1) - \dim G_2/P_2$.

Proof. This Theorem easily follows from Theorem 3.5. □

Remark 7.57 Polarized flag varieties with positive defect were first classified in [KM] using Theorem 3.1. For 'very big' polarizations this result can be deduced from Theorem 6.8. Indeed, the fundamental weights $\omega_1, \dots, \omega_k$ dual to the simple roots in $\Pi_{G/P}$ generate the lattice $\mathcal{P}_{G/P}$ isomorphic to $\mathrm{Pic}(G/P)$. We shall denote the line bundle L_{ω_i} by $\mathcal{L}_i$. Then any line bundle on G/P has the form

$$\mathcal{L} = \mathcal{L}_1^{\otimes n_1} \otimes \dots \otimes \mathcal{L}_r^{\otimes n_r}.$$

$\mathcal{L}$ is very ample if and only if all $n_i > 0$. Then the degree $\deg \Delta_{(X,\mathcal{L})}$ is a function of $n_1, \dots, n_r$. Let us introduce new non-negative integers $m_i = n_i - 1$. Then $\deg \Delta_{(X,\mathcal{L})} = f(m_1, \dots, m_r)$. All conditions of Theorem 6.8 are clearly satisfied, and therefore f is a non-trivial polynomial with non-negative coefficients. Assume that the weight λ satisfies the conditions $n_i \geq 2$ (λ is very strictly dominant). Then these results clearly imply that the dual variety $(G/P, \lambda)^*$ is a hypersurface.

Let us consider two examples. Let $V = \mathbb{C}^r$, $G = \mathrm{SL}(V)$ and let P be the stabiliser of the line in V. Then $G/P = \mathbb{P}(V)$. Take $\lambda = n\omega_1$, where $n > 0$. Then $\mathcal{L}_1 = \mathcal{O}(n)$. The degree of $\Delta_{(G/P,\lambda)}$ is the degree of the classical discriminant of a homogeneous form of degree n in r variables. Therefore by Example 6.4 we have

$$\deg \Delta_{(G/P,\lambda)} = r(n-1)^{r-1}.$$

In this case $f = rm^{r-1}$, and so it is a polynomial with nonnegative coefficients.

Let $G = SL_3$ and let $P = B$ be the Borel subgroup. The weight λ can be written as $\lambda = n_1\omega_1 + n_2\omega_2$. One can check using the Kleiman formula that

$$\deg \Delta_{(G/B,\lambda)} = 12(n_1^2 n_2 + n_1 n_2^2) - 6(n_1^2 + 4n_1n_2 + n_2^2) + 12(n_1 + n_2) - 6.$$

Substituting $m_1 = n_1 - 1$ and $m_2 = n_2 - 1$, we see that

$$f = 12(m_1^2 m_2 + m_1 m_2^2) + 6(m_1^2 + 4m_1m_2 + m_2^2) + 12(m_1 + m_2) + 6.$$

Again, this polynomial has nonnegative coefficients.

8

Dual Varieties of Homogeneous Spaces

Here we study dual varieties and discriminants of several special homogeneous spaces. We start in 8.1 by showing how to use standard results of representation theory such as the Borel–Weyl–Bott theorem, the BGG homomorphism, identities with Schur functors, and formulas of the Schubert calculus to find the codegree of Grassmannians or full and partial flag varieties. We give a list of formulas for the degree of hyperdeterminants and sketch the proof of a Theorem of Zak about varieties of codegree 3. In 8.2 we generalize the Theorem of Matsumura and Monsky about automorphisms of smooth hypersurfaces to automorphisms of smooth very ample divisors on flag varieties. In 8.3 we study commutative algebras without identities from the "discriminantal" point of view. As a corollary we prove that the algebra of diagonal matrices does not have quasiderivations. In 8.4 we study anticommutative algebras (nets of skewsymmetric forms). We show that they have beatiful geometric properties related to cubic surfaces, Del Pezzo surfaces, representation theory of S_5, etc. In 8.5 we show that the discriminant in a simple Lie algebra defined by analogy with the disciminant of a linear operator is equal to the discriminant of the minimal orbit, the so-called adjoint variety. Finally, in 8.6 we study related questions about schemes of zeros of irreducible homogeneous vector bundles. In particular, we address a question of classifying irreducible homogeneous vector bundles with a trivial line subbundle, find the maximal dimension of an isotropic subspace of a generic symmetric or skewsymmetric form, and study properties of the related Moore–Penrose involution.

8.1 Calculations of $\deg X^*$

8.1.1 Borel–Weyl–Bott Theorem

If P is an algebraic group then the exact sequence

$$\{e\} \to R_u(P) \to P \to P/R_u(P) \to \{e\}$$

always splits, i.e. there exists a reductive subgroup $L \subset P$ that maps isomorphically on $P/R_u(P)$. In other words, P is a semi-direct product $P = L \leftthreetimes R_u(P)$, called the *Levi decomposition*. L is called the *Levi subgroup*. A subgroup is a Levi subgroup if and only if it is a maximal reductive subgroup of P. All Levi subgroups are conjugated to each other.

If G is a reductive group, $P \subset G$ is a parabolic subgroup and $L \subset P$ is a Levi subgroup, then we abuse language and call L a Levi subgroup of G.

We choose the maximal torus $T \subset L$, and then T is also a maximal torus of G. Let $B \supset T$ be a Borel subgroup such that $B \cap P = B \cap L$. Then $B \cap L$ is a Borel subgroup in L. We choose the corresponding systems $\Delta_G \supset \Delta_L$ of roots, $\Delta_G^+ \supset \Delta_L^+$ of positive roots, and $\Pi_G \supset \Pi_L$ of simple roots in G and L. The semigroup $\mathcal{P}_L^+$ of dominant weights of L contains the semigroup $\mathcal{P}_G^+$ of dominant weights of G. For example, if $P = B$ is a Borel subgroup then $L = T$ and $\mathcal{P}_L^+ = \mathcal{X}(T)$.

We take any $\lambda \in \mathcal{P}_L^+$, and let U_λ be the corresponding irreducible L-module. We may consider U_λ as P-module with the trivial action of P_u. Now we define the fiber bundle $G \times_P U_\lambda$ as the quotient of $G \times U_\lambda$ by the action of P given by:

$$p \cdot (g, z) = (gp^{-1}, p \cdot z).$$

Projection onto the first factor induces the map $G \times_P U_\lambda \to G/P$, which is an equivariant vector bundle $\mathcal{E}_\lambda$ on G/P. ("Equivariant" means that G acts on $\mathcal{E}_\lambda$ compatibly with the action of G on the base G/P.)

The Borel–Weyl–Bott theorem shows how to calculate cohomology of $\mathcal{E}_\lambda$. Let V_λ be the irreducible G-module with the highest weight λ for any $\lambda \in \mathcal{P}_G^+$. Let ρ be the half-sum of positive roots of G. Then $(\rho, \alpha_i) = 1$ for any simple root $\alpha_i \in \Pi_G$. The element $z \in \mathfrak{t}_{\mathbb{R}}^\vee$ is called *singular* if there exists a root $\alpha \in \Delta_G$ such that $(z, \alpha) = 0$, i.e. if z belongs to the wall of a Weyl chamber. If z is not singular then there exists a unique element $w \in W_G$ of the Weyl group of G such that $w \cdot z$ belongs to the fixed Weyl chamber C^+.

Theorem 8.1 ([Bo])

- *If $\lambda \in \mathcal{P}_G^+$ then $H^0(G/P, \mathcal{E}_\lambda) = V_\lambda^\vee$ and $H^i(G/P, \mathcal{E}_\lambda) = 0$ for $i > 0$.*
- *If $\lambda + \rho$ is singular then $H^i(G/P, \mathcal{E}_\lambda) = 0$ for any i.*
- *If $\lambda + \rho$ is not singular, then, for the unique element $w \in W_G$,*

$$\lambda' = w(\lambda + \rho) - \rho \in \mathcal{P}_G^+.$$

Then $H^{l(w)}(G/P, \mathcal{E}_\lambda) = V_{\lambda'}^\vee$ and $H^i(G/P, \mathcal{E}_\lambda) = 0$ for $i \neq l(w)$.

Corollary 8.2

$$\chi(\mathcal{E}_\lambda) = \sum (-1)^i \dim H^i(G/P, \mathcal{E}_\lambda) = \prod_{\alpha \in \Delta_G^+} (\lambda + \rho, \alpha)/(\rho, \alpha).$$

Proof. Indeed, if $\lambda \in \mathcal{P}_G^+$, then this is a well-known Weyl formula for the dimension of V_λ; see [OV]. If λ is singular then this formula is obvious. Suppose that λ is not singular, and let $\lambda' = w(\lambda+\rho) - \rho \in \mathcal{P}_G^+$ for the unique element $w \in W_G$. Then

$$\chi(\mathcal{E}_\lambda) = (-1)^{l(w)} \dim V_{\lambda'} = (-1)^{l(w)} \prod_{\alpha\in\Delta_G^+} (\lambda'+\rho,\alpha)/(\rho,\alpha) =$$

$$(-1)^{l(w)} \prod_{\alpha\in\Delta_G^+} (w(\lambda+\rho),\alpha)/(\rho,\alpha) = (-1)^{l(w)} \prod_{\alpha\in\Delta_G^+} (\lambda+\rho, w^{-1}\alpha)/(\rho,\alpha) =$$

$$\prod_{\alpha\in\Delta_G^+} (\lambda+\rho,\alpha)/(\rho,\alpha).$$

The corollary is proved. □

8.1.2 Representation Theory of GL_n

We choose the standard maximal torus T of diagonal matrices and the Borel subgroup of upper-triangular matrices. The group of characters $\mathcal{X}(T)$ is freely generated by characters $\varepsilon_1, \ldots, \varepsilon_n$, where

$$\varepsilon_k(\mathrm{diag}\{x_1, \ldots, x_n\}) = x_k.$$

The roots are $\alpha_{ij} = \varepsilon_i - \varepsilon_j$. The root subspace of α_{ij} is $\mathbb{C}E_{ij}$. The simple roots are $\varepsilon_1 - \varepsilon_2, \ldots, \varepsilon_{n-1} - \varepsilon_n$. A weight $\lambda = \sum \lambda_i \varepsilon_i$ is dominant if and only if $\lambda_1 \geq \lambda_2 \geq \ldots \geq \lambda_n$. For example, the highest weight of $S^d\mathbb{C}^n$ is $d\varepsilon_1$ and the corresponding highest weight vector is e_1^d. Similarly, the highest weight of $\Lambda^d\mathbb{C}^n$ is $\varepsilon_1+\ldots+\varepsilon_d$ and the corresponding highest weight vector is $e_1\wedge\ldots\wedge e_d$.

One-dimensional representations of GL_n correspond to characters of GL_n, i.e. to functions $\det^p$, $p \in \mathbb{Z}$. The corresponding highest weights are $p\sum\varepsilon_i$. Therefore any irreducible representation, after tensoring by a suitable character, has the highest weight λ with the property $\lambda_n \geq 0$. We identify these dominant weights with the set $\mathcal{P}_n$ of all partitions

$$\lambda_1 \geq \ldots \geq \lambda_n \geq 0.$$

Let $\mathcal{P}_n^m \subset \mathcal{P}_n$ be the subset of partitions such that $\lambda_1 \leq m$. For any partition $\lambda \in \mathcal{P}_n$ we set

$$|\lambda| = \lambda_1 + \ldots + \lambda_n.$$

For any partition $\lambda \in \mathcal{P}_n^m$ we denote by $\lambda^t \in \mathcal{P}_m^n$ the transposed partition given by

$$\lambda_i^t = \max\{j \mid \lambda_j \geq i\}.$$

For any partition $\lambda \in \mathcal{P}_n$ (or just the dominant weight λ), we denote by $S_\lambda(\mathbb{C}^n)$ the irreducible representation of GL_n with the highest weight λ.

We will need the *Cauchy formula*

$$\Lambda^d(\mathbb{C}^n \otimes \mathbb{C}^m) = \bigoplus_{\lambda \in \mathcal{P}_n^m,\ |\lambda|=d} S_\lambda(\mathbb{C}^n) \otimes S_{\lambda^t}(\mathbb{C}^m),$$

the decomposition of $\mathrm{GL}_n \times \mathrm{GL}_m$-modules.

A parabolic subgroup $P \subset \mathrm{GL}_n$ consists of block triangular matrices

$$\begin{pmatrix} A_{11} & 0 & \dots & 0 \\ A_{21} & A_{22} & \ddots & \vdots \\ \vdots & \ddots & \ddots & 0 \\ A_{k1} & \dots & \dots & A_{kk} \end{pmatrix}$$

and its Levi subgroup L consists of block diagonal matrices with $A_{ij} = 0$ for $i > j$. Here $A_{ij} \in \mathrm{Mat}_{n_i,n_j}$ and $\sum n_i = n$. Dominant weights λ for L are given by sequences $\lambda_1, \dots, \lambda_n$ such that

$$\lambda_1 \geq \dots \geq \lambda_{n_1}, \quad \dots, \quad \lambda_{n-n_k+1} \geq \dots \geq \lambda_n.$$

By Corollary 8.2, the Euler characteristic $\chi(\mathcal{E}_\lambda)$ is given by the formula

$$\chi(\mathcal{E}_\lambda) = \prod_{1 \leq i < j \leq n} \frac{\lambda_i - \lambda_j + j - i}{j - i}.$$

8.1.3 Dual Variety of the Grassmannian

Let $X = \mathrm{Gr}(n_1, \mathbb{C}^{n_1+n_2})$ be the Grassmannian of n_1-dimensional subspaces in $\mathbb{C}^{n_1+n_2}$ in the Plücker embedding $X \subset \mathbb{P}(\mathbb{C}^N)$, where $\mathbb{C}^N = \Lambda^{n_1}\mathbb{C}^{n_1+n_2}$. We are going to calculate degree of the dual variety X^*.

Theorem 8.3 ([Las]) *Let $X = \mathrm{Gr}(n_1, \mathbb{C}^{n_1+n_2}) \subset \mathbb{P}(\Lambda^{n_1}\mathbb{C}^{n_1+n_2})$ be the Grassmannian of n_1-dimensional subspaces in $\mathbb{C}^{n_1+n_2}$ in the Plücker embedding. Then*

$$\deg \Delta_X = \sum_{\lambda \in \mathcal{P}_{n_1}^{n_2}, S} \frac{(-1)^S (S+1) \nabla(-\lambda_m^t, \dots, -\lambda_1^t, \lambda_1 - p + S, \dots, \lambda_n - p + S)}{\nabla(1, \dots, n_1 + n_2)},$$

where $S = n_1 n_2 - |\lambda|$ or $S = n_1 n_2 - |\lambda| - 1$, $p \in \mathbb{Z}$ is arbitrary, and

$$\nabla(l_1, \dots, l_{n_1+n_2}) = \prod_{1 \leq i < j \leq n_1+n_2} (l_i - l_j + j - i).$$

Proof. Let $\mathcal{L} = \mathcal{O}_X(1)$, and we have $\dim X = n_1 n_2$. By Theorem 6.13,

$$\deg \Delta_X = \chi\big(\Lambda^{n_1 n_2}([T_X^\vee] - 2[\mathcal{L}^\vee] + [\mathcal{O}_X])[\mathcal{L}]^p\big)$$

for any $p \in \mathbb{Z}$. We have

$$\Lambda^{n_1 n_2}([T_X^\vee] - 2[\mathcal{L}^\vee] + [\mathcal{O}_X]) = \sum_{i=0}^{n_1 n_2} \Lambda^i(T_X^\vee)\Lambda^{n_1 n_2 - i}([\mathcal{O}_X] - 2[\mathcal{L}^\vee]),$$

$$\Lambda^k([\mathcal{O}_X] - 2[\mathcal{L}^\vee]) = (-1)^k((k+1)[\mathcal{L}^\vee]^k - k[\mathcal{L}^\vee]^{k-1}).$$

Let $\mathcal{S}$ be the tautological vector bundle on X, and let $V/\mathcal{S}$ be the quotient tautological bundle. Then $T_X^\vee = \mathcal{S} \otimes (V/\mathcal{S})^\vee$ and by the Cauchy formula,

$$\Lambda^i(\mathcal{S} \otimes (V/\mathcal{S})^\vee) = \sum_{\lambda \in \mathcal{P}_{n_1}^{n_2},\ |\lambda| = i} S_\lambda(\mathcal{S}) \otimes S_{\lambda^t}((V/\mathcal{S})^\vee).$$

Therefore we have

$$\deg \Delta_X = \sum_{\lambda \in \mathcal{P}_{n_1}^{n_2}, S} (-1)^S (S+1) \chi\Big(S_\lambda(\mathcal{S}) \otimes S_{\lambda^t}((V/\mathcal{S})^\vee)[\mathcal{L}]^{p-S}\Big),$$

where $S = n_1 n_2 - |\lambda|$ or $S = n_1 n_2 - |\lambda| - 1$ and $p \in \mathbb{Z}$ is arbitrary.

By Corollary 8.2,

$$\chi\Big(S_\lambda(\mathcal{S}) \otimes S_{\lambda^t}((V/\mathcal{S})^\vee)[\mathcal{L}]^{p-S}\Big) =$$

$$\frac{\nabla(-\lambda_m^t, -\lambda_{m-1}^t, \ldots, -\lambda_1^t, \lambda_1 - p + S, \ldots, \lambda_n - p + S)}{\nabla(1, \ldots, n_1 + n_2)}.$$

The theorem is proved. □

If $m = 2$ then this degree was already calculated in Example 2.13. For small values of m and n the degree can be found from the following table.

(m,n)	$(3,3)$	$(3,4)$	$(3,5)$	$(3,6)$	$(3,7)$	$(3,8)$	$(4,4)$	$(4,5)$
d	4	7	16	120	640	3608	126	2943

Table 8.1.

8.1.4 Codegree of G/B

Let G be a simple algebraic group of rank r with a Borel subgroup B. The minimal equivariant projective embedding of G/B corresponds to the line bundle $\mathcal{L}_\rho$, where

$$\rho = \omega_1 + \ldots + \omega_r$$

is a half sum of positive roots. We shall follow [CW] here and calculate the degree of the corresponding discriminant.

Let $\{\beta_1, \ldots, \beta_N\} = \Delta_+$ be the collection of all positive roots. We consider the matrix

$$M = \left(\frac{\langle \beta_j \,|\, \beta_i \rangle}{\langle \rho \,|\, \beta_i \rangle} \right)_{i,j=1,\dots,N}.$$

Let $P_s(M)$ be the sum of all permanents of $s \times s$ submatrices of M. Recall that the permanent of a $n \times n$ matrix (a_{ij}) is equal to $\sum_{\sigma \in S_n} \prod_{i=1}^{n} a_{i,\sigma(i)}$.

Theorem 8.4 ([CW])

$$\deg \Delta_{(G/B,\mathcal{L}_\rho)} = \sum_{s=0}^{N} (s+1)! P_{N-s}(M).$$

Proof. Recall that by Theorem 6.2 we have

$$\deg \Delta_{(G/B,\mathcal{L}_\rho)} = \sum_{i=0}^{N} (i+1) \int_{G/B} c_{n-i}(T^\vee_{G/B}) \cdot c_1(\mathcal{L}_\rho)^i.$$

Therefore we need to calculate these integrals. The correspondence $\lambda \mapsto \mathcal{L}_\lambda$ identifies $\mathcal{P} \otimes \mathbb{Q}$ with $\mathrm{Pic}(G/B) \otimes \mathbb{Q}$. Therefore we have a homomorphism of commutative algebras

$$c \colon S^\bullet(\mathcal{P} \otimes \mathbb{Q}) \to A^\bullet(G/B),$$

where $S^\bullet(\mathcal{P} \otimes \mathbb{Q})$ is the symmetric algebra of $\mathcal{P} \otimes \mathbb{Q}$ and $A^\bullet(G/B)$ is the rational Chow ring. It is well-known (see e.g. [BGG]) that this homomorphism is surjective and its kernel is generated as an ideal by W-invariants of positive degree $(S^\bullet(\mathcal{P} \otimes \mathbb{Q}))^W_+$, where W is the Weyl group. For example,

$$c_1(\mathcal{L}_\rho) = c(\rho).$$

Using an appropriate filtration of the cotangent bundle $T^\vee_{G/B}$ and 'the splitting principle' it is easy to see that

$$c_{n-i}(T^\vee_{G/B}) = c(X_i), \quad X_i = \sum_{\substack{\Gamma \subset \Delta_+ \\ |\Gamma| = n-i}} \prod_{\alpha \in \Gamma} \alpha.$$

Therefore it suffices to compute $c(\rho^i X_i)$.

For any root α we define a linear function

$$D_\alpha(\lambda) = \langle \lambda \,|\, \alpha \rangle.$$

It extends by a Leibniz rule to the differential operator on $S^\bullet(\mathcal{P} \otimes \mathbb{Q})$. Consider the differential operator D given by

$$D = \frac{\prod_{\alpha > 0} D_\alpha}{\prod_{\alpha > 0} \langle \rho \,|\, \alpha \rangle}.$$

D decreases the degree by N, and so in particular we get a linear form

$$D : S^N(\mathcal{P} \otimes \mathbb{Q}) \to \mathbb{Q}.$$

One can show (see [CW]) that for any $x \in S^N(\mathcal{P} \otimes \mathbb{Q})$ we have

$$D(x) = \int_{G/B} c(x).$$

Now the claim of the theorem follows by an easy calculation. □

8.1.5 A Closed Formula

In most circumstances, known formulas for the degree of the discriminant depend on a certain set of discrete parameters. There are two possibilities for these parameters. First, we may fix a projective variety and vary its polarizations, see Theorem 6.8. The second possibility is to change a variety and to fix a polarization (in a certain sense). For example, consider the irreducible representation of SL_{n_0} with the highest weight λ. Then this weight can be considered as a highest weight of SL_n for any $n \geq n_0$ with respect to the natural embedding $\mathrm{SL}_n \subset \mathrm{SL}_{n+1} \subset \ldots$. As a result, we obtain a tower of flag varieties with "the same" polarization. The degree of the corresponding discriminants will be a function in n. This function can be very complicated, see e.g. Theorem 8.3.

However, sometimes this function has a closed expression – the Boole formula (6.4) is a formula of this kind. Another simple example is the formula (2.13) for the degree of the dual variety to $\mathrm{Gr}(2, n)$ in the Plücker embedding. The following theorem is a mixture of these two cases.

Theorem 8.5 ([T1]) *Let V be an irreducible SL_n-module with the highest weight $(a-1)\varphi_1 + \varphi_2$, where $a \geq 2$. Then the variety $X^* \subset \mathbb{P}(V^\vee)$ projectively dual to the projectivization X of the orbit of the highest vector is a hypersurface of degree*

$$\frac{(n^2-n)a^{n+1} - (n^2+n)a^{n-1} - 2n(-1)^n}{(a+1)^2}.$$

Proof. We have to calculate the degree of Δ_X. We use Kleiman's formula (6.2) for the degree of the dual variety. In our case $X = G/P$, where $G = \mathrm{SL}_n$ and $P \subset G$ is the parabolic subgroup of matrices

$$\begin{pmatrix} * & 0 & 0 & \ldots & 0 \\ * & * & 0 & \ldots & 0 \\ * & * & * & \ldots & * \\ \vdots & \vdots & \vdots & \ddots & \vdots \\ * & * & * & \ldots & * \end{pmatrix}. \tag{8.1}$$

Assume that $T \subset G$ is the diagonal torus, B is the Borel subgroup of lower-triangular matrices, $x_1, \ldots, x_n$ are the weights of the tautological representation, $X(T)$ is the lattice of characters of T, S is the symmetric algebra of $X(T)$ (over $\mathbb{Q}$), $W \simeq S_n$ is the Weyl group of G, and $W_P \simeq S_{n-2}$ is the Weyl group of P. It is well known (see [BGG]) that the map $c: X(T) \to \mathrm{Pic}(G/B)$ that assigns to λ the first Chern class of the invertible sheaf $\mathcal{L}_\lambda$ can be extended to a surjective homomorphism $c: S \to A^\bullet(G/B)$ in the (rational) Chow ring, and its kernel coincides with $S^W_+ S$. The projection $\alpha: G/B \to G/P$ induces an embedding $\alpha^*: A^\bullet(G/P) \to A^\bullet(G/B)$. The image coincides with the subalgebra of W_P-invariants. Hence $A^\bullet(G/P) = S^{W_P}/S^W_+ S^{W_P}$. We denote the homomorphism $S^{W_P} \to A^\bullet(G/P)$ by the same letter c.

To apply Kleiman's formula we need $c_1(\mathcal{L})$ (which is equal to $c(ax_1 + x_2)$) and the total Chern class of $T^\vee_Z$, which is equal to

$$c(T^\vee_Z) = c\left((1 - x_1 + x_2) \prod_{i=3,\ldots,n} (1 - x_1 + x_i) \prod_{i=3,\ldots,n} (1 - x_2 + x_i)\right).$$

(This can be shown using a filtration of $T^\vee_Z$ and the splitting principle.)

Let $\alpha_1, \ldots, \alpha_{n-2}$ be the elementary symmetric polynomials in $x_3, \ldots, x_n$. Then $S^{W_P} = \mathbb{Q}[x_1, x_2, \alpha_1, \ldots, \alpha_{n-2}]$, and the ideal $S^W_+ S^{W_P}$ is generated by

$$x_1 + x_2 + \alpha_1, \quad x_1x_2 + x_1\alpha_1 + x_2\alpha_1 + \alpha_2,$$
$$x_1x_2\alpha_1 + x_1\alpha_2 + x_2\alpha_2 + \alpha_3, \ \ldots, \ x_1x_2\alpha_{n-4} + x_1\alpha_{n-3} + x_2\alpha_{n-3} + \alpha_{n-2},$$
$$x_1x_2\alpha_{n-3} + x_1\alpha_{n-2} + x_2\alpha_{n-2}, \quad x_1x_2\alpha_{n-2}.$$

Hence $\alpha_i = (-1)^i(x_1^i + x_1^{i-1}x_2 + \ldots + x_2^i) \mod S^W_+ S^{W_P}$, and $A^\bullet(G/P)$ is isomorphic to the quotient ring $\mathbb{Q}[x_1, x_2]/\langle f_1, f_2\rangle$, where $f_1 = x_1^{n-1} + x_1^{n-2}x_2 + \ldots + x_2^{n-1}$ and $f_2 = x_1^n$. Note that f_1, f_2 is the Gröbner basis of the ideal $\langle f_1, f_2\rangle$ with respect to the ordering $x_2 > x_1$ (see [Berg]). Hence the set of X^iY^j is a basis of the quotient algebra, where $X = x_1 \mod \langle f_1, f_2\rangle$ and $Y = x_2 \mod \langle f_1, f_2\rangle$, $i = 1, \ldots, n-1$, $j = 1, \ldots, n-2$.

To calculate the degree of the discriminant by Kleiman's formula, we have to calculate $\int_Z c(X^{n-1}Y^{n-2})$. Let $\tilde{w}_0$ be the longest element in W, and let w_0 be the shortest element in $\tilde{w}_0 W_P$ with the reduced factorization

$$w_0 = \begin{pmatrix} 1 & 2 & 3 & 4 & \ldots & n \\ n & n-1 & 1 & 2 & \ldots & n-2 \end{pmatrix} =$$
$$(n-1, n)(n-2, n-1)\ldots(12) \cdot (n-1, n)(n-2, n-1)\ldots(23).$$

Let

$$A_{w_0} = A_{(n-1,n)}A_{(n-2,n-1)} \cdots A_{(12)}A_{(n-1,n)}A_{(n-2,n-1)} \cdots A_{(23)}$$

be the corresponding endomorphism of S of degree $-(2n-3)$, where

$$A_{(ij)} = \frac{\mathrm{Id} - s_{(ij)}}{x_i - x_j},$$

and $s_{(ij)}$ is the reflection that transposes x_i and x_j. Then

$$\int_Z c(X^{n-1}Y^{n-2}) = A_{w_0}(x_1^{n-1}x_2^{n-2})$$

(see [BGG]). It is obvious that

$$A_{w_0}(x_1^{n-1}x_2^{n-2}) = A_{\rho_1}A_{\rho_2}(x_1^{n-1}x_2^{n-2}) = \\ A_{\rho_1}(x_1^{n-1}A_{\rho_2}(x_2^{n-2})) = A_{\rho_1}(x_1)A_{\rho_2}(x_2),$$

where $\rho_k = (n-1,n)(n-2,n-1)\ldots(k,k+1)$. These factors are both equal to 1, since they are equal to $\int_{\mathbb{P}^{n-1}} c_1(\mathcal{O}(1))^{n-1}$ and $\int_{\mathbb{P}^{n-2}} c_1(\mathcal{O}(1))^{n-2}$, respectively. We finally see that $\int_Z c(X^{n-1}Y^{n-2}) = 1$.

It remains to calculate the polynomial

$$\sum_{i=0}^{2n-3} (i+1)c_{2n-3-i}(aX+Y)^i, \tag{8.2}$$

in the ring $\mathbb{Q}[X,Y]$ (with the basis X^iY^j, $i=1,\ldots,n-1$, $j=1,\ldots,n-2$, and relations $X^{n-1}+X^{n-2}Y+\ldots+Y^{n-1}=0$, $Y^n=0$ and $X^n=0$, which follow from the preceding relations), where c_k is the kth homogeneous component of the polynomial

$$(1-X+Y)\prod_{i=3}^{n}(1-X+x_i)\prod_{i=3}^{n}(1-Y+x_i),$$

in which the ith symmetric function of $x_3,\ldots,x_n$ must be replaced by $(-1)^i(X^i+X^{i-1}Y+\ldots+Y^i)$. By the previous discussion, the result of this calculation will be $\deg(\Delta_X)X^{n-1}Y^{n-2}$.

Note that the polynomial (8.2) is equal to

$$F'(T)|_{T=2X+Y} = (F_1F_2F_3F_4)'(T)|_{T=2X+Y},$$

where

$$F_1 = T, \quad F_2 = T-X+Y,$$
$$F_3 = \prod_{i=3}^{n}(T-X+x_i) = \sum_{i=0}^{n-2}(T-X)^{n-2-i}(-1)^i(X^i+\ldots+Y^i),$$
$$F_4 = \prod_{i=3}^{n}(T-Y+x_i) = \sum_{i=0}^{n-2}(T-Y)^{n-2-i}(-1)^i(X^i+\ldots+Y^i).$$

We have, further,

$$F_3(T) = \left(\sum_{i=0}^{n-2}(T-X)^{n-2-i}(-1)^i\frac{X^{i+1}-Y^{i+1}}{X-Y}\right) =$$

$$\frac{X(T-X)^{n-2}}{X-Y}\left(\sum_{i=0}^{n-2}(T-X)^{-i}(-1)^i X^i\right) -$$

$$\frac{Y(T-X)^{n-2}}{X-Y}\left(\sum_{i=0}^{n-2}(T-X)^{-i}(-1)^i Y^i\right) =$$

$$\frac{X\left((T-X)^{n-1}-(-X)^{n-1}\right)}{T(X-Y)} - \frac{Y\left((T-X)^{n-1}-(-Y)^{n-1}\right)}{(T-X+Y)(X-Y)}.$$

We obtain, likewise, that

$$F_4(T) = \left(\sum_{i=0}^{n-2}(T-Y)^{n-2-i}(-1)^i\frac{X^{i+1}-Y^{i+1}}{X-Y}\right) =$$

$$\frac{X(T-Y)^{n-2}}{X-Y}\left(\sum_{i=0}^{n-2}(T-Y)^{-i}(-1)^i X^i\right) -$$

$$\frac{Y(T-Y)^{n-2}}{X-Y}\left(\sum_{i=0}^{n-2}(T-Y)^{-i}(-1)^i Y^i\right) =$$

$$\frac{X\left((T-Y)^{n-1}-(-X)^{n-1}\right)}{(T+X-Y)(X-Y)} - \frac{Y\left((T-Y)^{n-1}-(-Y)^{n-1}\right)}{T(X-Y)}.$$

We deduce from the latter formula that

$$F'(T) = \frac{n(X-Y)(T-X)^{n-1}+(-X)^n-(-Y)^n}{X-Y} \times$$

$$\frac{(X-Y)(T-Y)^n+T(-X)^n-(T+X-Y)(-Y)^n}{T(T+X-Y)(X-Y)} +$$

$$\frac{(X-Y)(T-X)^n+(T-X+Y)(-X)^n-T(-Y)^n}{T(X-Y)} \times$$

$$\frac{\Big(n(T+X-Y)-(T-Y)\Big)(T-Y)^{n-1}+(-X)^n}{(T+X-Y)^2} -$$

$$\frac{(X-Y)(T-X)^n+(T-X+Y)(-X)^n-T(-Y)^n}{T(X-Y)} \times$$

$$\frac{(X-Y)(T-Y)^n+T(-X)^n-(T+X-Y)(-Y)^n}{T(T+X-Y)(X-Y)}.$$

Using the elementary formulas

$$\frac{\alpha^n(\gamma-\beta)+\beta^n(\alpha-\gamma)+\gamma^n(\beta-\alpha)}{(\alpha-\beta)(\beta-\gamma)(\gamma-\alpha)} = \sum_{i+j+k=n-2}\alpha^i\beta^j\gamma^k,$$

$$\frac{\alpha^n(\gamma-\beta)+\beta^n(\alpha-\gamma)+\gamma^n(\beta-\alpha)}{(\alpha-\beta)(\gamma-\beta)}=\sum_{i=0}^{n-2}\beta^i(\alpha^{n-1-i}-\gamma^{n-1-i}),$$

$$\frac{(n(\alpha-\beta)-\alpha)\alpha^{n-1}+\beta^n}{(\alpha-\beta)^2}=\sum_{i=0}^{n-2}(n-1-i)\alpha^{n-2-i}\beta^i,$$

we obtain

$$\begin{aligned}F'(T)=\Big(n(T-X)^{n-1}+(-1)^n\sum_{i=0}^{n-1}X^{n-1-i}Y^i\Big)\times\\ \Big(\sum_{i+j+k=n-2}(T-Y)^i(-X)^j(-Y)^k\Big)+\\ \Big(\sum_{i=0}^{n-2}(-X)^i((T-X)^{n-1-i}-(-Y)^{n-1-i})\Big)\times\\ \Big(\sum_{i=0}^{n-2}(n-1-i)(T-Y)^{n-2-i}(-X)^i\Big)-\\ \Big(\sum_{i=0}^{n-2}(-X)^i((T-X)^{n-1-i}-(-Y)^{n-1-i})\Big)\times\\ \Big(\sum_{i+j+k=n-2}(T-Y)^i(-X)^j(-Y)^k\Big).\end{aligned}$$

Putting $T=aX+bY$ in the latter formula, we obtain

$$\begin{aligned}\Big(\sum_{i=0}^{n-1}(1+(-1)^n n(a-1)^{n-1-i}b^i\binom{n-1}{i})X^{n-1-i}Y^i\Big)\times\\ \Big(\sum_{i=0}^{n-2}\begin{bmatrix}n-2\\ i\end{bmatrix}_{a,b-1}X^iY^{n-2-i}\Big)-\\ \Big(\sum_{i=0}^{n-1}(\begin{bmatrix}n-1\\ i\end{bmatrix}_{a-1,b}-\begin{bmatrix}n-2\\ i-1\end{bmatrix}_{a-1,b}-1)X^{n-1-i}Y^i\Big)\times\\ \Big(\sum_{i=0}^{n-2}(\begin{Bmatrix}n-2\\ i\end{Bmatrix}_{a,b-1}-\begin{Bmatrix}n-3\\ i\end{Bmatrix}_{a,b-1})X^{n-2-i}Y^i\Big)+\\ \Big(\sum_{i=0}^{n-1}(\begin{bmatrix}n-1\\ i\end{bmatrix}_{a-1,b}-\begin{bmatrix}n-2\\ i-1\end{bmatrix}_{a-1,b}-1)X^iY^{n-1-i}\Big)\times\\ \Big(\sum_{i=0}^{n-2}\begin{bmatrix}n-2\\ i\end{bmatrix}_{a,b-1}X^iY^{n-2-i}\Big),\end{aligned}$$

where

$$\begin{bmatrix}n\\ k\end{bmatrix}_{xy}=\sum_{p=0}^{k}\sum_{q=0}^{n-k}(-1)^{p+q}x^py^q\binom{p+q}{p},$$

$$\left\{ \begin{matrix} n \\ k \end{matrix} \right\}_{xy} = \sum_{p=0}^{k} \sum_{q=0}^{n-k} (p+q+1)(-1)^{p+q} x^p y^q \binom{p+q}{p}.$$

It remains to calculate the degree of the discriminant, that is, the difference between the coefficients of $X^{n-1}Y^{n-2}$ and $X^{n-2}Y^{n-1}$ in the expression for $F'(aX + bY)$. After some transformations we obtain

$$\sum_{i=0}^{n-1} \left((-1)^n n(a-1)^{n-1-i} b^i \binom{n-1}{i} + W_i \right) \times$$
$$\left(\begin{bmatrix} n-2 \\ i \end{bmatrix}_{a,b-1} - \begin{bmatrix} n-2 \\ i-1 \end{bmatrix}_{a,b-1} \right) -$$
$$\sum_{i=0}^{n-1} (W_i - 1) \times$$
$$\left(\left\{ \begin{matrix} n-2 \\ i \end{matrix} \right\}_{a,b-1} - \left\{ \begin{matrix} n-3 \\ i \end{matrix} \right\}_{a,b-1} - \left\{ \begin{matrix} n-2 \\ i-1 \end{matrix} \right\}_{a,b-1} + \left\{ \begin{matrix} n-3 \\ i-1 \end{matrix} \right\}_{a,b-1} \right),$$

where

$$W_i = \begin{bmatrix} n-1 \\ n-1-i \end{bmatrix}_{a-1,b} - \begin{bmatrix} n-2 \\ n-1-i \end{bmatrix}_{a-1,b}.$$

This is the degree of the discriminant of the irreducible SL_n-module with the highest weight $(a-b)\varphi_1 + b\varphi_2$.

Substituting $b = 1$ in the last formula, we obtain

$$\frac{(n^2 - n)a^{n+1} - (n^2 + n)a^{n-1} - 2n(-1)^n}{(a+1)^2}.$$

□

8.1.6 Degree of Hyperdeterminants

Consider the flag variety $\mathbb{P}^{k_1} \times \ldots \times \mathbb{P}^{k_r}$ of the group $\mathrm{SL}_{k_1+1} \times \ldots \times \mathrm{SL}_{k_r+1}$. The projectively dual variety of its 'minimal' equivariant projective embedding is related to a nice theory of hyperdeterminants initiated by Cayley and Schläfli [Ca1, Ca2, Schl].

Let $r \geq 2$ be an integer, and let $A = (a_{i_1 \ldots i_r})$, $0 \leq i_j \leq k_j$ be an r-dimensional complex matrix of *format* $(k_1 + 1) \times \ldots \times (k_r + 1)$. The hyperdeterminant of A is defined as follows; cf. Example 3.8. Consider the product $X = \mathbb{P}^{k_1} \times \ldots \times \mathbb{P}^{k_r}$ of several projective spaces embedded in $\mathbb{P}^{(k_1+1)\times \ldots \times (k_r+1)-1}$ via the Segre embedding. Let X^* be the projectively dual variety. If X^* is a hypersurface then it is defined by a corresponding discriminant Δ_X, which in this case is called the *hyperdeterminant* Det. Clearly $\mathrm{Det}(A)$ is a polynomial function in matrix entries of A invariant under the action of $\mathrm{SL}_{k_1+1} \times \ldots \times \mathrm{SL}_{k_r+1}$. If X^* is not a hypersurface then we set $\mathrm{Det} = 1$.

Example 3.8 shows that X^* is a hypersurface (and, hence, defines a hyperdeterminant) if and only if $2k_j \leq k_1 + \ldots + k_r$ for $j = 1, \ldots, r$. If for one of the j we have an equality $2k_j = k_1 + \ldots + k_r$ then the format is called *boundary*. Let $N(k_1, \ldots, k_r)$ be the degree of the hyperdeterminant of format $(k_1 + 1) \times \ldots \times (k_r + 1)$. The proof of the following theorem can be found in [GKZ2] or [GKZ3].

Theorem 8.6

(i) *The generating function for* $N(k_1, \ldots, k_r)$ *is given by*

$$\sum_{k_1,\ldots,k_r \geq 0} N(k_1, \ldots, k_r) z_1^{k_1} \ldots z_r^{k_r} = \frac{1}{\left(1 - \sum_{i=2}^{r} (i-1) e_i(z_1, \ldots, z_r)\right)^2}$$

where $e_i(z_1, \ldots, z_r)$ *is the i-th elementary symmetric polynomial.*

(ii) *The degree* $N(k_1, \ldots, k_r)$ *of the boundary format is given by (assuming that* $k_1 = k_2 + \ldots + k_r$*)*

$$N(k_2 + \ldots + k_r, k_2, \ldots, k_r) = \frac{(k_2 + \ldots + k_r + 1)!}{k_2! \ldots k_r!}.$$

(iii) *The degree of the hyperdeterminant of the cubic format is given by*

$$N(k, k, k) = \sum_{0 \leq j \leq k/2} \frac{(j + k + 1)!}{(j!)^3 (k - 2j)!} \cdot 2^{k-2j}.$$

(iv) *The exponential generating function for the degree* N_r *of the hyperdeterminant of format* $2 \times 2 \times \ldots \times 2$ *(*r *times) is given by*

$$\sum_{r \geq 0} N_r \frac{z^r}{r!} = \frac{e^{-2z}}{(1-z)^2}.$$

8.1.7 Varieties of Small Codegree

Let X be a projective variety. We define the *codegree* codeg X by

$$\operatorname{codeg} X = \deg X^*.$$

If X^* is a hypersurface, then codeg X is also called the *class* of X. This is a classical invariant playing an important role in enumerative geometry. If X^* is not a hypersurface and X' is a generic hyperplane section of X, then by Theorem 5.3 it is clear that $\operatorname{codeg} X' = \operatorname{codeg} X$.

With regard to codegree, the most simple nonsingular projective varieties are those whose codegree is small. The problem of classification of varieties of small codegree should be compared with that of classification of varieties of small degree. Much is known about this last problem. The case of varieties of degree two is classical (quadrics). The complete description of varieties of degree three was given by A. Weil in the anonymous publication [XXX]. Swinnerton–Dyer [Sw] classified all varieties of degree four. In the smooth case the classification up to degree 8 has been completed in several papers by Okonek [O1, O2, O3, O4] and Ionescu [I, I1, I2]. Fania and Livorni [FaL1, FaL2] classified varieties of degree 9, 10.

The same question can be asked for the codegree, but this problem is quite different. For example, in the case of varieties of small degree one can proceed by induction using the fact that the degree is stable with respect to passing to hyperplane sections, whereas there is no such inductive procedure for codegree.

Theorem 8.7 ([Z6]) *There exist exactly* 10 *non-degenerate nonsingular complex projective varieties of codegree three, namely*

- *the self-dual Segre threefold* $\mathbb{P}^1 \times \mathbb{P}^2 \subset \mathbb{P}^5$,
- *its hyperplane section* $F_1 \subset \mathbb{P}^4$ *obtained by blowing up a point in* $\mathbb{P}^2$ *by means of the map defined by the linear system of conics passing through this point,*
- *the four Severi varieties, i.e. the Veronese surface* $v_2(\mathbb{P}^2) \subset \mathbb{P}^5$, *the Segre variety* $\mathbb{P}^2 \times \mathbb{P}^2 \subset \mathbb{P}^8$, *the Grassmann variety* $Gr_2(\mathbb{C}^6) \subset \mathbb{P}^{14}$ *of lines in* $\mathbb{P}^5$ *and the* 16*-dimensional variety* $E \subset \mathbb{P}^{26}$ *corresponding to the orbit of highest weight vector for the standard representation of the algebraic group of type* E_6,
- *the four varieties obtained by projecting the Severi varieties from generic points of their ambient linear spaces.*

Sketch of the proof. If X^* is not a hypersurface, then we can either apply Weil's classification theorem to X^* or consider the intersection of X with a general hyperplane, thus reducing the problem to the case when X^* is a (singular) cubic in $(\mathbb{P}^N)^\vee$. Suppose that X^* is a hypersurface, and let $\Sigma = X^*_{sing}$. Then each point of Σ has multiplicity two and, since $\deg X^* = 3$, we conclude that $\mathrm{Sec}(\Sigma) \subseteq X^*$. Let pr_1 and pr_2 denote the projections of the conormal variety $I_X \subset \mathbb{P}^N \times (\mathbb{P}^N)^\vee$ on X and X^*. Let x be a generic point of X, let $\mathcal{P}_x = \mathrm{pr}_2(\mathrm{pr}_1^{-1}(x))$, and let

$$\Sigma_x = \{H \in \mathcal{P}_x \,|\, x \text{ is not a non-degenerate quadratic singularity of } H \cap X\}.$$

It is easy to see that $\Sigma_x \subset \Sigma \cap \mathcal{P}_x$ is either a hyperplane or a quadric in $\mathcal{P}_x$. In the first case one can show that Σ is a linear subspace in $(\mathbb{P}^N)^\vee$ from which it is possible to deduce that $X = F_1$. In the second case $\mathrm{Sec}(\Sigma) = X^*$ and one can show that Σ is nonsingular and, either $\dim \Sigma = n$ and X is a Severi variety, or $\dim \Sigma = n - 1$ and X is a nonsingular projection of a Severi variety from a point. □

The classification of smooth varieties of codegree 4 is still unknown. All known examples arise from Pyasetskii pairing (2.2). The conjectural list consists of

- The Segre embedding of $\mathbb{P}^1 \times Q$, where Q is a quadric hypersurface.
- The twisted cubic curve $v_3(\mathbb{P}^1)$.
- The Plücker embedding of the Grassmannian $\mathrm{Gr}_3(\mathbb{C}^6)$.
- The isotropic Grassmannian $\mathrm{Gr}_3^0(\mathbb{C}^6)$ of isotropic 3 dimensional subspaces in the symplectic space $\mathbb{C}^3$ in its minimal equivariant embedding.
- The spinor variety $\mathbb{S}_6$.
- 27-dimensional variety $E \subset \mathbb{P}^{55}$ corresponding to the orbit of highest weight vector for the minimal representation of the algebraic group of type E_7.
- The Segre embedding of $\mathbb{P}^1 \times \mathbb{P}^3$ (self-dual variety with defect 2), its non-singular hyperplane section and its non-singular section by two hyperplanes (which is either $\mathbb{P}^1 \times Q^1$ or the blow-up of a projective quadratic cone in its vertex).

8.2 Matsumura–Monsky Theorem

Let $D \subset \mathbb{P}^n$ be a smooth hypersurface of degree d. It was first proved in [MM] that the group of projective automorphisms preserving D is finite if $d > 2$. In fact, it was also proved that the group of biregular automorphisms of D is finite if $d > 2$ (except the cases $d = 3$, $n = 2$ and $d = 4$, $n = 3$). Though this generalization looks much stronger, actually it is an easy consequence of the "projective" version and the Bart Theorem [Ba].

More generally, let G/P be a flag variety of a simple Lie group and $D \subset G/P$ be a smooth ample divisor. Let $L_\lambda = \mathcal{O}(D)$ be the corresponding ample line bundle, where $\lambda \in \mathcal{P}^+$ is a dominant weight. Then one might expect that the normalizer $N_G(D)$ of D in G is finite if λ is big enough. Indeed, if $N_G(D)$ (or actually any linear algebraic group of transformations of D) contains a one-parameter subgroup of automorphisms of D then D is covered by rational curves. However, if λ is big enough then the canonical class K_D is nef by the adjunction formula and, therefore, D can not be covered by rational curves. Unfortunately, this transparent approach does not give strong estimates on λ. Much better estimates can be obtained generalizing the original proof of the Matsumura-Monsky theorem.

The problem can be reformulated as follows. Suppose that V_λ is an irreducible G-module with the highest weight λ. Let $\mathcal{D} \subset V_\lambda$ be the discriminant variety (the dual variety to the orbit of the highest weight vector in $V_\lambda^\vee$). We will show that if λ is big enough then any point $x \in V_\lambda \setminus \mathcal{D}$ has a finite stabilizer G_x and the orbit of x is closed, $Gx = \overline{Gx}$ (therefore x is a stable point of V_λ in the sense of Geometric Invariant Theory, see [MFK]). This result can be compared with the results of [AVE], where all irreducible modules of simple algebraic groups with infinite stabilizers of generic points were found (this classification was extended later in [El1] and [El2] to handle irreducible representations of semisimple groups and any representations of simple groups).

If $\mathcal{D}$ is not a hypersurface then an easy inspection using Theorem 7.56 shows that the stabilizer of any point is infinite. So from now on we shall assume that $\mathcal{D}$ is a hypersurface defined by vanishing of the discriminant Δ.

A dominant weight λ is called *self-dual* if V_λ is isomorphic to $V_\lambda^\vee$ as a G-module. Let $\mathcal{P}_S^+ \subset \mathcal{P}^+$ be the subcone of self-dual dominant weights. Let γ be the highest root.

Theorem 8.8 *Let V_λ be an irreducible representation of a simple algebraic group G with the highest weight λ such that $\mathcal{D}$ is a hypersurface. Suppose that $(\lambda-\gamma,\mu) > 0$ for any $\mu \in \mathcal{P}_S^+$. Let $x \in V_\lambda \backslash \mathcal{D}$. Then G_x is finite and $Gx = \overline{Gx}$. Moreover, $G_{[x]}$ is also finite, where $[x]$ is the line spanned by x.*

Proof. If G_x is finite and $G_{[x]}$ is infinite then $[x]\backslash\{0\} \subset Gx$. Therefore $0 \in \overline{Gx}$ and $\Delta(x) = \Delta(0) = 0$, and hence, $x \in \mathcal{D}$. The same argument shows that if $x \in V_\lambda \setminus \mathcal{D}$ then $\overline{Gx} \subset V_\lambda \setminus \mathcal{D}$. If Gx is not closed then G_y is infinite for any point $y \in \overline{Gx} \setminus Gx$. Therefore, in order to prove the theorem, it suffices to prove that G_x is finite for any $x \in V_\lambda \setminus \mathcal{D}$.

Suppose that G_x is infinite. Then G_y is infinite and reductive (see e.g. [PV]) for any point y from the closed orbit in $\overline{Gx}$. Therefore, it suffices to prove that if $S \subset G$ is a one-dimensional torus and $Sx = x$ then $x \in \mathcal{D}$.

Without loss of generality we may assume that $S \subset T$, where T is the fixed maximal torus, and $\mathfrak{t} = \operatorname{Lie} T$ is the Cartan subalgebra. Let $x = \sum_{\pi\in\mathcal{P}} x_\pi$ be the weight decomposition of x. Let $\operatorname{Supp}(x) = \{\pi \in \mathcal{P} \mid x_\pi \neq 0\}$ be the *support* of x. For any $\mu \in \mathcal{P}$ let H_μ denote the hyperplane of weights perpendicular to μ. Then there exists $\mu \in \mathcal{P}$ such that $\operatorname{Supp}(x) \subset H_\mu$. Using the action of the Weyl group we may assume that $\mu \in \mathcal{P}^+$.

For any $\lambda \in \mathcal{P}^+$ let $\lambda^\vee$ be the highest weight of the dual module $V_\lambda^\vee$. Then $\lambda^\vee = -w_0(\lambda)$, where w_0 is the longest element of the Weyl group.

Suppose that $(\mu, \lambda^\vee - \gamma) > 0$. Then for any positive root α we have $(\mu, \lambda^\vee - \alpha) > 0$. Therefore x is perpendicular to $[\mathfrak{g}, v_{\lambda^\vee}]$, where $v_{\lambda^\vee}$ is the highest weight vector of $V_\lambda^\vee$. It follows that $x \in \mathcal{D}$.

Suppose that $(\mu^\vee, \lambda^\vee - \gamma) > 0$. Then for any positive root α we have $(\mu^\vee, \lambda^\vee - \alpha) > 0$. Therefore $w_0(x)$ is perpendicular to $[\mathfrak{g}, v_{\lambda^\vee}]$. It follows that $w_0(x) \in \mathcal{D}$ and hence $x \in \mathcal{D}$.

Suppose now that $(\mu, \lambda^\vee - \gamma) \leq 0$ and $(\mu^\vee, \lambda^\vee - \gamma) \leq 0$. Then

$$(\mu + \mu^\vee, \lambda^\vee - \gamma) \leq 0.$$

But $\mu + \mu^\vee$ is a self-dual weight, and hence this contradicts assumptions of the theorem. □

Example 8.9 If $G = \mathrm{SL}_n$ and $\lambda = \sum n_i\omega_i$, where $\omega_1, \dots, \omega_n$ are the fundamental weights, then the conditions of the theorem are satisfied if and only if $\sum n_i > 2$; for example if $\lambda = n\omega_1$ and $n > 2$. In particular, we recover the original Matsumura–Monsky Theorem.

8.3 Discriminants of Commutative Algebras

8.3.1 Commutative Algebras Without Identities

Let $V = \mathbb{C}^n$. Consider the vector space $\mathcal{A} = S^2V^\vee \otimes V$ parametrizing bilinear commutative multiplications in V. In the sequel we identify points of $\mathcal{A}$ with the corresponding commutative algebras.

Definition 8.10 Let $A \in \mathcal{A}$. A non-zero element $v \in A$ is called a *quadratic nilpotent* if $v^2 = 0$. Let $\mathcal{D}_1 \subset \mathcal{A}$ be the subset of all algebras that contain quadratic nilpotents. A one-dimensional subalgebra $U \subset A$ is called *singular* if there exists linear independent vectors $u \in U$ and $v \in A$ such that

$$u^2 = \alpha u, \quad uv = \frac{\alpha}{2}v, \quad \text{where } \alpha \in \mathbb{C}.$$

Let $\mathcal{D}_2 \subset \mathcal{A}$ be the subset of all algebras that contain singular subalgebras.

Then the following theorem holds.

Theorem 8.11 ([T6])

(i) *$\mathcal{D}_1$ and $\mathcal{D}_2$ are irreducible hypersurfaces.*
(ii) *Let $A \in \mathcal{A}$. Then A contains a one-dimensional subalgebra.*
(iii) *Let $A \in \mathcal{A} \setminus (\mathcal{D}_1 \cup \mathcal{D}_2)$. Then A contains exactly $2^n - 1$ one-dimensional subalgebras; all these subalgebras are spanned by idempotents.*

Proof. Clearly $\mathcal{A}$ can be identified with a set of at most n-dimensional linear systems of quadrics in V. Namely, any linear function $f \in V^\vee$ defines a homogeneous quadratic function

$$v \to f(v^2).$$

Then $\mathcal{D}_1$ corresponds to the set of linear systems of quadrics with zero resultant. This proves that $\mathcal{D}_1$ is an irreducible hypersurface.

There exists, however, another useful identification. Any algebra A determines the n-dimensional linear system of *affine* quadrics in V. Namely, any linear form $f \in V^\vee$ defines an affine quadratic form

$$v \to f(v - v^2).$$

We can embed V into a projective space $\mathbb{P}$ as an affine chart. Then this linear system is naturally identified with a n-dimensional linear system of quadrics in $\mathbb{P}$. The base points of this linear system that do not lie on the infinite hyperplane coincide with idempotents of A. Infinite base points are the projectivizations of lines spanned by quadratic nilpotents. It is easy to see that quadrics from our linear system intersect transversally at 0. Moreover, quadrics intersect non-transversally at some point $v \neq 0$ if and only if the subalgebra spanned by v (if v is finite) or the subalgebra with the projectivization v (if v is infinite) is singular. Therefore (iii) follows from the Bezout theorem.

$\mathcal{D}_2$ is irreducible since $\mathcal{D}_2 = \mathrm{GL}_n \cdot \mathcal{D}_2'$, where $D_2' \subset \mathcal{A}$ is the linear subset of all algebras that have a fixed singular subalgebra and a fixed line spanned by the vector v from the definition of a singular subalgebra. It is also quite easy to check that D_2 is actually a closed hypersurface.

Since quadrics of our linear system intersect transversally at 0, it follows that there exist other base points, i.e. there exists at least one 1-dimensional subalgebra. □

Algebras that do not belong to discriminant varieties $\mathcal{D}_1$ and $\mathcal{D}_2$ are called *regular*. Both hypersurfaces $\mathcal{D}_1$ and $\mathcal{D}_2$ can be interpreted as standard discriminants. First, we can enlarge the symmetry group and consider $S^2(\mathbb{C}^n)^\vee \otimes \mathbb{C}^n$ as an $\mathrm{SL}_n \times \mathrm{SL}_n$-module. Then this module is irreducible and its discriminant variety (the dual variety of the projectivization of the highest weight vector orbit) coincides with $\mathcal{D}_1$. Now, consider $\mathcal{A} = S^2V^\vee \otimes V$ as an $\mathrm{SL}(V)$-module. Then this module is reducible, $\mathcal{A} = \mathcal{A}_0 + \tilde{\mathcal{A}}$, where $\mathcal{A}_0$ is a set of algebras with zero trace and $\tilde{\mathcal{A}}$ is isomorphic to $V^\vee$ as an $\mathrm{SL}(V)$-module. Consider the discriminant of $\mathcal{A}_0$ as a function on $\mathcal{A}$ (forgetting other coordinates). Then the corresponding hypersurface is exactly $\mathcal{D}_2$. If we consider the set of linear operators $\mathrm{Hom}(V, V) = V^\vee \otimes V$ instead of $\mathcal{A}$, then these constructions give rise to the determinant and the discriminant of a linear operator (see Example 2.11, Theorem 8.25).

8.3.2 Quasiderivations

Let $\mathfrak{g}$ be a Lie algebra with representation

$$\rho : \mathfrak{g} \to \mathrm{End}(V).$$

Consider any $v \in V$. A subalgebra

$$\mathfrak{g}_v = \{g \in \mathfrak{g} \,|\, \rho(g)v = 0\}$$

is called the *annihilator* of v. The subset

$$Q\mathfrak{g}_v = \{g \in \mathfrak{g} \,|\, \rho(g)^2 v = 0\}$$

is called the *quasi-annihilator* of v. Clearly, $\mathfrak{g}_v \subset Q\mathfrak{g}_v$. Of course, the quasi-annihilator is not a linear subspace in general, However, we have the following version of a Jordan decomposition:

Lemma 8.12 *Suppose that ρ is the differential of the representation of an algebraic group. Let $g \in Q\mathfrak{g}_v$. Consider the Jordan decomposition in $\mathfrak{g}$, $g = g_s + g_n$, where g_s is semisimple and g_n is nilpotent, and $[g_s, g_n] = 0$. Then $g_s \in \mathfrak{g}_v$ and $g_n \in Q\mathfrak{g}_v$.*

The proof is easy.

Now we apply this construction for $\mathfrak{g} = \mathfrak{gl}_n$ and ρ being the natural representation in the vector space $V^\vee \otimes V^\vee \otimes V$ that parametrizes algebras (bilinear multiplications in V). Let A be any algebra. Then $\mathfrak{g}_A$ is identified with the Lie algebra of derivations $\mathrm{Der}(A)$. Operators $D \in Q\mathfrak{g}_A$ are called *quasiderivations*. Of course, it is possible to write down explicit equations that determine $Q\,\mathrm{Der}(A) = Q\mathfrak{g}_A$ in $\mathrm{End}(A)$, but this formula is quite useless (see [Vi1]). We shall use a particular case of it that is quite easy to verify:

Lemma 8.13 *Let A be an algebra, $D \in \mathrm{End}(A)$, where $D^2 = 0$. Then $D \in Q\,\mathrm{Der}(A)$ if and only if*

$$D(x)D(y) = D(D(x)y) + D(xD(y)) \tag{8.3}$$

for any $x, y \in A$.

Example 8.14 It was conjectured in [Vi1] that all quasiderivations of the algebra of matrices Mat_n have the form $D(x) = ax+xb$, where $(a+b)^2 = [a,b]$. Let us give a counterexample. Consider the linear operator $D(x) = exe$, where $e^2 = 0$, $e \neq 0$. Then it is easy to check using (8.3) that D is a quasiderivation. However, D of course can not be written in the form $ax + xb$.

Quasiderivations can be used to define naive deformations. Namely, suppose that A is any algebra with the multiplication $u{\cdot}v$ and D is its quasiderivation. Consider the algebra A_D with the same underlying vector space but with the new multiplication

$$u \star v = u \cdot D(v) + D(u) \cdot v - D(u \cdot v).$$

Then, if A satisfies a polynomial identity, A_D satisfies this identity as well. More generally, let $\rho : \mathfrak{g} \to \mathrm{End}(V)$ be the differential of a representation of an algebraic group G, and let $v \in V$ and $D \in Q\mathfrak{g}_v$. Suppose that $H \subset V$ is a closed conical G-equivariant hypersurface and $v \in H$. Then $\rho(D)v$ also belongs to H. Indeed, since H is equivariant, $\exp(\lambda\rho(D))v$ belongs to H for any $\lambda \in \mathbb{C}$. Since D is a quasiderivation, $\exp(\lambda\rho(D))v = v + \lambda\rho(D)v$. Since H is conical, $v/\lambda + \rho(D)v$ belongs to H. Since H is closed, $\rho(D)v$ also belongs to H.

Now suppose that A is a regular commutative algebra. Then we claim that $Q\,\mathrm{Der}(A) = 0$.

Theorem 8.15 ([T6]) *Let $A \in \mathcal{A}$ and $A \notin \mathcal{D}_1 \cup \mathcal{D}_2$. Then $Q\,\mathrm{Der}(A) = 0$.*

Proof. By Lemma 8.12 it suffices to check that A has no semisimple derivations and no nilpotent quasiderivations. Suppose that $\mathrm{Der}(A)$ contains a nonzero semisimple element. Then the group $\mathrm{Aut}(A)$ contains a one-dimensional algebraic torus T. Let $\mathfrak{t}$ be its Lie algebra, $H \in \mathfrak{t}$, and $H \neq 0$. We way assume that the spectrum of H in A is integer-valued. Let $A_n \subset A$ be a weight space of weight n. Then $A = \oplus A_n$ is a $\mathbb{Z}$-grading. Let $v \in A$ be a homogeneous element of a maximal positive (or minimal negative) degree. Then $v^2 = 0$, and hence $A \in \mathcal{D}_1$.

Suppose now that E is a non-zero nilpotent quasiderivation. We can embed E in an $\mathfrak{sl}_2$-triple $\langle F, H, E\rangle \subset \mathrm{End}(A)$. Let

$$A = \oplus m_d R_d$$

be an $\mathfrak{sl}_2$-module decomposition, where R_d is an irreducible $(d+1)$-dimensional module. Consider also the weight decomposition

$$A = \oplus A^n \quad \text{and} \quad \mathcal{A} = \oplus \mathcal{A}^n$$

with respect to H. Let

$$J^n = A^n + A^{n+1} + A^{n+2} + \dots.$$

To avoid the abuse of notations, denote by $\alpha \in \mathcal{A}$ the point corresponding to the algebra A. Let Supp α be the support of α (i.e. all $n \in \mathbb{Z}$ such that $\alpha_n \neq 0$, where $\alpha = \sum \alpha_n$, $\alpha_n \in \mathcal{A}^n$). Since $E^2\alpha = 0$, the vector $E\alpha$ is a linear combination of highest weight vectors, and therefore Supp $\alpha \subset \{-1, 0, 1, 2, \dots\}$. This is equivalent to $J^n J^m \subset J^{n+m-1}$. In particular, if $v \in A$ is a weight vector of the weight n then $v^2 = 0$ since $n > 1$. Since A is regular, it follows that

$$A = m_0 R_0 \oplus m_1 R_1 = A^{-1} \oplus A^0 \oplus A^1.$$

Since $A^1 = J^1$, A^1 is a subalgebra in A. By Theorem 8.11 (ii), A^1 has a one-dimensional subalgebra U spanned by an idempotent u (since A has no quadratic nilpotents). Let $v \in A^{-1}$ be the unique vector such that $Ev = u$. Since $E^2 = 0$, we can apply formula (8.3) with $x = y = v$. We get $E(v)^2 = 2E(E(v)v)$, and hence $u = 2E(uv)$. Therefore $uv - \frac{1}{2}v \in J^0$. Notice that the operator of left multiplication by u preserves J^0. Since this operator has an eigenvector in A/J^0 with eigenvector $1/2$, it has such an eigenvector in A. Therefore U is a singular subalgebra in A. □

The following Corollary was proved in [An] using other methods.

Corollary 8.16 ([An]) *Let A be an n-dimensional semisimple commutative algebra, i.e. a direct sum of n copies of $\mathbb{C}$, i.e. the algebra of diagonal $n \times n$ matrices. Then A has no nonzero quasiderivations.*

Proof. It is sufficient to check that $A \notin \mathcal{D}_1 \cup \mathcal{D}_2$. Clearly, A has no nilpotents. Suppose that $U \subset A$ is a one-dimensional subalgebra spanned by an idempotent $e \in U$. Since the spectrum of the operator of left multiplication by e consists of 0 and 1, it follows that U is not singular. □

8.4 Discriminants of Anticommutative Algebras

8.4.1 Generic Anticommutative Algebras

If the set of certain objects is parametrized by an algebraic variety X then we can speak about generic objects. Namely, we say that a generic object satisfies some property if there exists a dense Zariski-open subset $X_0 \subset X$ such that objects parametrized by points of X_0 share this property. Sometimes it is possible to find a discriminant-type closed subvariety $Y \subset X$ and to study properties of 'regular' objects parametrized by points from $X \setminus Y$. For example, instead of studying generic hypersurfaces it is beneficial to study smooth hypersurfaces. In this section we implement this program for quite non-geometric objects, namely anticommutative algebras.

Let $V = \mathbb{C}^n$. We fix an integer k, $1 < k < n-1$. Let $\mathcal{A}_{n,k} = \Lambda^k V^\vee \otimes V$ be the vector space of k-linear anticommutative maps from V to V. We identify points of $\mathcal{A}_{n,k}$ with the corresponding algebras, that is, we assume that $A \in \mathcal{A}_{n,k}$ is the space V equipped with the structure of a k-argument anticommutative algebra. Subalgebras in generic algebras with $k = 2$ were studied in [T2]. The following theorem is a generalization of these results.

Theorem 8.17 ([T1]) *Let $A \in \mathcal{A}_{n,k}$ be a generic algebra. Then*

(i) *Every m-dimensional subspace is a subalgebra if $m < k$.*
(ii) *A contains no m-dimensional subalgebras with $k+1 < m < n$.*
(iii) *The set of k-dimensional subalgebras is a smooth irreducible $(k-1)(n-k)$-dimensional subvariety in the Grassmannian $\mathrm{Gr}(k, A)$.*
(iv) *There are finitely many $(k+1)$-dimensional subalgebras, and their number is*

$$\sum_{\substack{n-k-1\geq\mu_1\geq\ldots\geq\mu_{k+1}\geq 0\\ n-k-1\geq\lambda_1\geq\ldots\geq\lambda_{k+1}\geq 0\\ \mu_1\leq\lambda_1,\ldots,\mu_{k+1}\leq\lambda_{k+1}}} (-1)^{|\mu|}\frac{(\lambda_1+k)!(\lambda_2+k-1)!\ldots\lambda_{k+1}!}{(\mu_1+k)!(\mu_2+k-1)!\ldots\mu_{k+1}!}(|\lambda|-|\mu|)!\times$$

$$\left|\frac{1}{(i-j+\lambda_j-\mu_i)!}\right|^2_{i,j=1,\ldots,k+1},$$

where $|\lambda| = \lambda_1 + \ldots + \lambda_{k+1}$, $|\mu| = \mu_1 + \ldots + \mu_{k+1}$, $1/N! = 0$ if $N < 0$.
(v) *A contains a $(k+1)$-dimensional subalgebra.*
(vi) *If $k = n-2$, then the number of $(k+1)$-dimensional subalgebras is equal to*

$$\frac{2^n - (-1)^n}{3.}$$

Proof. The GL_n-module $\mathcal{A}_{n,k}$ is a sum of two irreducible submodules:

$$\mathcal{A}_{n,k} = \mathcal{A}^0_{n,k} \oplus \tilde{\mathcal{A}}_{n,k}. \tag{8.4}$$

Here $\tilde{\mathcal{A}}_{n,k}$ is isomorphic to $\Lambda^{k-1}V^\vee$: we assign to every $(k-1)$-form ω the algebra with multiplication

$$[v_1, \ldots, v_k] = \sum_{i=1}^{k}(-1)^{i-1}\omega(v_1, \ldots, \hat{v}_i, \ldots, v_k)v_i.$$

Note that every subspace of this algebra is a subalgebra. Hence the lattice of subalgebras of $A \in \mathcal{A}_{n,k}$ coincides with the lattice of subalgebras of A^0, where $A \mapsto A^0$ is the GL_n-equivariant projector on the first summand in (8.4). Algebras in $\mathcal{A}^0_{n,k}$ will be called *zero trace algebras*, since $A \in \mathcal{A}^0_{n,k}$ if and only if the $(k-1)$-form $\mathrm{Tr}[v_1, \dots, v_{k-1}, \cdot]$ is equal to zero. Hence the theorem will be proved once we have proved it for generic algebras in $\mathcal{A}^0_{n,k}$.

We choose a basis $\{e_1, \dots, e_n\}$ in V, identify GL_n with the group of matrices, consider the standard diagonal maximal torus T, and take for B and B_- the subgroups of upper- and lower-triangular matrices. We fix an $m \geq k$. Consider the parabolic subgroup of matrices

$$P = \begin{pmatrix} A & 0 \\ * & B \end{pmatrix},$$

where B is an $m \times m$ matrix and A is an $(n-m) \times (n-m)$ matrix. Then G/P coincides with $\mathrm{Gr}(m, V)$. Consider the vector bundle

$$\mathcal{L} = \Lambda^k \mathcal{S}^\vee \otimes V/\mathcal{S}$$

on G/P, where $\mathcal{S}$ is the tautological bundle and $V/\mathcal{S}$ is the factor-tautological bundle. Then the assumptions of Theorem 8.27 are fulfilled since $\mathcal{L} = \mathcal{L}_\lambda$, where λ is the highest weight of $\mathcal{A}^0_{n,k}$. Therefore

$$\mathcal{A}^0_{n,k} = H^0(\mathrm{Gr}(m, V), \Lambda^k \mathcal{S}^\vee \otimes V/\mathcal{S}).$$

Let $A \in \mathcal{A}^0_{n,k}$, and let s_A be the corresponding global section. Then the scheme of zeros $(Z_{s_A})_{red}$ coincides with the variety of m-dimensional subalgebras of A.

Let us return to the theorem. Statement (i) is obvious. (ii) follows from Theorem 8.27 (i). Theorem 8.27 (ii) implies that if every k-argument anti-commutative algebra A contains a k-dimensional subalgebra, then the variety of k-dimensional subalgebras of a generic algebra is a smooth unmixed $(k-1)(n-k)$-dimensional subvariety in $\mathrm{Gr}(k, A)$. We claim that any $(k-1)$-dimensional subspace U can be included in a k-dimensional subalgebra. The multiplication in the algebra defines a linear map from V/U to V/U. Let $v + U$ be a non-zero eigenvector. It is obvious that $\mathbb{C}v \oplus U$ is a k-dimensional subalgebra.

We have only to prove that the variety of k-dimensional subalgebras is irreducible. Assume that a section s of the bundle $\mathcal{L} = \Lambda^k \mathcal{S}^\vee \otimes V/\mathcal{S}$ over $\mathrm{Gr}(k, V)$ corresponding to A has transversal intersection with the zero section. Then the Koszul complex

$$0 \to \Lambda^{n-k}\mathcal{L}^\vee \xrightarrow{s} \dots \xrightarrow{s} \Lambda^2\mathcal{L}^\vee \xrightarrow{s} \mathcal{L}^\vee \xrightarrow{s} \mathcal{O} \to \mathcal{O}_{Z(s)} \to 0$$

is exact by Theorem 5.17.

Note that $\Lambda^p \mathcal{L}^\vee$ is isomorphic to the bundle $S^p \Lambda^k \mathcal{S} \otimes \Lambda^p (V/\mathcal{S})^\vee$. This is a homogeneous bundle over G/P of the form $\mathcal{L}_\mu$, where

$$\mu = -\varepsilon_1 - \ldots - \varepsilon_p + \varepsilon_{n-k+1} + \ldots + \varepsilon_n$$

and ε_i are the weights of the diagonal torus in the tautological representation. Note that the weight $\mu + \rho$ (where ρ is the half-sum of the positive roots) is singular (belongs to the wall of the Weyl chamber) for any p, $1 \leq p \leq n-k$. By the Borel–Weyl–Bott Theorem 8.1, $H^\bullet(\mathrm{Gr}(k,V), \Lambda^p\mathcal{L}^\vee) = 0$ for $1 \leq p \leq n-k$. Hence $H^0(Z(s), \mathcal{O}_{Z(s)}) = H^0(\mathrm{Gr}(k,V), \mathcal{O}) = \mathbb{C}$, as was to be shown.

We postpone the proof of (v) till the end of this section and consider (iv) and (vi), that is, we shall calculate the highest Chern class of the bundle $\Lambda^k\mathcal{S}^\vee \otimes V/\mathcal{S}$ over $\mathrm{Gr}(k+1,V)$.

We use the standard notation, facts, and formulae from the Schubert calculus (cf. [Ful]. The letters λ and μ always denote Young diagrams in the rectangle with $k+1$ rows and $n-k-1$ columns. It is well known that such diagrams parametrize the basis of the Chow ring of $\mathrm{Gr}(k+1,V)$. The cycle corresponding to λ is denoted by σ_λ, where $\sigma_\lambda \in A^{|\lambda|}(\mathrm{Gr}(k+1,V))$. (We grade the Chow ring by the codimension, $|\lambda| = \lambda_1 + \ldots + \lambda_{k+1}$, where λ_i is the length of the ith row of λ.) We need the total Chern class of the bundles $\mathcal{S}$ and $V/\mathcal{S}$:

$$\begin{aligned} c(\mathcal{S}) &= 1 - \sigma_1 + \sigma_{1,1} - \ldots + (-1)^{k+1}\sigma_{1,\ldots,1}, \\ c(V/\mathcal{S}) &= 1 + \sigma_1 + \sigma_2 + \ldots + \sigma_{n-k-1}. \end{aligned}$$

We begin by calculating the total Chern class of $\mathcal{S} \otimes V/\mathcal{S}$. The standard formula for the total Chern class of the tensor products of two bundles implies that

$$c(\mathcal{S} \otimes V/\mathcal{S}) = \sum_{\mu \subset \lambda} d_{\lambda\mu} \Delta_{\tilde{\mu}}(c(\mathcal{S}))\Delta_{\lambda'}(c(V/\mathcal{S})), \tag{8.5}$$

where

$$d_{\lambda\mu} = \left| \binom{\lambda_i + k + 1 - i}{\mu_j + k + 1 - j} \right|_{1 \leq i,j \leq k+1}, \qquad \Delta_\lambda(c) = |c_{\lambda_i + j - i}|,$$

$\lambda' = (n-k-1-\lambda_{k+1}, n-k-1-\lambda_k, \ldots, n-k-1-\lambda_1)$, and $\tilde{\mu}$ is the diagram obtained from μ by transposition. We shall use the fact that $\Delta_{\tilde{\mu}}(c(\mathcal{S})) = \Delta_\mu(s(\mathcal{S}))$, where $s(E)$ is the Segre class of E. Another fact is that $\Delta_\lambda(c(V/\mathcal{S})) = \sigma_\lambda$ and $\Delta_\lambda(s(\mathcal{S})) = (-1)^{|\lambda|}\sigma_\lambda$ (since $s(\mathcal{S}) = 1 - \sigma_1 + \sigma_2 + \ldots + (-1)^{n-k-1}\sigma_{n-k-1}$). Therefore (8.5) can be written as

$$c(\mathcal{S} \otimes V/\mathcal{S}) = \sum_{\mu \subset \lambda} d_{\lambda\mu}(-1)^{|\mu|}\sigma_\mu\sigma_{\lambda'}. \tag{8.6}$$

To calculate the highest Chern class of $\mathcal{L} = \Lambda^k\mathcal{S}^\vee \otimes V/\mathcal{S}$, we use the formula $\mathcal{L} = \Lambda^{k+1}\mathcal{S}^\vee \otimes (\mathcal{S} \otimes V/\mathcal{S})$. The total Chern class of the first factor is equal to $c(\Lambda^{k+1}\mathcal{S}^\vee) = 1 + \sigma_1$. The total Chern class of the second is given by (8.6). Hence the highest Chern class of $\mathcal{L}$ is

$$c_{top}(\mathcal{L}) = \sum_{\mu \subset \lambda} d_{\lambda\mu}(-1)^{|\mu|}\sigma_\mu \sigma_{\lambda'} \sigma_1^{|\lambda|-|\mu|}. \tag{8.7}$$

The last result of the Schubert calculus that we need is the exact formula for the degree of a product of two cycles. In our case this can be written as

$$\sigma_\mu \sigma_{\lambda'} \sigma_1^{|\lambda|-|\mu|} = \deg(\sigma_\mu \sigma_{\lambda'}) = (|\lambda| - |\mu|)! \left| \frac{1}{(i-j+\lambda_j-\mu_i)!} \right|_{i,j=1,\dots,n}, \tag{8.8}$$

where $1/N! = 0$ if $N < 0$. The formula in part (iv) of the theorem can be obtained from (8.7) and (8.8) by a slight modification of the determinant in the formula for $d_{\lambda\mu}$.

It remains to verify the formula in part (vi). Let $k = n - 2$. Then the formula in (iv) can be written as

$$\sum_{0 \le i \le j \le k+1} (-1)^i \frac{(k+1)!k!\dots(k-j+2)!(k-j)!\dots 0!}{(k+1)!k!\dots(k-i+2)!(k-i)!\dots 0!}(j-i)! \det{}^2 A, \tag{8.9}$$

where

$$A = \begin{pmatrix} X & 0 & 0 \\ * & Y & 0 \\ * & * & Z \end{pmatrix},$$

X is an $i \times i$ matrix, Y is a $(j-i) \times (j-i)$ matrix, and the format of Z is a $(k+1-j) \times (k+1-j)$ matrix. X and Z are lower-triangular matrices with 1s on the diagonal and Y is given by

$$Y = \begin{pmatrix} 1 & 1 & 0 & 0 & \cdots & 0 \\ 1/2! & 1 & 1 & 0 & \cdots & 0 \\ 1/3! & 1/2! & 1 & 1 & \cdots & 0 \\ 1/4! & 1/3! & 1/2! & 1 & \cdots & 0 \\ \vdots & \vdots & \vdots & \vdots & \ddots & \vdots \\ 1/(j-i)! & 1/(j-i-1)! & 1/(j-i-2)! & 1/(j-3)! & \cdots & 1 \end{pmatrix}.$$

It is easy to verify that $\det Y = 1/(j-i)!$, which enables us to rewrite (8.9) as

$$\sum_{i \le j} (-1)^i \frac{(1+(k-i))!\dots(1+(k-j+1))!}{(k-i)!\dots(k-j+1)!} / (j-i)! =$$

$$\sum_{i \le j} (-1)^i \binom{k+1-i}{j-i} = \sum_i (-1)^i 2^{k+1-i} = \frac{2^{k+2} - (-1)^{k+2}}{3} = \frac{2^n - (-1)^n}{3},$$

as was to be shown.

It remains to prove part (v) of the theorem. We have to prove that every $A \in \mathcal{A}_{n,k}$ has a $(k+1)$-dimensional subalgebra. We fix a basis $\{e_1, \dots, e_n\}$ in V and consider the subspace $U = \langle e_{n-k}, \dots, e_n \rangle$. Let $M \subset \mathcal{A}_{n,k}$ be the subspace that consists of the algebras for which U is a $(k+1)$-dimensional subalgebra. We claim that $\mathcal{A}_{n,k} = \mathrm{GL}_n \cdot M$. It is sufficient to prove that the differential of the canonical morphism $\phi : \mathrm{GL}_n \times M \to \mathcal{A}_{n,k}$ is surjective at a point (e, A). Consider the algebra $A \in M$ in which $[e_{n-k}, \ldots, \hat{e}_i, \ldots, e_n] = e_i$ for all $n-k \leq i \leq n$ and the other products are zero. We claim that $d\phi$ is surjective at (e, A). Consider the map $\pi : \mathcal{A}_{n,k} \to \mathcal{A}_{n,k}/M$. It is sufficient to verify that $\pi \circ d\phi_{(e,A)}(\mathfrak{gl}_n, 0) = \mathcal{A}_{n,k}/M \simeq \Lambda^k U^\vee \otimes V/U$. But this is obvious, since multiplication in $d\phi_{(e,A)}(E_{ji}, 0)$ for $n-k \leq i \leq n$, $1 \leq j \leq n-k-1$ (E_{ji} is the matrix identity) is given by $[e_{n-k}, \ldots, \hat{e}_i, \ldots, e_n] = e_j$ with the other products equal to zero. This completes the proof of the theorem. □

8.4.2 Regular Algebras

It is an essential drawback of Theorem 8.17 that we cannot use it to study the structure of subalgebras of any particular algebra. To correct this, we introduce an explicit class of 'regular' algebras instead of the implicit class of generic algebras. The natural way to remove degeneracies is to consider discriminants.

Let $(\mathcal{A}^0_{n,k})^*$ be the GL_n-module dual to $\mathcal{A}^0_{n,k}$, and let $S_D \subset (\mathcal{A}^0_{n,k})^*$ be the closure of the orbit of the highest vector. Let $PS_D \subset \mathbb{P}(\mathcal{A}^0_{n,k})^*$ be its projectivization, let $P\mathcal{D} \subset \mathbb{P}\mathcal{A}^0_{n,k}$ be the subvariety projectively dual to the subvariety PS_D, and let $\mathcal{D} \subset \mathcal{A}^0_{n,k}$ be the cone over it. We call $\mathcal{D}$ the *D-discriminant subvariety.* Algebras $A \in \mathcal{D}$ are called *D-singular*, and algebras $A \notin \mathcal{D}$ are called *D-regular.*

Theorem 8.18 ([T1])

(i) *$\mathcal{D}$ is a hypersurface.*
(ii) *If A is D-regular then k-dimensional subalgebras of A form a smooth irreducible $(k-1)(n-k)$-dimensional variety.*
(iii) *Let $k = n-2$. Then the degree of $\mathcal{D}$ is equal to*

$$\frac{(3n^2 - 5n)2^n - 4n(-1)^n}{18}. \tag{8.10}$$

It follows that D-singularity of A is equivalent to vanishing of the SL_n-invariant polynomial D called the *D-discriminant.*

Proof. The fact that $\mathcal{D}$ is a hypersurface follows immediately from the results of Section 7.4. (ii) can be deduced from the corresponding assertion of Theorem 8.17 by an easy calculation with differentials. Namely, let A be a D-regular algebra. We use the arguments in the proof of Theorem 8.27 (ii) and Theorem 8.17 (iii). According to these calculations, it is sufficient to verify that, if U is a k-dimensional subalgebra of A, then the map

$$\psi : \mathfrak{gl}_n \to \Lambda^k U^\vee \otimes A/U,$$

$$\psi(g)(v_1 \wedge \ldots \wedge v_k) = g[v_1, \ldots, v_k] - [gv_1, \ldots, v_k] - \ldots - [v_1, v_2, \ldots, gv_k] + U$$

is surjective. Assume the opposite. Then there is a hyperplane $H \supset U$ such that the image of ψ lies in $\Lambda^k U^\vee \otimes H/U$. Consider a non-zero algebra $\tilde{A}$ in $S_D \subset (\mathcal{A}^0_{n,k})^*$ such that $[U^\perp, V^\vee, \ldots, V^\vee] = 0$, $[V^\vee, \ldots, V^\vee] \subset H^\perp$, where $U^\perp$ and $H^\perp$ are the annihilators of U and H in $(\mathcal{A}^0_{n,k})^*$, and square brackets denote multiplication in the algebra. $\tilde{A}$ is defined by these conditions uniquely up to a scalar. Then $\tilde{A}$ annihilates $[\mathfrak{gl}_n, A]$, which is equivalent to the fact that A annihilates $[\mathfrak{gl}_n, \tilde{A}]$, that is, the tangent space to S_D at $\tilde{A}$. This means that A lies in $\mathcal{D}$, i.e. A is a D-singular algebra.

Finally, (iii) follows from Theorem 8.5. □

We will also define the E-discriminant and E-regularity but only for $(n-2)$-argument n-dimensional anticommutative algebras. Let $\mathcal{A} = \mathcal{A}^0_{n,n-2}$. Consider the projection $\pi : \mathrm{Gr}(n-1, V) \times P\mathcal{A} \to P\mathcal{A}$ on the second summand and the incidence subvariety $Z \subset \mathrm{Gr}(n-1, V) \times P\mathcal{A}$ that consists of pairs $S \subset PA$, where S is a subalgebra in A. Let $\tilde{\pi} = \pi|_Z$. By Theorem 8.17, we have $\tilde{\pi}(Z) = \mathcal{A}$. Let $\tilde{\mathcal{E}} \subset Z$ be the set of critical points of $\tilde{\pi}$, let $P\mathcal{E} = \tilde{\pi}(\tilde{\mathcal{E}})$ be the set of critical values of $\tilde{\pi}$, and let $\mathcal{E} \subset \mathcal{A}$ be the cone over $P\mathcal{E}$. Then $\mathcal{E}$ is called the *E-discriminant subvariety*. The algebras $A \in \mathcal{E}$ are said to be *E-singular*. The algebras $A \notin \mathcal{E}$ are said to be *E-regular*.

Theorem 8.19 ([T1])

(i) *$\mathcal{E}$ is an irreducible hypersurface.*
(ii) *Let A be an E-regular algebra. Then A has precisely*

$$\frac{2^n - (-1)^n}{3}$$

$(n-1)$-dimensional subalgebras.
(iii) *The map $\tilde{\pi} : \tilde{\mathcal{E}} \to P\mathcal{E}$ is birational.*

Hence the E-singularity of A is determined by the vanishing of the SL_n-invariant polynomial that defines $\mathcal{E}$. This polynomial is called the *E-discriminant.* (iii) can be formulated as follows: a generic E-singular algebra has precisely one "critical" $(n-1)$-dimensional subalgebra.

Proof. The proof of (ii) is similar to the proof of Theorem 8.17 (iv). We claim that (i) follows from (iii). We choose a basis $\{f_1, \ldots, f_n\}$ in $V^\vee$ dual to the basis $\{e_1, \ldots, e_n\}$ in V. Let $U \in \mathrm{Gr}(n-1, V)$ be the hyperplane $f_1 = 0$. It is clear that $\tilde{\mathcal{E}} = \mathrm{GL}_n \cdot M_0$, where $M_0 = \tilde{\mathcal{E}} \cap (U, P\mathcal{A})$. Moreover, $M = Z \cap (U, P\mathcal{A})$ is the linear subspace of the algebras for which U is an $(n-1)$-dimensional subalgebra. Let P be the parabolic subgroup of matrices $\begin{pmatrix} A & 0 \\ * & B \end{pmatrix}$, and let $\mathfrak{u}$ be the Lie algebra of matrices $\begin{pmatrix} 0 & X \\ 0 & 0 \end{pmatrix}$, where B is an $(n-1) \times (n-1)$ matrix and X is an $1 \times (n-1)$ matrix. Every algebra $A \in M$ defines a linear map $u \to \Lambda^{n-1}U^\vee \otimes V/U$. Since $A \in M_0$ if and only if this map is degenerate, we have $\operatorname{codim}_M M_0 = 1$. It is obvious that M_0 is irreducible, since M_0 is the spreading of the subspace

$$M_0^1 = \{A \in M_0 \mid \text{the algebra } E_{12}A \text{ has a subalgebra } U\}$$

by the group P. Hence $\tilde{\mathcal{E}}$ is an irreducible divisor in Z, and (i) follows from (iii).

We prove (iii). We say that an $(n-1)$-dimensional subalgebra U' of $A \in \mathcal{E}$ is *critical* if (U', A) lies in $\tilde{\mathcal{E}}$. In this case there is an $(n-2)$-dimensional subspace $W' \subset U'$ such that, if $V = U' \oplus \mathbb{C}e$ and $v \in \mathfrak{gl}_n$ is a non-zero linear operator such that $v(V) \subset \mathbb{C}e$ and $v(W') = 0$, then U' is a subalgebra of vA. To prove (iii) it suffices to prove that in generic algebras in M_0^1 the subalgebra U is the unique critical subalgebra. Let $N \subset M_0^1$ be the subvariety of all algebras that have another critical subalgebra.

Note that M_0^1 is normalized by the parabolic subgroup

$$Q = \begin{pmatrix} * & 0 & 0 & \dots & 0 \\ * & * & 0 & \dots & 0 \\ * & * & * & \dots & * \\ \vdots & \vdots & \vdots & \ddots & \vdots \\ * & * & * & \dots & * \end{pmatrix}.$$

Then N is the spreading of the subvarieties N_2 and N_3 by the group Q, where $A \in N_i$ if and only if the hyperplane $f_i = 0$ is a critical subalgebra. In turn, N_2 is the spreading of the vector spaces N_2^1 and N_2^3, and N_3 is the spreading of the subspaces N_3^1, N_3^2, and N_3^4, where $N_i^j \subset N_i$ is the subspace of all algebras such that the $(n-2)$-dimensional subspace W' (mentioned above) is given by $f_i = f_j = 0$. Let $Q_i^j \subset Q$ be the subgroup that normalizes the flag $f_i \subset \langle f_i, f_j \rangle$. It is easy to verify that

$$\operatorname{codim}_Q Q_2^1 = 1,\ \operatorname{codim}_Q Q_2^3 = n-1,\ \operatorname{codim}_Q Q_3^1 = n-1,$$

$$\operatorname{codim}_Q Q_3^2 = n,\ \operatorname{codim}_Q Q_3^4 = 2n-3.$$

On the other hand,

$$\operatorname{codim}_{M_0^1} N_2^1 = n,$$

$$\operatorname{codim}_{M_0^1} N_2^3 = \operatorname{codim}_{M_0^1} N_3^1 = \operatorname{codim}_{M_0^1} N_3^2 = \operatorname{codim}_{M_0^1} N_3^4 = 2n-2.$$

Hence $\operatorname{codim}_{M_0^1} QN_i^j \geq 1$ in all cases, which completes the proof. □

8.4.3 Regular 4-dimensional Anticommutative Algebras

An $(n-2)$-argument n-dimensional anticommutative algebra is called *regular* if it is both D-regular and E-regular. In this subsection we consider 2-argument 4-dimensional algebras. The corresponding generic algebras were studied in [T2].

Theorem 8.20 ([T1]) *Let A be a 4-dimensional regular anticommutative algebra. Then*

(i) *A has precisely five 3-dimensional subalgebras. The set of these subalgebras is a generic configuration of five hyperplanes. In particular, A has a pentahedral normal form, that is, it can be reduced by a transformation that belongs to GL_4 to an algebra such that the set of its five subalgebras is a Sylvester pentahedron $x_1 = 0$, $x_2 = 0$, $x_3 = 0$, $x_4 = 0$, $x_1+x_2+x_3+x_4 = 0$.*
(ii) *A has neither one- nor two-dimensional ideals.*
(iii) *The set of two-dimensional subalgebras of A is a Del Pezzo surface of degree 5 (a blowing up of $\mathbb{P}^2$ at four generic points).*
(iv) *A has precisely 10 fans, that is, flags $V_1 \subset V_3$ of 1-dimensional and 3-dimensional subspaces such that every intermediate subspace U, $V_1 \subset U \subset V_3$, is a two-dimensional subalgebra.*

Proof. We begin with (i). We have to prove that, if A is a 4-dimensional regular anticommutative algebra with zero trace, then the set of its three-dimensional subalgebras $S_1, S_2, \ldots, S_5$ is a generic configuration of hyperplanes, that is, the intersection of any three of them is one-dimensional and the intersection of any four of them is zero-dimensional. Indeed, assume, for example, that $U = S_1 \cap S_2 \cap S_3$ is two-dimensional. Let $v \in U$ and $v \neq 0$. Then $[v, \cdot]$ induces a linear operator on A/U since U is a subalgebra. S_1/U, S_2/U, and S_3/U are one-dimensional eigenspaces. Since $\dim A/U = 2$, the operator is a dilation. Since this is true for any $v \in U$, any three-dimensional subspace that contains U is a three-dimensional subalgebra, which contradicts the fact that there are precisely five such subalgebras.

Now assume that $U = S_1 \cap S_2 \cap S_3 \cap S_4$ is one-dimensional, and let $v \in U$, $v \neq 0$. Then the operator $[v, \cdot]$ induces an operator on A/U. This operator has four two-dimensional eigenspaces $S_1/U, \ldots, S_4/U$ of which any three have zero intersection. Hence this operator is a dilation. Let $W \supset U$ be an arbitrary two-dimensional subspace, and let $w \in W$ be a vector that is not proportional to v. Then the operator $[w, \cdot]$ induces a linear operator on A/W. Let z be a non-zero eigenvector. Then $\langle v, w, z\rangle$ is a three-dimensional subalgebra. Therefore every vector can be included in a three-dimensional subalgebra, which contradicts the fact that there are only five such subalgebras.

This argument also shows that A has no one-dimensional ideals. Since every three-dimensional subspace that contains a two-dimensional ideal is a subalgebra, there are no two-dimensional ideals, which completes the proof of (ii).

To prove (iii), we consider the subvariety X of two-dimensional subalgebras in A. Then $X \subset \mathrm{Gr}(2,4)$. Consider the Plücker embedding $\mathrm{Gr}(2,4) \subset \mathbb{P}^5 = P(\Lambda^2\mathbb{C}^4)$. First we claim that the embedding $X \subset \mathbb{P}^5$ is non-degenerate, that is, the image is contained in no hyperplane. Let S_1, S_2, S_3, S_4 be four three-dimensional subalgebras. Since it is a generic configuration, we can choose a basis $\{e_1, e_2, e_3, e_4\}$ in A such that $S_i = \langle e_1, \ldots, \hat{e}_i, \ldots, e_4\rangle$. Since the intersection of three-dimensional subalgebras is a two-dimensional subalgebra, A has six subalgebras $\langle e_i, e_j\rangle$, $i \neq j$. The set of corresponding bivectors $e_i \wedge e_j$ is a basis in $\Lambda^2\mathbb{C}^4$. Therefore they can lie in no hyperplane. Simple calculation with Koszul complexes (cf. Theorem 5.17) shows that $H^0(X, \mathcal{O}_X(1))^\vee = \Lambda^2\mathbb{C}^4$.

To prove that X is a Del Pezzo surface of degree five, we have only to verify that $\mathcal{O}_X(1)$ coincides with the anticanonical sheaf (see [Ma]). Since $Y = \mathrm{Gr}(2,4)$ is a quadric in $\mathbb{P}^5$, we have

$$\omega_Y = \mathcal{O}_Y(2-5-1) = \mathcal{O}_Y(-4).$$

The set X is a non-singular subvariety of codimension 2 in Y. Therefore $\omega_X = \omega_Y \otimes \Lambda^2\mathcal{N}_{X/Y}$, where $\mathcal{N}_{/Y}$ is the normal sheaf. Further, X is the scheme of zeros of a regular section of the vector bundle $\mathcal{L} = \Lambda^2\mathcal{S}^\vee \otimes V/\mathcal{S}$, and hence $\mathcal{N}_{X/Y} = \mathcal{L}|_Y$ and $\Lambda^2\mathcal{N}_{X/Y} = \mathcal{O}_X(3)$, since $c_1(\mathcal{L}) = 3H$. We obtain that $\omega_X = \mathcal{O}_X(-4) \otimes \mathcal{O}_X(3) = \mathcal{O}_X(-1)$, as was to be shown.

It remains to prove (iv). Since X is a del Pezzo surface of degree five, it contains ten straight lines. Since the embedding $X \subset P(\Lambda^2\mathbb{C}^4)$ is anticanonical, these straight lines are ordinary straight lines in $P(\Lambda^2\mathbb{C}^4)$ that lie in $\mathrm{Gr}(2,4)$. It remains to establish a bijection between these straight lines and fans. If $b \in \Lambda^2\mathbb{C}^4$, then b belongs to the cone over $\mathrm{Gr}(2,4)$ if and only if $b \wedge b = 0$. If b_1 and b_2 belong to this cone, then the straight line that joins them belongs to this cone if and only if $b_1 \wedge b_2 = 0$, which coincides with the fan condition. □

8.4.4 Dodecahedral Section

Let us start with some definitions. Let X be an irreducible G-variety (a variety with an action of algebraic group G), $S \subset X$ be an irreducible subvariety. Then S is called a *section* of X if $\overline{G \cdot S} = X$. The section S is called a *relative section* if the following condition holds: there exists a dense Zariski-open subset $U \subset S$ such that if $x \in U$ and $gx \in S$ then $g \in H$, where $H = N_G(S) = \{g \in G \,|\, gS \subset S\}$ is the normalizer of S in G (see [PV]). In this case for any invariant function $f \in \mathbb{C}(X)^G$ the restriction $f|_S$ is well-defined and the map

$$\mathbb{C}(X)^G \to \mathbb{C}(S)^H, \quad f \mapsto f|_S,$$

is an isomorphism. Any relative section defines a G-equivariant rational map $\psi : X \to G/H$: if $g^{-1}x \in S$ then $x \mapsto gH$. Conversely, any G-equivariant rational map $\psi : X \to G/H$ with irreducible fibers defines the relative section $\overline{\psi^{-1}(eH)}$.

We are going to apply Theorem 8.20 and construct a relative section in the SL_4-module $\mathcal{A}_0$ (the module of 4-dimensional anticommutative algebras with zero trace). The action of SL_4 on 'Sylvester pentahedrons' is transitive with finite stabilizer H (which is the central extension of the permutation group S_5). In the sequel the Sylvester pentahedron will always mean the standard configuration formed by the hyperplanes

$$x_1 = 0,\ x_2 = 0,\ x_3 = 0,\ x_4 = 0,\ x_1 + x_2 + x_3 + x_4 = 0.$$

Let $S \subset \mathcal{A}_0$ be a linear subspace formed by all algebras such that the hyperplanes of Sylvester pentahedron are their subalgebras. Then Theorem 8.20 implies that S is a 5-dimensional linear relative section of SL_4-module $\mathcal{A}_0$.

It is easy to see that the multiplication in algebras from S is given by formulas

$$[e_i\,, e_j] = a_{ij}e_i + b_{ij}e_j \quad (1 \leq i < j \leq 4),$$

where a_{ij} and b_{ij} satisfy a certain set of linear conditions. Consider 6 algebras $A_1, \ldots, A_6$ with the following structure constants:

	A_1	A_2	A_3	A_4	A_5	A_6
a_{12}	0	1	−1	1	0	−1
b_{12}	1	0	1	−1	−1	0
a_{13}	1	1	−1	0	−1	0
b_{13}	0	−1	0	1	1	−1
a_{14}	1	0	0	1	−1	−1
b_{14}	−1	1	−1	0	0	1
a_{23}	−1	−1	0	1	0	1
b_{23}	1	0	−1	0	1	−1
a_{24}	0	−1	−1	0	1	1
b_{24}	−1	1	0	1	−1	0
a_{34}	−1	1	1	−1	0	0
b_{34}	0	0	−1	1	−1	1

Table 8.2.

Then it is easy to see that $A_i \in S$ for any i. Moreover, the algebras A_i satisfy the unique linear relation $A_1 + \ldots + A_6 = 0$. It follows that any $A \in S$ can be written uniquely in the form $\alpha_1 A_1 + \ldots + \alpha_6 A_6$, where $\alpha_1 + \ldots + \alpha_6 = 0$. The coordinates α_i are called *dodecahedral coordinates* and S is called the *dodecahedral section* (this name will be clear later).

The stabilizer of the standard Sylvester pentahedron in PGL_4 is isomorphic to $\mathbb{S}_5$ represented by permutations of its hyperplanes. The group S_5 is generated by the transposition (12) and the cycle (12345). The preimages of these elements in GL_4 are given by matrices

$$\sigma = \begin{pmatrix} 0 & 1 & 0 & 0 \\ 1 & 0 & 0 & 0 \\ 0 & 0 & 1 & 0 \\ 0 & 0 & 0 & 1 \end{pmatrix} \qquad \tau = \begin{pmatrix} -1 & -1 & -1 & -1 \\ 1 & 0 & 0 & 0 \\ 0 & 1 & 0 & 0 \\ 0 & 0 & 1 & 0 \end{pmatrix}.$$

The preimage of S_5 in SL_4 is the group H of 480 elements. The representation of H in S induces the projective representation of S_5 in $\mathbb{P}^4$. We have the following

Proposition 8.21 *This projective representation is the projectivization of the 5-dimensional irreducible representation of S_5 (either of two possible).*

Proof. Recall certain 'folklore' facts about the representation theory of S_5. It is well-known that S_5 admits exactly two embeddings in S_6 up to conjugacy. One is standard via permutations of the first five elements of the six-element set permuted by S_6. The other one can be obtained from the first by taking the composition with the unique (up to conjugacy) outer involution of S_6. S_5 has exactly two irreducible 5-dimensional representations, which have the same projectivizations. One of these representations has the following model. One takes the tautological 5-dimensional irreducible representation of S_6 and considers its composition with the non-standard embedding $S_5 \subset S_6$.

Now let us return to our projective representation. The action of σ and τ on algebras A_i is given by the formulas

$$\begin{pmatrix} \sigma A_1 \\ \sigma A_2 \\ \sigma A_3 \\ \sigma A_4 \\ \sigma A_5 \\ \sigma A_6 \end{pmatrix} = \begin{pmatrix} -A_2 \\ -A_1 \\ -A_4 \\ -A_3 \\ -A_6 \\ -A_5 \end{pmatrix} \quad \text{and} \quad \begin{pmatrix} \tau A_1 \\ \tau A_2 \\ \tau A_3 \\ \tau A_4 \\ \tau A_5 \\ \tau A_6 \end{pmatrix} = \begin{pmatrix} A_1 \\ A_6 \\ A_2 \\ A_3 \\ A_4 \\ A_5 \end{pmatrix}. \tag{8.11}$$

Therefore S_5 permutes the lines spanned by the A_i. Moreover, the induced embedding is clearly non-standard: transposition in S_5 maps to the composition of 3 independent transpositions in S_6. It is clear that that the corresponding projective representation of S_5 is isomorphic to the projectivization of the 5-dimensional irreducible representation in the model described above. □

Remark 8.22 The section S is called dodecahedral for the following reason. Though the surjection $H \to S_5$ does not split, the alternating group A_5 can be embedded in H. The induced representation of A_5 in S has the following description. $\mathbb{A}_5$ can be realised as a group of rotations of the dodecahedron. Let $\{\Gamma_1, \ldots, \Gamma_6\}$ be the set of pairs of opposite faces of the dodecahedron. Consider the vector space of functions

$$f : \{\Gamma_1, \ldots, \Gamma_6\} \to \mathbb{C}, \quad \sum_{i=1}^{6} f(\Gamma_i) = 0.$$

Then this vector space is an A_5-module. It is easy to see that this module is isomorphic to S via the identification $A_i \mapsto f_i$, where

$$f_i(\Gamma_i) = 5, \quad f_i(\Gamma_j) = -1, \quad j \neq i.$$

The following proposition follows from the discussion above

Theorem 8.23 *The restriction of invariants induces an isomorphism of invariant fields*

$$\mathbb{C}(\mathcal{A}_0)^{GL_4} \simeq \mathbb{C}(\mathbb{C}^5)^{\mathbb{C}^* \times \mathbb{S}_5},$$

where $\mathbb{C}^*$ *acts on* $\mathbb{C}^5$ *by homotheties and* $\mathbb{S}_5$ *acts via any of two* 5*-dimensional irreducible representations.*

Remark 8.24 The Sylvester pentahedron also naturally arises in the theory of cubic surfaces. The SL_4-module of cubic forms $S^3(\mathbb{C}^4)^\vee$ admits the relative section (the so-called Sylvester section, or Sylvester normal form). Namely, a generic cubic form in a suitable system of homogeneous coordinates $x_1, \ldots, x_5$, $x_1 + \ldots + x_5 = 0$, can be written as a sum of 5 cubes $x_1^3 + \ldots + x_5^3$. The Sylvester pentahedron can be recovered from a generic cubic form f in a very interesting way: its 10 vertices coincide with 10 singular points of a quartic surface $\det \mathrm{Hes}(f)$. The Sylvester section has the same normalizer H as our dodecahedral section. It can be proved [Bek] that in this case the restriction of invariants induces an isomorphism

$$\mathbb{C}(S^3(\mathbb{C}^4)^\vee)^{GL_4} \simeq \mathbb{C}(\mathbb{C}^5)^{\mathbb{C}^* \times \mathbb{S}_5},$$

where $\mathbb{C}^*$ acts via homotheties and $\mathbb{S}_5$ via permutations of coordinates (i.e. via the *reducible* 5-dimensional representation). Other applications of the Sylvester pentahedron to moduli varieties can be found in [Bar].

These results were used in [T2] in order to prove that the field of invariant functions of the 5-dimensional irreducible representation of S_5 is rational (is isomorphic to the field of invariant functions of a vector space). From this result it is easy to deduce that in fact the field of invariant functions of any representation of S_5 is rational (see [T5]).

8.5 Adjoint Varieties

For the adjoint representation of $\mathrm{SL}_n = \mathrm{SL}(V)$ there is a natural notion of the discriminant defined as follows. For any operator $A \in \mathfrak{sl}(V)$ let

$$P_A = \det(t\,\mathrm{Id} - A)$$

be the characteristic polynomial. This is a polynomial in one variable t and its coefficients are homogeneous forms in the matrix entries of A. The discriminant $D(A) = D(P_A)$ of this polynomial is a homogeneous form on $\mathfrak{sl}(V)$ of degree $n^2 - n$. Clearly $D(A) \neq 0$ if and only if all eigenvalues of A are distinct, in other words, if A is a *regular semisimple operator.*

This discriminant can be defined for other simple Lie algebras as well. Before doing this, notice that $D(A)$ can be also defined as follows. Consider the characteristic polynomial

$$Q_A = \det(t\,\mathrm{Id} - \mathrm{ad}(A)) = \sum_{i=0}^{n^2-1} t^i D_i(A)$$

of the adjoint operator $\mathrm{ad}(A) = [A, \cdot]$. If $\lambda_1, \ldots, \lambda_n$ are the eigenvalues of A (counted with multiplicities) then the set of eigenvalues of $\mathrm{ad}(A)$ consists of $n-1$ zeros and $n(n-1)$ differences $\lambda_i - \lambda_j$, $i,j = 1,\ldots,n$, $i \neq j$. Therefore $D_0(A) = \ldots = D_{n-2}(A) = 0$ and $D_{n-1}(A)$ coincides with $D(A)$ up to a non-zero scalar.

Suppose now that $\mathfrak{g}$ is a simple Lie algebra of rank r. For any $x \in \mathfrak{g}$ let

$$Q_x = \det(t\,\mathrm{Id} - \mathrm{ad}(x)) = \sum_{i=0}^{\dim \mathfrak{g}} t^i D_i(x)$$

be the characteristic polynomial of the adjoint operator $\mathrm{ad}(x) = [x, \cdot]$. Then $D(x) = D_r(x)$ is called the *discriminant* of x. Clearly, D is a homogeneous Ad-invariant polynomial on $\mathfrak{g}$ of degree $n - r$. Since the dimension of the centralizer $\mathfrak{g}_x$ of any element $x \in \mathfrak{g}$ is greater or equal to r, it follows that $D_0 = \ldots = D_{r-1} = 0$, and therefore $D(x) = 0$ if and only if $\mathrm{ad}(x)$ has the eigenvalue 0 with multiplicity $> r$. We claim that actually $D(x) \neq 0$ if and only if x is regular semisimple (recall that x is called *regular* if $\dim \mathfrak{g}_x = r$). Indeed, if x is semisimple then $\mathrm{ad}(x)$ is a semisimple operator, and therefore $D(x) = 0$ if and only if the dimension of the centralizer $\mathfrak{g}_x$ is greater than r, i.e. x is not regular. If x is not semisimple then we take the Jordan decomposition $x = x_s + x_n$, where x_s is semisimple, x_n is nilpotent, and $[x_s, x_n] = 0$. Then x_s is automatically not regular, therefore since $Q_x = Q_{x_s}$ we have $D(x) = D(x_s) = 0$ (of course, this last equality also follows from the Ad-invariance of D and the fact that the Ad-orbit of x contains the Ad-orbit of x_s in its closure).

To study $D(x)$ further, we can use the Chevalley restriction theorem (see [PV]) $\mathbb{C}[\mathfrak{g}]^G = \mathbb{C}[\mathfrak{t}]^W$, where $\mathfrak{t} \subset \mathfrak{g}$ is any Cartan subalgebra and W is the Weyl group. Let $\Delta \subset \mathfrak{t}^\vee$ be the root system, $|\Delta| = n - r$. Let $x \in \mathfrak{t}$. Then $D(x) = 0$ if and only if x is not regular if and only if $\alpha(x) = 0$ for some $\alpha \in \Delta$. Since $\deg D = n - r$ and $D|_\mathfrak{t}$ is W-invariant, it easily follows that

$$D|_\mathfrak{t} = \prod_{\alpha \in \Delta} \alpha.$$

The Weyl group acts transitively on the set of roots of the same length. Therefore $D|_\mathfrak{t}$, and hence D, is irreducible if and only if all roots in Δ have the same length, i.e. Δ is of type A, D, or E. If Δ is of type B, C, F, or G, we have $\Delta = \Delta_s \cup \Delta_l$, where Δ_s is the set of short roots and Δ_l is the set of long roots. Then we have $D = D_l D_s$, where D_l and D_s are irreducible polynomials and

$$D_l|_\mathfrak{t} = \prod_{\alpha \in \Delta_l} \alpha, \quad D_s|_\mathfrak{t} = \prod_{\alpha \in \Delta_s} \alpha.$$

In the $A-D-E$ case we also set $D_l = D$ to simplify notations.

We are going to show that D_l is also the discriminant in our usual sense. The adjoint representation $\mathrm{Ad} : G \to \mathrm{GL}(\mathfrak{g})$ is irreducible. In $A-D-E$ case let $\mathcal{O} = \mathcal{O}_l$ be the Ad-orbit of any root vector. In $B-C-F-G$ case let $\mathcal{O}_l \subset \mathfrak{g}$ (resp. $\mathcal{O}_s \subset \mathfrak{g}$) be the Ad-orbit of any long root vector (resp. any short root vector). Then $\mathcal{O}_l$ is the orbit of the highest weight vector. Its projectivization is called the adjoint variety. Both orbits $\mathcal{O}_l$ and $\mathcal{O}_s$ are conical. Let $X_l = \mathbb{P}(\mathcal{O}_l)$ and $X_s = \overline{\mathbb{P}(\mathcal{O}_s)}$.

Theorem 8.25 *D_l is the discriminant of X_l. D_s is the discriminant of X_s.*

Proof. We identify $\mathfrak{g}$ and $\mathfrak{g}^\vee$ via the Killing form. Let $\alpha \in \Delta$ be any root, and $e_\alpha \in \mathfrak{g}$ the corresponding root vector. Since $[\mathfrak{g}, e_\alpha]^\perp = \mathfrak{g}_{e_\alpha}$, we have

$$\left(\overline{\mathbb{P}(\mathrm{Ad}(G)\cdot e_\alpha)}\right)^* = \overline{\mathbb{P}(\mathrm{Ad}(G)\cdot \mathfrak{g}_{e_\alpha})}.$$

Let

$$\mathfrak{t}_\alpha = \{x \in \mathfrak{t} \mid \alpha(x) = 0\}, \quad \hat{\mathfrak{t}}_\alpha = \mathfrak{t}_\alpha + e_\alpha.$$

Then, clearly, $\hat{\mathfrak{t}}_\alpha \in \mathfrak{g}_{e_\alpha}$. Moreover, if α is long then $D_l|_{\hat{\mathfrak{t}}_\alpha} = D_l|_{\mathfrak{t}_\alpha} = 0$. If α is short then $D_s|_{\hat{\mathfrak{t}}_\alpha} = D_s|_{\mathfrak{t}_\alpha} = 0$. Therefore, in order to prove the theorem, it suffices to check that

$$\dim \mathrm{Ad}(G)\cdot \hat{\mathfrak{t}}_\alpha = n - 1. \tag{8.12}$$

Clearly, for generic $x \in \hat{\mathfrak{t}}_\alpha$ we have

$$\dim \mathrm{Ad}(G)\cdot \hat{\mathfrak{t}}_\alpha = \dim G + \dim \hat{\mathfrak{t}}_\alpha - \dim \mathrm{Tran}(x, \hat{\mathfrak{t}}_\alpha), \tag{8.13}$$

where

$$\mathrm{Tran}(x, \hat{\mathfrak{t}}_\alpha) = \{g \in G \mid \mathrm{Ad}(g)x \in \hat{\mathfrak{t}}_\alpha\}.$$

Let $x = y + e_\alpha \in \hat{\mathfrak{t}}_\alpha$, where $y \in \mathfrak{t}_\alpha$ is such that $\beta(y) \neq 0$ for any $\beta \in \Delta \setminus \{\pm\alpha\}$. Suppose that $\mathrm{Ad}(g)x \in \hat{\mathfrak{t}}_\alpha$. Since $[e_\alpha, \mathfrak{t}_\alpha] = 0$, it easily follows that $\mathrm{Ad}(g)y \in \mathfrak{t}_\alpha$ and $\mathrm{Ad}(g)e_\alpha = e_\alpha$. Under our assumptions on y this means that

$$\mathrm{Tran}(x, \hat{\mathfrak{t}}_\alpha) = (N_G(\mathfrak{t}_\alpha))_{e_\alpha},$$

where $N_G(\mathfrak{t}_\alpha)$ is the normalizer of t_α in G. Therefore, in order to prove (8.12) using (8.13), it suffices to check that

$$\dim (\mathfrak{g}_{\mathfrak{t}_\alpha})_{e_\alpha} = r, \tag{8.14}$$

where $\mathfrak{g}_{\mathfrak{t}_\alpha}$ is the centralizer of t_α in $\mathfrak{g}$. Now $\mathfrak{g}_{\mathfrak{t}_\alpha}$ is a Levi subalgebra equal to $\mathfrak{t} + \mathbb{C}e_\alpha + \mathbb{C}e_{-\alpha}$. Therefore $(\mathfrak{g}_{\mathfrak{t}_\alpha})_{e_\alpha} = \mathfrak{t}_\alpha + \mathbb{C}e_\alpha$ and (8.14) follows. □

Remark 8.26 In A-D-E cases one can prove by inspection the following interesting formula:

$$2\dim \mathfrak{g} = \mathrm{rank}\, \mathfrak{g}(\dim \mathcal{O} + 4).$$

8.6 Homogeneous Vector Bundles

8.6.1 Zeros of Generic Global Sections

Both resultants and discriminants are related to the following general construction. Suppose that X is a smooth projective variety and E is a vector bundle on X generated by global sections. Let $Z(s)$ denote the scheme of zeros of any global section $s \in H^0(X, E)$. One might expect that for generic s the scheme $Z(s)$ is a smooth variety of codimension $\dim E$. Then we can define the degeneration variety $D \subset H^0(X, E)$ parametrizing all global sections s such that $Z(s)$ is not smooth of expected codimension. For example, if E is a very ample line bundle then D is a cone over the dual variety. If E is a very ample vector bundle and $\dim E = \dim X + 1$ then D is the resultant variety. If E is a very ample vector bundle and $\dim E = \dim X$ then the corresponding homogeneous polynomial can be called Bézoutian. Indeed, if $X = \mathbb{P}^n$ and $E = \mathcal{O}(d_1) \oplus \ldots \oplus \mathcal{O}(d_n)$ then D parametrizes sets of homogeneous forms of degrees $d_1, \ldots, d_n$ such that the Bézout theorem is not applicable.

In general, in order to make the theory consistent, it is necessary to impose very strong conditions like ampleness on the vector bundle E. These conditions are not always satisfied even in the case of homogeneous vector bundles on flag varieties. Even the question whether or not a generic zero scheme is non-empty can be quite difficult.

Assume that G is a connected reductive group, T is a fixed maximal torus, B is a fixed Borel subgroup, $T \subset B \subset G$, B_- is the opposite Borel subgroup, P is a parabolic subgroup containing B_-, $\mathbf{X}(T)$ is the lattice of characters of T, and $\lambda \in \mathbf{X}(T)$ is the dominant weight. Consider the homogeneous vector bundle $\mathcal{L}_\lambda = G \times_P U_\lambda$ over G/P, where U_λ is the irreducible P-module with highest weight λ. By the Borel–Weyl–Bott Theorem 8.1, $V_\lambda = H^0(G/P, \mathcal{L}_\lambda)$ is an irreducible G-module with highest weight λ.

Theorem 8.27 *Let $s \in V_\lambda$ be a generic global section. Then*

(i) *If $\dim U_\lambda > \dim G/P$, the scheme of zeros Z_s is empty.*
(ii) *If $\dim U_\lambda \leq \dim G/P$, either Z_s is empty or s intersects the zero section of $\mathcal{L}_\lambda$ transversally and Z_s is a smooth unmixed subvariety of expected codimension $\dim U_\lambda$.*
(iii) *If $\dim U_\lambda = \dim G/P$, the geometric number of points in Z_s is equal to the top Chern class of $\mathcal{L}_\lambda$.*

Proof. (i) is obvious and follows by an easy dimension count. (ii) and (iii) follow from the fact that $\mathcal{L}_\lambda$ is generated by global sections. Let us recall this argument here. Let $\dim U_\lambda \leq \dim G/P$, and assume that every global section has a zero. We have to prove that a generic global section s intersects the zero section of $\mathcal{L}_\lambda$ transversally. This will imply, in particular, that Z_s is a smooth unmixed subvariety of codimension $\dim U_\lambda$. For simplicity we suppress the index λ. Consider the incidence variety

$$Z \subset G/P \times V, \quad Z = \{(x,s) \mid x \in (Z_s)_{red}\}.$$

Since G/P is homogeneous and Z is invariant, it follows that Z is obtained by spreading the fiber $Z_e = \{s \in V \mid s(eP) = 0\}$ by the group G. Since U is irreducible, we have $\dim Z_e = \dim V - \dim U$. Hence Z is a smooth irreducible subvariety of dimension $\dim V + \dim G/P - \dim U$.

Let $\pi : Z \to V$ be the restriction to Z of the projection of $G/P \times V$ on the second summand. By assumption, π is a surjection. By Sard's lemma for algebraic varieties (see [Mum1]), the differential $d\pi_{(x,s)}$ is surjective for a generic point $s \in V$ and any point (x,s) in $\pi^{-1}s$. We claim that s has the transversal intersection with the zero section. Indeed, Z can be regarded as a subbundle of the trivial bundle $G/P \times V$. Then $\mathcal{L} = (G/P \times V)/Z$ and the zero section of $\mathcal{L}$ is identified with $Z \mod Z$. The section s is identified with $(G/P \times \{s\}) \mod Z$. Hence, it is sufficient to prove that $G/P \times \{s\}$ is transversal to Z, which is equivalent to the following claim: $d\pi_{(x,s)}$ is surjective for all $(x,s) \in Z$. Now statement (iii) of the theorem follows from the standard intersection theory (see [Ful]). □

Example 8.28 It should be noted that Theorem 8.27 cannot be strengthened to the point where the non-emptiness and the irreducibility of the scheme of zeros in Theorem 8.27 could be established apriori, as the following example shows. Consider the vector bundle $S^2\mathcal{S}^\vee$ on $\mathrm{Gr}(k, 2n)$, where $\mathcal{S}$ is the tautological bundle. The dimension of a fiber does not exceed the dimension of the Grassmannian if $k \le \dfrac{4n-1}{3}$, but a generic section (that is, a non-degenerate quadratic form in $\mathbb{C}^{2n}$) has a zero (that is, a k-dimensional isotropic subspace) only if $k \le n$. For $k = n$ the scheme of zeros is a reducible variety of dimension $\dfrac{n(n-1)}{2}$ with two irreducible components (spinor varieties) that correspond to two families of maximal isotropic subspaces on an even-dimensional quadric.

Consider the P-submodule

$$M_\lambda = \{s \in H^0(G/P, \mathcal{L}_\lambda) \mid s(P) = 0\}$$

in V_λ. It is easy to see that M_λ can be characterized as the unique maximal proper P-submodule of V_λ. Clearly $V_\lambda/M_\lambda \simeq U_\lambda$ as P-modules. Consider the map

$$\Psi : G \times M_\lambda \to V_\lambda, \quad \Psi(g,v) = gv.$$

Then generic global sections of $\mathcal{L}$ have zeros iff Ψ is dominant iff the differential of Ψ at a generic point is surjective. We write $\mathfrak{g}$ and $\mathfrak{p}$ for Lie algebras of G and P, respectively. The natural "orbital" map $\mathfrak{g} \times V_\lambda \to V_\lambda$ defines the P-equivariant map $\psi : \mathfrak{g}/\mathfrak{p} \times M_\lambda \to U_\lambda$. Then Ψ is dominant iff $\psi(\cdot, x)$ is surjective at a generic point $x \in M_\lambda$. Let $Z_\lambda = \psi^{-1}(0)_{red} \subset \mathfrak{g}/\mathfrak{p} \times M_\lambda$ be the incidence variety. Then the following proposition follows from an easy dimension count:

Proposition 8.29 *Suppose that*

$$\dim Z_\lambda = \dim \mathfrak{g}/\mathfrak{p} + \dim M_\lambda - \dim U_\lambda.$$

Then generic global sections of $\mathcal{L}_\lambda$ have zeros.

Denote by π the restriction on Z_λ of the projection of $\mathfrak{g}/\mathfrak{p} \times M_\lambda$ to the first factor. The variety Z_λ could be rather complicated. Say, it is not irreducible in general. We shall use the fact that fibers $\pi^{-1}(x)$ are linear spaces. So consider the function $l_\lambda(x) = \dim \pi^{-1}(x)$, $x \in \mathfrak{g}/\mathfrak{p}$.

Conjecture 8.30 *There exists an algebraic stratification $\mathfrak{g}/\mathfrak{p} = \bigsqcup_{i=1}^{r} X_i$ such that for any λ the function $l_\lambda(x)$ is constant along each X_i:*

$$l_\lambda(x) = l_\lambda^i, \quad x \in X_i.$$

If such stratification exists then

$$\dim Z_\lambda = \max_i (\dim X_i + l_\lambda^i). \tag{8.15}$$

If P acts on $\mathfrak{g}/\mathfrak{p}$ with a finite number of orbits then we can take these orbits as X_i (since ψ is P-equivariant). Unfortunately, this class of parabolic subgroups is very small. The dual P-module $(\mathfrak{g}/\mathfrak{p})^\vee$ is isomorphic to the representation of P on the unipotent radical $\mathfrak{p}_n$ of $\mathfrak{p}$, and hence by the Pyasetskii Theorem 2.4 the action $P : \mathfrak{g}/\mathfrak{p}$ has a finite number of orbits iff the action $P : \mathfrak{p}_n$ satisfies this property. Such actions were studied in [PR], and the complete classification for classical groups was obtained in [HRo]. But even in these cases the problem of the complete description of the orbital decomposition seems very messy. In the next section we shall introduce another class of parabolic subgroups that satisfy the Conjecture 8.30.

The most wonderful class of parabolic subgroups is the class of parabolic subgroups P with abelian unipotent radical (aura) $\mathfrak{p}_n$; see Section 2.3.2. Let P^- be an opposite parabolic subgroup, let $L = P \cap P^-$ be a Levi subgroup with Lie algebra $\mathfrak{l}$, and let $\mathfrak{p}_n^-$ be a unipotent radical of $\mathfrak{p}^-$. Then $\mathfrak{p}_n$ is abelian iff the decomposition $\mathfrak{g} = \mathfrak{p}_n^- \oplus \mathfrak{l} \oplus \mathfrak{p}_n$ is a $\mathbb{Z}$-grading: $\mathfrak{g} = \mathfrak{g}_{-1} \oplus \mathfrak{g}_0 \oplus \mathfrak{g}_1$. Notice that $\mathfrak{g}/\mathfrak{p}$ is isomorphic to $\mathfrak{p}_n^-$ as a L-module. It is well-known that the action of L on $\mathfrak{p}_n^-$ has a finite number of orbits in this case. Take the orbital decomposition $\mathfrak{p}_n^- = \bigcup_{i=1}^{r} Lf_i$, for whose detailed description see Theorem 2.10. Then (8.15) takes the form

$$\dim Z_\lambda = \max_i (\dim Lf_i + l_\lambda(f_i)). \tag{8.16}$$

It remains to calculate $l_\lambda(f_i)$. If $f_i = 0$ then $l_\lambda(f_i) = \dim M_\lambda$. In the opposite case we can include f_i in a $\mathfrak{sl}_2$-triple $\langle f_i, h_i, e_i \rangle$. Evidently we can assume that $h_i \in \mathfrak{l}$, $e_i \in \mathfrak{p}_n$. These h_i were also written down explicitly in Theorem 2.10. We shall use the following easy lemma from the $\mathfrak{sl}_2$-theory:

Lemma 8.31 *Let V be a finite-dimensional $\mathfrak{sl}_2$-module, with $\mathfrak{sl}_2 = \langle f, h, e\rangle$. Let $\mathfrak{b} = \langle h, e\rangle$. Suppose that $M \subset V$ is a $\mathfrak{b}$-submodule such that $eV \subset M$. Then*

$$\dim \operatorname{Coker}(M \subset V \xrightarrow{f} V \to V/M) = \dim (V/M)^h,$$

where $(V/M)^h = \{v \in V/M \mid hv = 0\}$.

Applying this Lemma to the representation of $\mathfrak{sl}_2$ in V_λ we get

$$l_\lambda(f_i) = \dim \operatorname{Ker} \psi(f_i, \cdot) = \dim M_\lambda - \dim U_\lambda + \dim \operatorname{Coker} \psi(f_i, \cdot) = \dim M_\lambda - \dim U_\lambda + \dim U_\lambda^{h_i}.$$

Notice that this formula remains valid for $f_i = 0$ if we assume that $h_i = 0$ in this case. Joining this formula with (8.16) and applying Lemma 8.31 we get the following result.

Theorem 8.32 ([T3]) *Suppose that for any i*

$$\dim U_\lambda^{h_i} \leq \operatorname{codim}_{\mathfrak{g}/\mathfrak{p}} Lf_i.$$

Then generic global sections of the bundle $\mathcal{L}_\lambda$ have zeros.

8.6.2 Isotropic Subspaces of Forms

All ingredients of the formula in Theorem 8.32 can be easily computed in many particular cases. We will apply Theorem 8.32 for the proof of the following

Theorem 8.33 ([T3]) *Let $w \in S^d V^\vee$, resp. $w \in \Lambda^d V^\vee$, be a generic symmetric or skew-symmetric form. Then V contains a k-dimensional isotropic subspace of w if and only if*

$$n \geq \frac{\binom{d+k-1}{d}}{k} + k, \quad \text{resp. } n \geq \frac{\binom{k}{d}}{k} + k, \tag{8.17}$$

with the following exceptions: V contains a k-dimensional isotropic subspace if and only if

- *$n \geq 2k$ for $w \in S^2V^\vee$ or $w \in \Lambda^2 V^\vee$.*
- *$k \leq n-2$ for $w \in \Lambda^{n-2} V^\vee$, n is even.*
- *$k \leq 4$ for $w \in \Lambda^3 V^\vee$, $n = 7$.*

Remark 8.34 The variety of 4-dimensional isotropic subspaces of a generic skew-symmetric 3-form in $\mathbb{C}^7$ is a smooth 8-dimensional Fano variety. Moreover, this variety is a compactification of the unique symmetric space of the simple algebraic group G_2.

Proof. We choose a basis $\{e_1, \dots, e_n\}$ in V and identify GL_n with the group of non-singular matrices. Let T, B, B^- be the subgroups of diagonal, upper- and lower-triangular matrices. We fix an integer k. Consider the parabolic subgroup

$$P = \begin{pmatrix} A & 0 \\ * & B \end{pmatrix},$$

where B is a $k \times k$ matrix. Then G/P is the Grassmannian $\mathrm{Gr}(k, V)$. Consider the vector bundle $\mathcal{L} = S^d \mathcal{S}^\vee$ (resp. $\mathcal{L} = \Lambda^d \mathcal{S}^\vee$, but only in the case $k \geq d$) on G/P, where $\mathcal{S}$ is the tautological bundle. Then $\mathcal{L} = \mathcal{L}_\lambda$, where λ is the highest weight of the GL_n-module $S^d V^\vee$ (resp. $\Lambda^d V^\vee$). Let $w \in S^d V^\vee$ (resp. $w \in \Lambda^d V^\vee$), and let s_w be the corresponding global section. It is easy to see that $(Z_{s_w})_{red}$ coincides with the variety of k-dimensional isotropic subspaces of w. Notice that the inequalities (8.17) are equivalent to the condition $\dim U_\lambda \leq \dim G/P$. In the sequel we suppose that these inequalities hold.

We take

$$f_i = E_{1,n} + E_{2,n-1} + \ldots + E_{i,n+1-i}, \quad i = 0, \dots, r = \min(k, n-k).$$

The unipotent radical $\mathfrak{p}_n^-$ can be identified with matrices of the shape $k \times (n-k)$, and then the orbit Lf_i is identified with the variety of matrices of rank i. Therefore

$$\mathrm{codim}_{\mathfrak{g}/\mathfrak{p}}\, Lf_i = (k-i)(n-k-i).$$

We have

$$h_i = (E_{n,n} - E_{1,1}) + (E_{n-1,n-1} - E_{2,2}) + \ldots + (E_{n+1-i,n+1-i} - E_{i,i}),$$

and hence

$$\dim U_\lambda^{h_i} = \binom{d+k-i-1}{d} \text{ if } V_\lambda = S^d V^\vee;$$

$$\dim U_\lambda^{h_i} = \binom{k-i}{d} \text{ if } V_\lambda = \Lambda^d V^\vee.$$

Applying Theorem 8.32 we get

Proposition 8.35

(i) *Suppose that for any $i = 0, \dots, r$ we have*

$$\binom{d+k-i-1}{d} \leq (k-i)(n-k-i).$$

Then a generic form $w \in S^d V^\vee$ has a k-dimensional isotropic subspace.

(ii) *Suppose that for any $i = 0, \dots, r$ we have*

$$\binom{k-i}{d} \leq (k-i)(n-k-i).$$

Then a generic form $w \in \Lambda^d V^\vee$ has a k-dimensional isotropic subspace.

It remains to clarify when the conditions of Proposition 8.35 follow from formulas (8.17).

A. Symmetric Case.

Clearly if $i = k$ then the conditions of Proposition 8.35 are satisfied. Therefore it suffices to find out when the inequality

$$n \geq \frac{\binom{d+k-i-1}{d}}{k-i} + k + i$$

follows from the inequality

$$n \geq \frac{\binom{d+k-1}{d}}{k} + k$$

as $i = 1, \dots, \min(k-1, n-k)$. If $d = 1$ then Theorem 8.33 is obvious. If $d = 2$ it reduces to a well-known result about isotropic subspaces of quadratic form. So assume that $d \geq 3$. We use the following lemma that can be easily verified by induction:

Lemma 8.36 *Let $d \geq 3$ and $\alpha \geq 2$. Then* $\frac{\binom{d+\alpha-1}{d}}{\alpha} \geq \frac{\binom{d+\alpha-2}{d}}{\alpha-1} + 1$.

It follows from Lemma 8.36 that

$$n \geq \frac{\binom{d+k-1}{d}}{k} + k \geq \frac{\binom{d+k-2}{d}}{k-1} + k + 1 \geq$$
$$\frac{\binom{d+k-3}{d}}{k-2} + k + 2 \geq \ldots \geq \frac{\binom{d+k-1-(k-1)}{d}}{k-(k-1)} + k + k - 1.$$

□

B. Skew-symmetric Case.

Clearly if $i > k - d$ then the conditions of Proposition 8.35 are satisfied. Therefore it suffices to check when the inequality

$$n \geq \frac{\binom{k-i}{d}}{k-i} + k + i$$

follows from the inequality

$$n \geq \frac{\binom{k}{d}}{k} + k$$

as $i = 1, \dots, \min(k-d, n-k)$. If $d = 1$ then Theorem 8.33 is obvious, if $d = 2$ it reduces to a well-known result about isotropic subspaces of a 2-form. So assume that $d \geq 3$. We use a following Lemma that can be easily verified by induction:

Lemma 8.37 *Let $3 \leq d \leq \alpha - 2$. Then* $\frac{\binom{\alpha}{d}}{\alpha} \geq \frac{\binom{\alpha-1}{d}}{\alpha-1} + 1$.

It follows from Lemma 8.37 that

$$n \geq \frac{\binom{k}{d}}{k} + k \geq \frac{\binom{k-1}{d}}{k-1} + k + 1 \geq$$
$$\frac{\binom{k-2}{d}}{k-2} + k + 2 \geq \ldots \geq \frac{\binom{k-(k-3-d)}{d}}{k-(k-3-d)} + k + (k-3-d).$$

It remains to consider 3 cases: $i = k-d-2$, $i = k-d-1$, $i = k-d$ as $i \leq n-k$.

Let $i = k-d-2$, $n \geq 2k-d-2$. We want to deduce the inequality $n \geq \frac{\binom{d+2}{d}}{d+2} + 2k - d - 2 = 2k - \frac{d-1}{2} - 2$ from $n \geq \frac{\binom{k}{d}}{k} + k$. It suffices to check $\binom{k}{d} \geq k(k - \frac{d-1}{2} - 2)$. Since $d \geq 3$ and $k - d = i + 2 \geq 3$ we have $\binom{k}{d} \geq \binom{k}{3} \geq k(k-3) \geq k(k - \frac{d-1}{2} - 2)$.

Let $i = k-d-1$, $n \geq 2k-d-1$. We want to deduce the inequality $n \geq \frac{\binom{d+1}{d}}{d+1} + 2k - d - 1 = 2k - d$ from $n \geq \frac{\binom{d+1}{d}}{d+1} + 2k - d - 1 = 2k - d$. It suffices to check $\binom{k}{d} \geq k(k-d)$. Since $d \geq 3$ and $k - d = i + 1 \geq 2$ we have $\binom{k}{d} \geq \binom{k}{3} \geq k(k-3) \geq k(k-d)$ as $k - d \geq 3$. If $k - d = 2$ then $\binom{k}{d} = \frac{(d+2)(d+1)}{2} \geq 2(d+2) \geq k(k-d)$.

Let $i = k - d$, $n \geq 2k - d$. We need to deduce the inequality $n \geq \frac{\binom{d}{d}}{d} + 2k - d = 2k - d + \frac{1}{d}$ from $n \geq \frac{\binom{k}{d}}{k} + k$. All exceptions are defined by the following system of equalities and inequalities:

$$n = 2k - d, \quad \binom{k}{d} \leq k(k-d), \quad k - d \geq 1.$$

There exist only two possibilities: either $k = d+1$, $n = d+2$ or $d = 3$, $k = 5$, $n = 7$.

In the first case we need to clarify whether a generic form $w \in \Lambda^{n-2}V^\vee$ has a $(n-1)$-dimensional isotropic subspace or not. This holds iff a generic 2-form in $V^\vee$ has a non-zero kernel. It is well-known that this is true only for odd n.

To complete the proof it remains to find out whether a generic form $w \in \Lambda^3(\mathbb{C}^7)^\vee$ has a 5-dimensional isotropic subspace. In suitable coordinates this isotropic subspace will coincide with the linear span $\langle e_1, \ldots, e_5 \rangle$. Therefore $w = \sum\limits_{1 \leq i < j < k \leq 7} \alpha_{ijk} x_i \wedge x_j \wedge x_k$, where $\alpha_{ijk} = 0$ as $k \leq 5$. Consider the one-parameter subgroup

$$H(t) = \mathrm{diag}(t^2, t^2, t^2, t^2, t^2, t^{-5}, t^{-5})$$

in SL_7. Clearly, $\lim\limits_{t \to 0} H(t)w = 0$. Therefore w belongs to the null-cone of the action $\mathrm{SL}_7 : \Lambda^3(\mathbb{C}^7)^\vee$. It follows that any non-constant homogeneous invariant of this action vanishes at w. But it is known that this action has non-constant invariants (see [PV]), for example, the discriminant is well-defined in this case. Therefore generic forms do not admit 5-dimensional isotropic subspaces. □

Remark 8.38 A 3-form in the 7-dimensional vector space has a 5-dimensional isotropic subspace if and only if its discriminant is equal to zero.

8.6.3 Moore–Penrose Inverse and Applications

The nice notion of a generalized inverse of an arbitrary matrix (possibly singular or even non-square) has been discovered independently by Moore [Mo] and Penrose [Pe]. The following definition belongs to Penrose (Moore's definition is different but equivalent):

Definition 8.39 A matrix A^+ is called an *a MP-inverse* of a matrix A if

$$AA^+A = A, \quad A^+AA^+ = A^+,$$

and AA^+, A^+A are Hermitian matrices.

It is quite surprising, but the MP-inverse always exists and is unique. Since the definition is symmetric with respect to A and A^+ it follows that $(A^+)^+ = A$. If A is a non-singular square matrix then A^+ coincides with an ordinary inverse matrix A^{-1}. The theory of MP-inverses and their numerous modifications becomes now a separate subfield of Linear Algebra [CM] with various applications. Here we show that this notion quite naturally arises in the theory of shortly graded simple Lie algebras, and we give applications. To explain this connection let us first give another definition of the MP-inverse.

Let $A \in \mathrm{Mat}_{n,m}(\mathbb{C})$. Then it is easy to see that a matrix $A^+ \in \mathrm{Mat}_{m,n}(\mathbb{C})$ is an MP-inverse of A if and only if there exist Hermitian matrices

$$B_1 \in \mathrm{Mat}_{n,n}(\mathbb{C}) \text{ and } B_2 \in \mathrm{Mat}_{m,m}(\mathbb{C})$$

such that the following matrices form an $\mathfrak{sl}_2$-triple in $\mathfrak{sl}_{n+m}(\mathbb{C})$:

$$E = \begin{pmatrix} 0 & A \\ 0 & 0 \end{pmatrix}, \quad H = \begin{pmatrix} B_1 & 0 \\ 0 & B_2 \end{pmatrix}, \quad F = \begin{pmatrix} 0 & 0 \\ A^+ & 0 \end{pmatrix}.$$

By an $\mathfrak{sl}_2$-triple $\langle e, h, f\rangle$ in a Lie algebra $\mathfrak{g}$ we mean a collection of (possibly zero) vectors such that

$$[e, f] = h, \quad [h, e] = 2e, \quad [h, f] = -2f.$$

In other words, an $\mathfrak{sl}_2$-triple is a homomorphic image of canonical generators of $\mathfrak{sl}_2$ with respect to some homomorphism of Lie algebras $\mathfrak{sl}_2 \to \mathfrak{g}$.

This definition admits an immediate generalization. Suppose that $\mathfrak{g}$ is a simple complex Lie algebra, and G is a corresponding simple simply-connected Lie group. Suppose further that P is a parabolic subgroup of G with abelian unipotent radical (with *aura*). Then $\mathfrak{g}$ admits a short grading

$$\mathfrak{g} = \mathfrak{g}_{-1} \oplus \mathfrak{g}_0 \oplus \mathfrak{g}_1$$

with only three nonzero parts. Here $\mathfrak{p} = \mathfrak{g}_0 \oplus \mathfrak{g}_1$ is a Lie algebra of P and $\exp \mathfrak{g}_1$ is the abelian unipotent radical of P. Let $\mathfrak{k}_0$ be a compact real form of $\mathfrak{g}_0$. In this section we shall permanently consider compact real forms of reductive subalgebras of simple Lie algebras. These subalgebras will always be Lie algebras of algebraic reductive subgroups of a corresponding simple complex algebraic group. Their compact real forms will always be understood as Lie algebras of compact real forms of corresponding algebraic groups. For example, a Lie algebra of an algebraic torus has a unique compact real form.

Suppose now that $e \in \mathfrak{g}_1$. It is well-known that there exists a *homogeneous* $\mathfrak{sl}_2$-triple $\langle e, h, f\rangle$ such that $h \in \mathfrak{g}_0$ and $f \in \mathfrak{g}_{-1}$.

Definition 8.40 An element $f \in \mathfrak{g}_{-1}$ is called *an MP-inverse* of $e \in \mathfrak{g}_1$ if there exists a homogeneous $\mathfrak{sl}_2$-triple $\langle e, h, f\rangle$ with $h \in i\mathfrak{k}_0$.

MP-inverses of elements $f \in \mathfrak{g}_{-1}$ are defined in the same way. It is clear that if f is an MP-inverse of e then e is an MP-inverse of f.

Example 8.41 Suppose that $G = \mathrm{SL}_{n+m}$ and $P \subset G$ is a maximal parabolic subgroup of block triangular matrices of the form

$$\begin{pmatrix} B_1 & A \\ 0 & B_2 \end{pmatrix}, \quad \text{where} \quad B_1 \in \mathrm{Mat}_{n,n},\ A \in \mathrm{Mat}_{n,m},\ B_2 \in \mathrm{Mat}_{m,m}\,.$$

The graded components of the correspondent grading consist of matrices of the following form:

$$\mathfrak{g}_{-1} = \begin{pmatrix} 0 & 0 \\ A' & 0 \end{pmatrix}, \quad \mathfrak{g}_0 = \begin{pmatrix} B_1 & 0 \\ 0 & B_2 \end{pmatrix}, \quad \mathfrak{g}_1 = \begin{pmatrix} 0 & A \\ 0 & 0 \end{pmatrix},$$

where $A' \in \mathrm{Mat}_{m,n}$, $B_1 \in \mathrm{Mat}_{n,n}$, $B_2 \in \mathrm{Mat}_{m,m}$, and $A \in \mathrm{Mat}_{n,m}$. One can take $\mathfrak{k}_0$ to be the real Lie algebra of block diagonal skew-Hermitian matrices with zero trace. Then $i\mathfrak{k}_0$ is a vector space of block diagonal Hermitian matrices with zero trace. Therefore in this case we return to a previous definition of a Moore–Penrose inverse.

Theorem 8.42 ([T4]) *For any $e \in \mathfrak{g}_1$ there exists a unique Moore–Penrose inverse $f \in \mathfrak{g}_{-1}$.*

The proof is similar to the proof of Theorem 8.44 below.

It obviously follows that for any non-zero $f \in \mathfrak{g}_{-1}$ there exists a unique MP-inverse $e \in \mathfrak{g}_1$. So taking an MP-inverse is a well-defined involutive operation. In general, it is not equivariant with respect to a Levi subgroup $L \subset P$ with Lie algebra $\mathfrak{g}_0$, but only with respect to its maximal compact subgroup $K_0 \subset L$.

Let us give an intrinsic description of the Moore-Penrose inverse in all cases arising from short gradings of classical simple Lie algebras. Exceptional cases may be found in [T4].

Linear Maps.

This is, of course, the classical Moore–Penrose inverse. Let us recall its intrinsic description. Suppose that $\mathbb{C}^n$ and $\mathbb{C}^m$ are vector spaces equipped with standard Hermitian scalar products. The Moore-Penrose inverse of a linear map $F : \mathbb{C}^n \to \mathbb{C}^m$ is a linear map $F^+ : \mathbb{C}^m \to \mathbb{C}^n$ defined as follows. Let $\operatorname{Ker} F \subset \mathbb{C}^n$ and $\operatorname{Im} F \subset \mathbb{C}^m$ be a kernel and an image of F. Let $\operatorname{Ker}^\perp F \subset \mathbb{C}^n$ and $\operatorname{Im}^\perp F \subset \mathbb{C}^m$ be their orthogonal complements with respect to the Hermitian scalar products. Then F defines via restriction a bijective linear map $\tilde{F} : \operatorname{Ker}^\perp F \to \operatorname{Im} F$. Then $F^+ : \mathbb{C}^m \to \mathbb{C}^n$ is the unique linear map such that $F^+|_{\operatorname{Im}^\perp F} = 0$ and $F^+|_{\operatorname{Im} F} = \tilde{F}^{-1}$. This MP-inverse corresponds to short gradings of $\mathfrak{sl}_{n+m}$.

Symmetric and Skew-symmetric Bilinear Forms.

Let $V = \mathbb{C}^n$ be a vector space equipped with a standard Hermitian scalar product. The Moore-Penrose inverse of a symmetric (resp. skew-symmetric) bilinear form ω on V is a symmetric (resp. skew-symmetric) bilinear form ω^+ on $V^\vee$ defined as follows. Let $\operatorname{Ker} \omega \subset V$ be the kernel of ω. Then ω induces a non-degenerate bilinear form $\tilde{\omega}$ on $V/\operatorname{Ker} \omega$. Let $\operatorname{Ann}(\operatorname{Ker} \omega) \subset V^\vee$ be an annihilator of $\operatorname{Ker} \omega$. Then $\operatorname{Ann}(\operatorname{Ker} \omega)$ is canonically isomorphic to the dual of $V/\operatorname{Ker} \omega$. Therefore the form $\tilde{\omega}^{-1}$ on $\operatorname{Ann}(\operatorname{Ker} \omega)$ is well-defined. The form ω^+ is defined as the unique form whose its restriction on $\operatorname{Ann}(\operatorname{Ker} \omega)$ coincides with $\tilde{\omega}^{-1}$ and whose kernel is $\operatorname{Ann}(\operatorname{Ker} \omega)^\perp$, the orthogonal complement with respect to a standard Hermitian scalar product on $V^\vee$. This MP-inverse corresponds to the short grading of $\mathfrak{sp}_{2n+2}$ (resp. $\mathfrak{so}_{2n+2}$).

Vectors in a Vector Space With the Scalar Product.

Suppose that $V = \mathbb{C}^n$ is a vector space with the standard bilinear scalar product $(\cdot,\cdot)$. The Moore-Penrose inverse $v^\vee$ of a vector $v \in V$ is again a vector in V defined as follows:

$$v^\vee = \begin{cases} \dfrac{2v}{(v,v)}, & \text{if } (v,v) \neq 0 \\ \dfrac{\overline{v}}{(\overline{v},v)}, & \text{if } (v,v) = 0,\ v \neq 0 \\ 0, & \text{if } v = 0. \end{cases}$$

This MP-inverse corresponds to the short grading of $\mathfrak{so}_{n+2}$.

It is quite natural to ask whether it is possible to extend the notion of the Moore–Penrose inverse from parabolic subgroups with aura to arbitrary parabolic subgroups. It is also interesting to consider the "non-graded" situation. Let us start with this situation. Suppose that G is a simple connected simply-connected Lie group with Lie algebra $\mathfrak{g}$. We fix a compact real form $\mathfrak{k} \subset \mathfrak{g}$.

Definition 8.43 A nilpotent orbit $\mathcal{O} \subset \mathfrak{g}$ is called *a Moore–Penrose orbit* if for any $e \in \mathcal{O}$ there exists an $\mathfrak{sl}_2$-triple $\langle e, h, f\rangle$ such that $h \in i\mathfrak{k}$.

It turns out that it is quite easy to find all Moore-Penrose orbits. Recall that the height $\mathrm{ht}(\mathcal{O})$ of a nilpotent orbit $\mathcal{O} = \mathrm{Ad}(G)e$ is equal to the maximal integer k such that $\mathrm{ad}(e)^k \neq 0$. Clearly $\mathrm{ht}(\mathcal{O}) \geq 2$.

Theorem 8.44 *$\mathcal{O}$ is a Moore–Penrose orbit if and only if* $\mathrm{ht}(\mathcal{O}) = 2$. *In this case for any $e \in \mathcal{O}$ there exists a unique $\mathfrak{sl}_2$-triple $\langle e, h, f\rangle$ such that $h \in i\mathfrak{k}$.*

Proof. Let $x \to \overline{x}$ denotes a complex conjugation in $\mathfrak{g}$ with respect to the compact form $\mathfrak{k}$. Therefore $x = \overline{x}$ iff $x \in \mathfrak{k}$ and $x = -\overline{x}$ iff $x \in i\mathfrak{k}$. Let $B(x, y) = \mathrm{Tr}\,\mathrm{ad}(x)\,\mathrm{ad}(y)$ be the Killing form of $\mathfrak{g}$. Finally, let $H(x, y) = -B(x, \overline{y})$ be a positive-definite Hermitian form on $\mathfrak{g}$.

Lemma 8.45 *We fix a nilpotent element $e \in \mathfrak{g}$. Suppose that $\langle e, h, f\rangle$ is an $\mathfrak{sl}_2$-triple in $\mathfrak{g}$ such that $h \in i\mathfrak{k}$. Then for any other $\mathfrak{sl}_2$-triple $\langle e, h', f'\rangle$ we have $H(h, h) < H(h', h')$. In particular, if there exists an $\mathfrak{sl}_2$-triple $\langle e, h, f\rangle$ with $h \in i\mathfrak{k}$ then the $\mathfrak{sl}_2$-triple with this property is unique.*

Proof. Recall that if $\langle e, h, f\rangle$ is an $\mathfrak{sl}_2$-triple then h is called *a characteristic* of e. Consider the subset $\mathcal{H} \subset \mathfrak{g}$ consisting of all possible characteristics of e. It is well-known that $\mathcal{H}$ is an affine subspace in $\mathfrak{g}$ such that the corresponding linear subspace is precisely the unipotent radical $\mathfrak{z}_{\mathfrak{g}}^u(e)$ of the centralizer $\mathfrak{z}_{\mathfrak{g}}(e)$ in $\mathfrak{g}$ of the element e. Since $H(h', h')$ is a strongly convex function on $\mathcal{H}$, there exists a unique element $h_0 \in \mathcal{H}$ such that $H(h_0, h_0) < H(h', h')$ for any $h' \in \mathcal{H}$, $h' \neq h_0$. We need to show that $h_0 = h$. It is clear that an element $h_0 \in \mathcal{H}$ minimizes $H(h, h)$ on $\mathcal{H}$ iff $H(h_0, \mathfrak{z}_{\mathfrak{g}}^u(e)) = 0$ iff $B(\overline{h}_0, \mathfrak{z}_{\mathfrak{g}}^u(e)) = 0$. If $h \in \mathcal{H} \cap i\mathfrak{k}_0$ then $\overline{h} = -h$ and we have

$$B(\overline{h}, \mathfrak{z}_{\mathfrak{g}}^u(e)) = -B(h, \mathfrak{z}_{\mathfrak{g}}^u(e)) = -B([e, f], \mathfrak{z}_{\mathfrak{g}}^u(e)) = B(f, [e, \mathfrak{z}_{\mathfrak{g}}^u(e)]) = 0.$$

Therefore $h = h_0$. □

Suppose that $\langle e, h, f\rangle$ is an $\mathfrak{sl}_2$-triple in $\mathfrak{g}$. Consider the grading $\mathfrak{g} = \bigoplus_k \mathfrak{g}_k$ such that $x \in \mathfrak{g}_k$ iff $[h, x] = kx$. Let $\mathfrak{n}_+ = \bigoplus_{k>0} \mathfrak{g}_k$ and $\mathfrak{n}_- = \bigoplus_{k<0} \mathfrak{g}_k$. It is well known that $\mathfrak{z}_{\mathfrak{g}}^u(e) \subset \mathfrak{n}_+$.

Lemma 8.46 *Suppose that $\mathfrak{z}_{\mathfrak{g}}^u(e) = \mathfrak{n}_+$. Then $\mathcal{O} = \mathrm{Ad}(G)e$ is a Moore–Penrose orbit.*

Proof. We need to prove that for any element $e' \in \mathcal{O}$ there exists an $\mathfrak{sl}_2$-triple $\langle e', h', f'\rangle$ such that $h' \in i\mathfrak{k}$, where $\mathfrak{k}$ is a *fixed* compact real form of $\mathfrak{g}$. Clearly it is sufficient to prove that for an *arbitrary* compact real form $\mathfrak{k}$ there exists an $\mathfrak{sl}_2$-triple $\langle e, h, f\rangle$ with $h \in i\mathfrak{k}$. According to the proof of Lemma 8.45 we should choose h to be a unique characteristic such that $B(\overline{h}, \mathfrak{z}_{\mathfrak{g}}^u(e)) = 0$, where $x \to \overline{x}$ denotes a complex conjugation in $\mathfrak{g}$ with respect to the compact form $\mathfrak{k}$. It remains to prove that $h \in i\mathfrak{k}$. Since B is a non-degenerate ad-invariant scalar product on $\mathfrak{g}$ and $\mathfrak{z}^u_{\mathfrak{g}}(e) = \mathfrak{n}_+$, it follows that $\overline{h} \in \mathfrak{p}$, where $\mathfrak{p} = \mathfrak{g}_0 \oplus \mathfrak{n}_+$. Let $\mathfrak{l}$ be some "standard" compact real form of $\mathfrak{g}$ such that $h \in i\mathfrak{l}$ and $\tilde{\mathfrak{n}}_\pm = \mathfrak{n}_\mp$, where $x \to \tilde{x}$ denotes a complex conjugation in $\mathfrak{g}$ with respect to the compact form $\mathfrak{l}$. Let $P \subset G$ be a parabolic subgroup of G with the Lie algebra $\mathfrak{p}$, and let $H \subset P$ be its Levi subgroup with the Lie algebra $\mathfrak{g}_0$. Since all compact real forms of a semisimple complex Lie algebra are conjugated by elements of any fixed Borel subgroup it follows that there exists $g \in P$ such that $\mathrm{Ad}(g)\mathfrak{k} = \mathfrak{l}$. Therefore

$$\widetilde{\mathrm{Ad}(g)h} = \mathrm{Ad}(g)\overline{h} \subset \mathrm{Ad}(g)(\mathfrak{p}) = \mathfrak{p}.$$

We can express g as a product uz, where $u \in \exp(\mathfrak{n}_+)$ and $\mathrm{Ad}(z)h = h$. Then $\widetilde{\mathrm{Ad}(g)h} = \widetilde{\mathrm{Ad}(u)h}$. If u is not the identity element of G then $\mathrm{Ad}(u)h = h + \xi$, where $\xi \in \mathfrak{n}_+$ and $\xi \neq 0$. Therefore $\widetilde{\mathrm{Ad}(u)h} = -h + \tilde{\xi}$. But $\tilde{\xi} \in \mathfrak{n}_-$ and hence $\widetilde{\mathrm{Ad}(u)h} \notin \mathfrak{p}$: a contradiction. Therefore u is trivial and since $\mathrm{Ad}(z)h = h$ we finally get

$$\overline{h} = \tilde{h} = -h.$$

□

Now we shall try to reverse this argument.

Lemma 8.47 *Suppose that $\mathcal{O} = \mathrm{Ad}(G)e$ is a Moore–Penrose orbit. Then $\mathfrak{z}^u_{\mathfrak{g}}(e) = \mathfrak{n}_+$.*

Proof. We choose a standard compact real form $\mathfrak{l}$ as in the proof of Lemma 8.46. Clearly, $\mathfrak{z}^u_{\mathfrak{g}}(e)$ is a graded subalgebra of $\mathfrak{n}_+ = \bigoplus_{k>0} \mathfrak{g}_k$. Suppose, on the contrary, that $\mathfrak{z}^u_{\mathfrak{g}}(e) \neq \mathfrak{n}_+$. Let $\xi \in \mathfrak{g}_p$, $p > 0$, be a homogeneous element that does not belong to $\mathfrak{z}^u_{\mathfrak{g}}(e)$. Let $u = \exp(\xi)$ and $e' = \mathrm{Ad}(u)e$. We claim that all characteristics of e' do not belong to $i\mathfrak{l}$. Indeed, all characteristics of e' have the form $\mathrm{Ad}(u)h + \mathrm{Ad}(u)x$, where $x \in \mathfrak{z}^u_{\mathfrak{g}}(e)$. Suppose that for some x we have $\mathrm{Ad}(u)h + \mathrm{Ad}(u)x \in i\mathfrak{l}$. Since $h \in i\mathfrak{l}$, $\tilde{\mathfrak{n}}_\pm = \mathfrak{n}_\mp$, and $\mathrm{Ad}(u)(h+x) - h \in \mathfrak{n}_+$ it follows that $\mathrm{Ad}(u)(h+x) = h$. In $\mathfrak{n}_+$ modulo $\bigoplus_{k>p} \mathfrak{g}_k$ we obtain the equation $[\xi, h] + x = 0$, but $[h, \xi] = p\xi$ and therefore $\xi \in \mathfrak{z}^u_{\mathfrak{g}}(e)$. This is the required contradiction. □

Now we can finish the proof of Theorem 8.44. Combining previous lemmas we see that $\mathcal{O}$ is a Moore–Penrose orbit if and only if $\mathfrak{z}^u_{\mathfrak{g}}(e) = \mathfrak{n}_+$. It follows from the $\mathfrak{sl}_2$-theory that $\dim \mathfrak{z}^u_{\mathfrak{g}}(e) = \dim \mathfrak{g}_1 + \dim \mathfrak{g}_2$. Therefore $\mathfrak{z}^u_{\mathfrak{g}}(e) = \mathfrak{n}_+$ if and only if $\mathfrak{g}_p = 0$ for $p > 2$. Clearly, this is precisely equivalent to $\mathrm{ht}(\mathcal{O}) = 2$. In this case for any $e \in \mathcal{O}$ there exists a unique $\mathfrak{sl}_2$-triple $\langle e, h, f \rangle$ such that $h \in i\mathfrak{k}$ by Lemma 8.45. □

Now let us turn to the graded situation. Suppose that $\mathfrak{g}$ is a $\mathbb{Z}$-graded simple Lie algebra with $\mathfrak{g} = \bigoplus_{k \in \mathbb{Z}} \mathfrak{g}_k$. Let $P \subset G$ be a parabolic subgroup with Lie algebra $\mathfrak{p} = \bigoplus_{k \geq 0} \mathfrak{g}_k$. Let $L \subset P$ be a Levi subgroup with Lie algebra $\mathfrak{g}_0$. We choose a compact real form $\mathfrak{k}_0$ of $\mathfrak{g}_0$. Suppose now that $e \in \mathfrak{g}_k$. It is well-known that there exists a homogeneous $\mathfrak{sl}_2$-triple $\langle e, h, f \rangle$ with $h \in \mathfrak{g}_0$ and $f \in \mathfrak{g}_{-k}$.

Definition 8.48 Take any $k > 0$ and any L-orbit $\mathcal{O} \subset \mathfrak{g}_k$. Then $\mathcal{O}$ is called *a Moore-Penrose orbit* if for any $e \in \mathcal{O}$ there exists a homogeneous $\mathfrak{sl}_2$-triple $\langle e, h, f \rangle$ such that $h \in i\mathfrak{k}_0$. In this case f is called an MP-inverse of e. A grading is called *a Moore–Penrose grading in degree* $k > 0$ if all L-orbits in $\mathfrak{g}_k$ are Moore-Penrose. A grading is called *a Moore–Penrose grading* if it is a Moore–Penrose grading in any positive degree. A parabolic subgroup $P \subset G$ is called *a Moore–Penrose parabolic subgroup* if there exists a Moore–Penrose grading $\mathfrak{g} = \bigoplus_{k \in \mathbb{Z}} \mathfrak{g}_k$ such that $\mathfrak{p} = \bigoplus_{k \geq 0} \mathfrak{g}_k$ is a Lie algebra of P.

One should be careful comparing graded and non-graded situations: if $\mathcal{O} \subset \mathfrak{g}_k$ is a Moore–Penrose L-orbit then $\mathrm{Ad}(G)\mathcal{O} \subset \mathfrak{g}$ is not necessarily a Moore–Penrose G-orbit. Let us give a criterion for an L-orbit to be Moore–Penrose. Suppose that $\mathcal{O} = \mathrm{Ad}(L)e \subset \mathfrak{g}_k$. Take any homogeneous $\mathfrak{sl}_2$-triple $\langle e, h, f \rangle$. Then h defines a grading $\mathfrak{g}_0 = \bigoplus_{n \in \mathbb{Z}} \mathfrak{g}_0^n$, such that $\mathrm{ad}(h)|_{\mathfrak{g}_0^n} = n \cdot \mathrm{Id}$.

Theorem 8.49 ([T4]) *$\mathcal{O}$ is a Moore–Penrose orbit if and only if* $\mathrm{ad}(e) g_0^n = 0$ *for any $n > 0$. In this case for any $e' \in \mathcal{O}$ there exists a unique homogeneous $\mathfrak{sl}_2$-triple $\langle e', h', f' \rangle$ such that $h' \in i\mathfrak{k}_0$.*

The proof is similar to the proof of Theorem 8.44.

Example 8.50 Suppose that G is a simple group of type G_2. We fix a root decomposition. There are two simple roots α_1 and α_2 such that α_1 is short and α_2 is long. There are 3 proper parabolic subgroups: the Borel subgroup B and two maximal parabolic subgroups P_1 and P_2 such that a root vector of α_i belongs to a Levi subgroup of P_i. Then the following is an easy application of Theorem 8.49. B is a Moore–Penrose parabolic subgroup (actually Borel subgroups in all simple groups are Moore–Penrose parabolic subgroups with respect to any grading). P_1 is not Moore–Penrose, but it is a Moore–Penrose parabolic subgroup in degree 2 (with respect to the natural grading of height 2). P_2 is a Moore–Penrose parabolic subgroup.

Example 8.51 Suppose that $G = \mathrm{SL}_n$. We fix positive integers $d_1, \ldots, d_k$ such that $n = d_1 + \ldots + d_k$. We consider the parabolic subgroup $P(d_1, \ldots, d_k) \subset \mathrm{SL}_n$ that consists of all upper-triangular block matrices with sizes of blocks equal to $d_1, \ldots, d_k$. We take a standard grading. Then $\mathfrak{g}_1$ is identified with the linear space of all tuples of linear maps $\{f_1, \ldots, f_k\}$,

$$\mathbb{C}^{d_1} \xleftarrow{f_1} \mathbb{C}^{d_2} \xleftarrow{f_2} \ldots \xleftarrow{f_{k-1}} \mathbb{C}^{d_k},$$

$\mathfrak{g}_{-1}$ is identified with the linear space of all tuples of linear maps $\{g_1, \ldots, g_k\}$,

$$\mathbb{C}^{d_1} \xrightarrow{g_1} \mathbb{C}^{d_2} \xrightarrow{g_2} \ldots \xrightarrow{g_{k-1}} \mathbb{C}^{d_k},$$

and Levi subgroup $L(d_1, \ldots, d_k)$ is just the group

$$(A_1, \ldots, A_k) \in \mathrm{GL}_{d_1} \times \ldots \times \mathrm{GL}_{d_k} \quad \text{such that} \quad \det(A_1) \cdot \ldots \cdot \det(A_k) = 1.$$

$L(d_1,\ldots,d_k)$ acts on these spaces of linear maps in an obvious way. The most important among L-orbits are varieties of complexes. To define them, let us fix in addition non-negative integers $m_1,\ldots,m_{k-1}$ such that $m_{i-1}+m_i \leq d_i$ (we set $m_0 = m_k = 0$), and consider the subvariety of all tuples $\{f_1\ldots,f_{k-1}\}$ as above such that $\operatorname{rk} f_i = m_i$ and $f_{i-1}\circ f_i = 0$ for any i. These tuples form a single L-orbit $\mathcal{O}$ called a variety of complexes. For any tuple $\{f_1,\ldots,f_{k-1}\} \in \mathcal{O}$ consider the tuple $\{f_1^+,\ldots,f_{k-1}^+\} \in \mathfrak{g}_{-1}$, where f_i^+ is a classical "matrix" Moore–Penrose inverse of f_i. It is easy to see that this new tuple is again a complex and, moreover, this complex is a Moore–Penrose inverse (in our latest meaning of this word) of an original complex. In particular, orbits of complexes are Moore–Penrose orbits.

At first glance only few parabolic subgroups are Moore–Penrose. But this is scarcely true. For example, we have the following theorem:

Theorem 8.52 ([T4]) *Any parabolic subgroup in SL_n is Moore–Penrose.*

To explain our interest in Moore–Penrose parabolic subgroups let us recall Conjecture 8.30 from the previous section. Suppose once again that G is a simple connected simply-connected Lie group, P is its parabolic subgroup, and $\mathfrak{p} \subset \mathfrak{g}$ are their Lie algebras. We take any irreducible G-module V. There exists a unique proper P-submodule M_V of V. We have the inclusion $i : M_V \to V$, the projection $\pi : V \to V/M_V$ and the map $R_V : \mathfrak{g} \to \operatorname{End}(V)$ defining the representation. Therefore we have a linear map $\tilde{R}_V : \mathfrak{g} \to \operatorname{Hom}(M_V, V/M_V)$, namely $\tilde{R}_V(x) = \pi \circ R_V(x) \circ i$. Clearly $\mathfrak{p} \subset \operatorname{Ker} \tilde{R}_V$. Finally, we have a linear map $\Psi_V : \mathfrak{g}/\mathfrak{p} \to \operatorname{Hom}(M_V, V/M_V)$. The Conjecture 8.30 states that there exists an algebraic stratification $\mathfrak{g}/\mathfrak{p} = \bigsqcup_{i=1}^{n} X_i$ such that for any V the function $\operatorname{rk} \Psi_V$ is constant along each X_i.

The following theorem shows the connection of this problem with the Moore–Penrose inverse.

Theorem 8.53 ([T4]) *Suppose that a grading $\mathfrak{g} = \bigoplus_{k\in\mathbb{Z}} \mathfrak{g}_k$ is a Moore–Penrose grading in all positive degrees except at most one. Then the Conjecture 8.30 is true for the corresponding parabolic subgroup P.*

Proof. Suppose that G is a connected reductive group with a Lie algebra $\mathfrak{g}$. For any elements $x_1,\ldots,x_r \in \mathfrak{g}$ let $\langle x_1,\ldots,x_r\rangle_{alg}$ denote the minimal algebraic Lie subalgebra of $\mathfrak{g}$ that contains $x_1,\ldots,x_r$ (algebraic subalgebras are the Lie algebras of algebraic subgroups). By a theorem of Richardson [Ri1] $\langle x_1,\ldots,x_r\rangle_{alg}$ is reductive if and only if an orbit of the r-tuple $(x_1,\ldots,x_r)$ in $\mathfrak{g}^r$ is closed with respect to the diagonal action of G. Suppose now that $h_1,\ldots,h_r$ are semi-simple elements of $\mathfrak{g}$. Consider the closed subvariety $\hat{\mathcal{O}} = (\operatorname{Ad}(G)h_1,\ldots,\operatorname{Ad}(G)h_r) \subset \mathfrak{g}^r$. For any closed G-orbit $\mathcal{O} \subset \hat{\mathcal{O}}$ let us denote by $G(\mathcal{O})$ the conjugacy class of the reductive subalgebra $\langle x_1,\ldots,x_r\rangle_{alg}$ for $(x_1,\ldots,x_r) \in \mathcal{O}$.

Lemma 8.54 *There are only finitely many conjugacy classes $G(\mathcal{O})$.*

Proof. We shall use induction on $\dim \mathfrak{g}$. Suppose that the claim of the Lemma is true for all reductive groups H with $\dim H < \dim G$. Let $\mathfrak{z} \subset \mathfrak{g}$ be the center of $\mathfrak{g}$, and $\mathfrak{g}' \subset \mathfrak{g}$ be its commutant. Consider two canonical homomorphisms

$$\mathfrak{g} \xrightarrow{\pi} \mathfrak{g}' \quad \text{and} \quad \mathfrak{g} \xrightarrow{\pi'} \mathfrak{z}.$$

We take any closed G-orbit $\mathcal{O} \subset \hat{\mathcal{O}}$. Let $(x_1, \dots, x_r) \in \mathcal{O}$, $y_i = \pi(x_i)$ for $i = 1, \dots, r$. Then $\langle y_1, \dots, y_r \rangle_{alg} = \pi(\langle x_1, \dots, x_r \rangle_{alg})$ and is therefore reductive. Let us consider two cases.

Suppose first that $\langle y_1, \dots, y_r \rangle_{alg} = \mathfrak{g}'$. Then $\mathfrak{g}'$ is a commutor subgroup of $\langle x_1, \dots, x_r \rangle_{alg}$ and, therefore, $\langle x_1, \dots, x_r \rangle_{alg} = \langle \pi'(h_1), \dots, \pi'(h_r) \rangle \oplus \mathfrak{g}'$. In this case we get one conjugacy class.

Suppose next that $\langle y_1, \dots, y_r \rangle_{alg} \neq \mathfrak{g}'$. Then $\langle y_1, \dots, y_r \rangle_{alg}$ is contained in some maximal reductive Lie subalgebra of $\mathfrak{g}'$. It is well-known (and not difficult to prove) that in a semisimple Lie algebra there are only finitely many conjugacy classes of maximal reductive subalgebras. Let $\mathfrak{h}'$ be one of them, $\mathfrak{h} = \mathfrak{z} \oplus \mathfrak{h}' \subset \mathfrak{g}$. Let H be a corresponding reductive subgroup of G. It is sufficient to prove that, for any closed G-orbit $\mathcal{O}$ of $\hat{\mathcal{O}}$ that meets $\mathfrak{h}^r$, there are only finitely many possibilities for $G(\mathcal{O})$. It easily follows from Richardson's Lemma [Ri] that for any i the intersection $\mathrm{Ad}(G)h_i \cap \mathfrak{h}$ is a union of finitely many closed H-orbits, say $\mathrm{Ad}(H)h_i^1, \dots, \mathrm{Ad}(H)h_i^{s_i}$. It remains to prove that if, for some r-tuple $(x_1, \dots, x_r) \in \mathrm{Ad}(H)h_1^{k_1} \times \dots \mathrm{Ad}(H)h_r^{k_r}$, the corresponding subalgebra $\langle x_1, \dots, x_r \rangle_{alg}$ is reductive then there are only finitely many possibilities for its conjugacy class. But this is precisely the claim of the lemma for the group H, which is true by the induction hypothesis. □

Suppose that $\mathfrak{k}$ is a compact real form of $\mathfrak{g}$.

Lemma 8.55 *If an r-tuple $(x_1, \dots, x_r)$ belongs to $(i\mathfrak{k})^r$, then its G-orbit is closed in $\mathfrak{g}^r$.*

Proof. Indeed, let B be a non-degenerate ad-invariant scalar product on $\mathfrak{g}$, which is negative-definite on $\mathfrak{k}$. Let $H(x) = -B(\overline{x}, x)$ be a positive-definite $\mathfrak{k}$-invariant Hermitian quadratic form on $\mathfrak{g}$, where the complex conjugation is taken with respect to $\mathfrak{k}$. Let H^r be a corresponding Hermitian quadratic form on $\mathfrak{g}^r$. More precisely, $H^r(x_1, \dots, x_r) = H(x_1) + \dots + H(x_r)$. By a Kempf–Ness criterion [PV], in order to prove that the G-orbit of $(x_1, \dots, x_r)$ is closed, it is sufficient to prove that the real function $H^r(\cdot)$ has a critical point on this orbit. Let us show that $(x_1, \dots, x_r)$ is this critical point. Indeed, for any $g \in \mathfrak{g}$

$$-B(\overline{x}_1, [g, x_1]) - \dots - B(\overline{x}_r, [g, x_r]) = B(x_1, [g, x_1]) + \dots + B(x_r, [g, x_r]) = 0.$$

□

Now let G be a simple simply-connected Lie group, and let $\mathfrak{g}$ be its Lie algebra with a $\mathbb{Z}$-grading $\mathfrak{g} = \bigoplus_{k \in \mathbb{Z}} \mathfrak{g}_k$. Let r be a maximal integer such that

$\mathfrak{g}_r \neq 0$. We denote the non-positive part of the grading $\underset{k \leq 0}{\oplus} \mathfrak{g}_k$ by $\mathfrak{p}$. Let $P \subset G$ be a parabolic subgroup with the Lie algebra $\mathfrak{p}$. We shall identify $\mathfrak{g}/\mathfrak{p}$ with $\underset{k>0}{\oplus} \mathfrak{g}_k$. Let $L \subset G$ be a connected reductive subgroup with Lie algebra $\mathfrak{g}_0$. Let V be an irreducible G-module. If we choose a Cartan subalgebra $\mathfrak{t} \subset \mathfrak{g}_0$ then the grading of $\mathfrak{g}$ originates from some $\mathbb{Z}$-grading on $\mathfrak{t}^\vee$. Therefore there exists a $\mathbb{Z}$-grading $V = \underset{k \in \mathbb{Z}}{\oplus} V_k$ such that $\mathfrak{g}_i V_j \subset V_{i+j}$. Let R be a maximal integer such that $V_R \neq 0$. It is easy to see that $M_V = \underset{k<R}{\oplus} V_k$ (notice that V_R is an irreducible L-module).

Now we can finish the proof of Theorem 8.53. It is sufficient to prove that there exists a finite set of points $\{x_1, \ldots, x_N\} \subset \mathfrak{g}/\mathfrak{p}$ such that for any $x \in \mathfrak{g}/\mathfrak{p}$ and for any V we have $\operatorname{rk} \Psi_V(x) = \operatorname{rk} \Psi_V(x_i)$ for some i. Recall that L has finitely many orbits on each $\mathfrak{g}_k$; see Theorem 2.6. We pick some L-orbit $\mathcal{O}_i$ in each $\mathfrak{g}_i$. Then it is sufficient to find a finite set of points as above only for points $x \in \mathfrak{g}/\mathfrak{p}$ of a form $x = x_1 + \ldots + x_r$, where $x_i \in \mathcal{O}_i$. For any orbit $\mathcal{O}_i$ let $\mathcal{H}_i$ denote the set of all possible homogeneous characteristics of all elements from $\mathcal{O}_i$. Clearly $\mathcal{H}_i$ is a closed $\operatorname{Ad}(L)$-orbit. Let $\hat{\mathcal{O}} = \mathcal{H}_1 \times \ldots \times \mathcal{H}_r \subset \mathfrak{g}_0^r$. Then by Lemma 8.54 the set of conjugacy classes of subgroups $G(\mathcal{O})$ for closed L-orbits $\mathcal{O}$ in $\hat{\mathcal{O}}$ is finite. Let us show that for any r-tuple $(x_1, \ldots, x_r) \in \mathcal{O}_1 \times \ldots \times \mathcal{O}_r$ there exists an r-tuple $(h_1, \ldots, h_r) \in \hat{\mathcal{O}}$ such that h_i is a homogeneous characteristic of x_i and an L-orbit $\operatorname{Ad}(L)(h_1, \ldots, h_r)$ is closed. Indeed, after simultaneous conjugation of elements x_i by some element $g \in L$ we may suppose that any x_i has a homogeneous characteristic $h_i \in i\mathfrak{k}_0$ (in all degrees except at most one, no conjugation is needed because of Moore–Penrose property; for one degree this is obvious). Then by Lemma 8.55 an orbit $\operatorname{Ad}(L)(h_1, \ldots, h_r)$ is closed. Since all functions $\operatorname{rk} \Psi_V$ are L-invariant, we may restrict ourselves to the points $x = \sum_i x_i \in \mathfrak{g}/\mathfrak{p}$ such that $x_i \in \mathcal{O}_i$, any x_i has a homogeneous characteristic $h_i \in i\mathfrak{k}_0$, and a conjugacy class of $\langle h_1, \ldots, h_r \rangle_{alg}$ is fixed. We claim that any function $\operatorname{rk} \Psi_V$ is constant along the set of these points. Moreover, we shall prove that

$$\operatorname{rk} \Psi_V(x) = \dim V_R - \dim V_R^{\langle h_1, \ldots, h_r \rangle_{alg}}. \tag{8.18}$$

Indeed,

$$\operatorname{rk} \Psi_V(x) = \dim \sum_i \operatorname{Im} \left(\operatorname{ad}(x_i)|_{V_{R-i}} \right).$$

Clearly V_R is $\operatorname{ad}(h_i)$-invariant and is killed by $\operatorname{ad}(e_i)$, therefore from the $\mathfrak{sl}_2$-theory we find that $V_R = \underset{k \geq 0}{\oplus} V_R^k$, where $\operatorname{ad}(h_i)|_{V_R^k} = k \cdot \mathrm{Id}$. Moreover,

$$\operatorname{Im} \left(\operatorname{ad}(x_i)|_{V_{R-i}} \right) = \underset{k>0}{\oplus} V_R^k.$$

Let H be a contravariant Hermitian form on V_R with respect to the compact form $\mathfrak{k}_0$ of $\mathfrak{g}_0$. Since $h_i \in i\mathfrak{k}_0$ and H is a contravariant form we find that $\underset{k>0}{\oplus} V_R^k = (V_R^0)^\perp$. Therefore

$$\sum_i \operatorname{Im}\left(\operatorname{ad}(x_i)|_{V_{R-i}}\right) = \left(\cap_i V_R^{h_i}\right)^{\perp} = \left(V_R^{\langle h_1,\dots,h_r\rangle_{alg}}\right)^{\perp}.$$

The formula (8.18) follows. □

For example, combining Theorem 8.53 and Theorem 8.52 we get the following corollary:

Corollary 8.56 *Conjecture 8.30 is true for any parabolic subgroup in* SL_n.

Though the conditions of Theorem 8.53 are not always satisfied, it seems that one can prove Conjecture 8.30 for any simple group using the same ideas.

9

Self-dual Varieties

In this chapter we study self-dual varieties, i.e. varieties isomorphic to their projectively dual variety. In Sect. 9.1 we consider smooth self-dual varieties. The complete list of these varieties is (conjecturally) surprisingly short. All known varieties are flag varieties so we start by considering this case, where everything follows from the classification of flag varieties of positive defect. After the brief introduction to the Hartshorne conjecture we sketch the proof of the amazing theorem of Ein that gives the complete list of self-dual varieties in the range that is allowed by the Hartshorne conjecture. We also prove a finiteness theorem of Munoz that uses distribution of primes to give restrictions on the Beilinson spectral sequence. We finish in Sect. 9.2 by describing results of Popov about self-dual nilpotent orbits.

9.1 Smooth Self-dual Varieties

9.1.1 Self-dual Flag Varieties

A projective variety $X \subset \mathbb{P}^n$ is called *self-dual* if X is isomorphic to X^* as an embedded projective variety, i.e. there exists an isomorphism $f : \mathbb{P}^N \to (\mathbb{P}^n)^\vee$ such that $f(X) = X^*$. Many examples are provided by the Pyasetskii pairing 2.4: If G is a connected reductive algebraic group acting on a vector space V then the dual action $G : V^\vee$ can be described as follows. Having passed to a finite covering of G, one may assume that there is an involution $\theta \in \mathrm{Aut}(G)$ (the so-called *involution of a maximal rank*) such that $\theta(t) = t^{-1}$ for any $t \in T$, where T is the fixed maximal torus in G. Along with the action of G on V we may consider the twisted action given by $g \star v = \omega(g)v$. Then it is easy to see that the twisted action is isomorphic to the dual action $G : V^\vee$. Therefore we may canonically identify G-orbits in V and in $V^\vee$. In particular, if G acts on V with finitely many orbits then we may consider the Pyasetskii pairing (hence the projective duality of projectivizations) as a pairing between G-orbits in V.

For example, let L be a simple algebraic group and P a parabolic subgroup with Abelian unipotent radical. In this case $\mathfrak{l} = \operatorname{Lie} L$ admits a short $\mathbb{Z}$-grading with only three non-zero parts:

$$\mathfrak{l} = \mathfrak{l}_{-1} \oplus \mathfrak{l}_0 \oplus \mathfrak{l}_1.$$

Here $\mathfrak{l}_0 \oplus \mathfrak{l}_1 = \operatorname{Lie} P$ and $\exp(\mathfrak{l}_1)$ is the Abelian unipotent radical of P. Let $G \subset L$ be a reductive subgroup with Lie algebra $\mathfrak{l}_0$. Recall that by Theorem 2.10 G has finitely many orbits in $\mathfrak{l}_1$, naturally labelled by integers from the segment $[0, r]$ such that $\mathcal{O}_k$ corresponds to $\mathcal{O}_{r-k}$ via Pyasetskii pairing (here we use the above identifications of G-orbits in $\mathfrak{l}_1$ and $\mathfrak{l}_{-1}$). Therefore if r is even then the closure of the projectivization of $\mathcal{O}_{\frac{r}{2}}$ is a self-dual projective variety.

However, it is well-known that the closure of the projectivization of $\mathcal{O}_i$ is smooth if and only if $i = 1$ or $i = r$ (the last case being not interesting: $\overline{\mathbb{P}(\mathcal{O}_r)} = \mathbb{P}(\mathfrak{l}_1)$); see e.g. [Pan] for the equivariant resolution of singularities of $\overline{\mathcal{O}_i}$. Therefore this construction gives a smooth self-dual projective variety iff $r = 2$. It is worthwhile to write down all arising cases. We use the notation from Sect. 2.3.2.

Example 9.1 Consider the short gradings of $\mathfrak{l} = \mathfrak{sl}_{n+2}$. Then $\mathcal{O}_1$ is the variety of $n \times 2$ matrices of rank 1. The projectivization of $\mathcal{O}_1$ is identified with $X = \mathbb{P}^{n-1} \times \mathbb{P}^1$ in the Segre embedding.

Example 9.2 Consider the short grading of $\mathfrak{l} = \mathfrak{so}_{n+2}$ that corresponds to β being the first simple root. The projectivization of $\mathcal{O}_1$ is identified with the quadric hypersurface in $\mathbb{P}^{n-1}$. The quadric hypersurface in $\mathbb{P}^5$ also arises from the short grading of $\mathfrak{l} = \mathfrak{so}_8 = D_4$ that corresponds to $\beta = \alpha_1$, $\beta = \alpha_3$, or $\beta = \alpha_4$ (Plücker quadric).

Example 9.3 Consider the short grading of $\mathfrak{l} = \mathfrak{so}_{10} = D_5$ that corresponds to $\beta = \alpha_4$ or $\beta = \alpha_5$. The projectivization of $\mathcal{O}_1$ is identified with $X = \operatorname{Gr}(2, 5)$ in the Plücker embedding.

Example 9.4 The short grading of E_6 gives the spinor variety $\mathbb{S}_5$.

Remarkably, there are no other known examples of smooth non-linear self-dual varieties. Moreover, it is widely expected that this list is complete.

Theorem 9.5 ([Sn]) *Let $X = G/P \subset \mathbb{P}^N$ be a non-linear polarized flag variety. If $\dim X = \dim X^*$ the X is one of the following:*

- *A quadric hypersurface.*
- *The Segre embedding of $\mathbb{P}^n \times \mathbb{P}^1$.*
- *The Plücker embedding of $\operatorname{Gr}(2, 5)$.*
- *The 10-dimensional spinor variety $\mathbb{S}_5$.*

Proof. If $\dim X = \dim X^*$ then either X is a hypersurface or $\operatorname{def} X > 0$. In the first case X is necessarily a quadric hypersurface. Indeed, let $Y \subset (\mathbb{P}^N)^\vee$ be the projectivization of the highest weight vector orbit of the dual representation. Then Y is isomorphic to X as an embedded projective variety. The intersection $Y^* \cap X$ is non-empty and G-invariant, and therefore $Y^* = X$ and X is a self-dual smooth hypersurface. Then X is a quadric hypersurface by Example 6.3. An alternative proof follows from the fact that the closure of the highest weight vector orbit is always cut out by quadrics [Li].

If $\operatorname{def} X > 0$ then the claim easily follows from Theorem 7.56. □

Theorem 9.5 was used in [Sn] to recover the classification of homogeneous real hypersurfaces in a complex projective space due to [Ta].

Theorem 9.6 ([Ta, Sn]) *Let M be a homogeneous complete real hypersurface embedded equivariantly in $\mathbb{P}^N$. Then M is a tube over a linear projective space or one of the 4 self-dual homogeneous spaces $X \subset \mathbb{P}^N$ listed in Theorem 9.5.*

Sketch of the proof. Let $M = K/L$, where K is a compact Lie group. The main idea is to use the fact that M is necessarily a tube over a *complex* submanifold $X \subset \mathbb{P}^N$ called a focal submanifold; see e.g. [CR]. Then X is easily seen to be homogeneous, and therefore X is a flag variety, $X = G/P$, where G is the complexification of K. Moreover, G acts transitively on the normal directions to X. It follows that the conormal variety $\mathbb{P}(N_X\mathbb{P}^N)$ is also homogeneous, and therefore the dual variety X^* is also homogeneous being the image of the conormal variety. Hence X^* is smooth and therefore $\dim X = \dim X^*$ by Theorem 4.25. Now we can apply Theorem 9.5. □

9.1.2 Hartshorne Conjecture

In [Ha2] R. Hartshorne has suggested a number of conjectures related to the geometry of projective varieties of small codimension. This work stimulated a remarkable research, for example Zak's theorems on linear normality and on Severi varieties. Undoubtedly, the most famous conjecture from this paper is

Conjecture 9.7 ([Ha2]) *If X is a smooth n-dimensional projective variety in $\mathbb{P}^N$ and* $\operatorname{codim} X < N/3$, *then X is a complete intersection.*

This conjecture is still very far from being solved. Only very partial results are known, for example the following Landsberg's theorem on complete intersections. The proof is based on the technique of Sect. 3.2.

Theorem 9.8 ([Lan1]) *Let $X \subset \mathbb{P}(V) = \mathbb{P}^N$ be a smooth irreducible projective variety cut out by quadrics. If* $\operatorname{codim} X < \dfrac{N+3}{4}$, *then X is a complete intersection.*

If the Hartshorne conjecture is true then any smooth projective variety X in $\mathbb{P}^N$ such that $\operatorname{codim} X < N/3$ is a complete intersection, and therefore X^* is a hypersurface by Theorem 5.11. It is not known whether this conjectural corollary is true. In particular, if X is a smooth self-dual variety then, up to the Hartshorne conjecture, either X is a quadric hypersurface or $\operatorname{codim} X \geq N/3$. Quite remarkably, in this range all self-dual varieties are known:

9.1.3 Ein's Theorem

Theorem 9.9 ([E1]) *Let X be a nonlinear smooth projective variety in $\mathbb{P}^N$. Assume that $\operatorname{codim} X \geq N/3$. Suppose that $\dim X = \dim X^*$. Then X is one of the following varieties:*

(i) *X is a hypersurface in $\mathbb{P}^2$ or $\mathbb{P}^3$.*
(ii) *X is the Segre embedding of $\mathbb{P}^1 \times \mathbb{P}^{n-1}$ in $\mathbb{P}^{2n-1}$.*
(iii) *X is the Plücker embedding of $\mathrm{Gr}(2,5)$ in $\mathbb{P}^9$.*
(iv) *X is the 10-dimensional spinor variety $\mathbb{S}_5$ in $\mathbb{P}^{15}$.*

We start with two preliminary results. Let $L = (X \cap H)_{sing}$ be a generic contact locus.

Theorem 9.10 ([E1]) *If $K_X = \mathcal{O}_X(l)$ for $l \in \mathbb{Z}$ and $n < 3d + 2$, then*

(i) $H^0(N^{\vee}_{L,X}(a)) = 0$ *for* $a \leq 0$.
(ii) $H^d(N^{\vee}_{L,X}(a)) = 0$ *for* $a \geq -d$.
(iii) $H^i(N^{\vee}_{L,X}(a)) = 0$ *if* $0 < i < d$ *and* $a \geq (n - 3d)/2$.
(iv) $H^i(N^{\vee}_{L,X}(a)) = 0$ *if* $0 < i < d$ *and* $a \leq (d - n)/2$.

Proof. Let $\tilde{X}$ be a blow up of X along L. Let $p : \tilde{X} \to X$ be the corresponding map. Denote by E the exceptional divisor and denote by F the proper transform of $H \cap X$. Therefore we have a following diagram

$$\begin{array}{ccccccc} \mathbb{P}(N_{L,X}) & = & E & \subset & \tilde{X} & \supset & F \\ & & \downarrow & & \downarrow & & \downarrow \\ & & L & \subset & X & \supset & H \cap X. \end{array}$$

We denote by $\mathcal{O}_{\tilde{X}}(a,b)$ the line bundle $p^*\mathcal{O}_X(a) \otimes \mathcal{O}_{\tilde{X}}(-bE)$. For example, $\mathcal{O}_{\tilde{X}}(F) = \mathcal{O}_{\tilde{X}}(1,2)$.

By 7.11, we have $l = (-n-d-2)/2$. Since $\tilde{X}$ is the blowing up of X along L, $K_{\tilde{X}} = \mathcal{O}_{\tilde{X}}(l, -n+d+1)$.

Let $f : \tilde{X} \to \mathbb{P}^{N-1-d}$ be the projection with center L. Let $Y = f(\tilde{X})$. The hyperplane section $H \cap X$ will correspond to a hyperplane section D of Y. Then $f^{-1}(D) = E + F$ and $f^*\mathcal{O}_Y(1) = \mathcal{O}_{\tilde{X}}(1,1)$. Let $y \in Y \setminus D$ and $Z = f^{-1}(y)$. Suppose that $\dim Z \geq 1$. Since $Z \cap (E \cup F) = \emptyset$, p maps Z isomorphically to a variety in X. Therefore $\mathcal{O}_{\tilde{X}}(1,0)|_Z$ is non-trivial. But $\mathcal{O}_{\tilde{X}}(1,1)|_Z = f^*\mathcal{O}_Y(1)|_Z$ is trivial. So $\mathcal{O}_{\tilde{X}}(0,1)$ is non-trivial. Hence $Z \cap E \neq \emptyset$, which is a contradiction. Therefore, all positive-dimensional fibers of f belong to $E \cap F$, and in particular $\dim Y = \dim X$. Now,

$$\mathcal{O}_{\tilde{X}}(a,1) = K_{\tilde{X}} \otimes f^*\mathcal{O}_Y(n-d) \otimes \mathcal{O}_{\tilde{X}}\left(a - (n-3d-2)/2, 0\right).$$

It follows from the Grauert–Rimenschneider vanishing theorem [GR] that

$$H^i(\mathcal{O}_{\tilde{X}}(a,1)) = 0 \quad \text{if} \quad i > 0 \quad \text{and} \quad a \geq (n-3d-2)/2. \tag{9.1}$$

Similarly,

$$H^i(\mathcal{O}_{\tilde{X}}(a,2)) = 0 \quad \text{if} \quad i > 0 \quad \text{and} \quad a \geq (n-3d)/2. \tag{9.2}$$

Consider the exact sequence

$$0 = H^0(\mathcal{O}_{\tilde{X}}(0,1)) \to H^0(\mathcal{O}_E(0,1)) \to H^1(\mathcal{O}_{\tilde{X}}(0,2)).$$

Since $H^1(\mathcal{O}_{\tilde{X}}(0,2)) = 0$, we have $H^0(\mathcal{O}_E(0,1)) \simeq H^0(N^\vee_{L,X}) = 0$. Hence $H^0(N^\vee_{L,X}(a)) = 0$ for $a \leq 0$ and we have (i). Recall that $N_{L,X} = N^\vee_{L,X}(1)$. So (ii) follows from (i) and Serre duality.

Consider the exact sequence

$$H^i(\mathcal{O}_{\tilde{X}}(a,1)) \to H^i(\mathcal{O}_E(a,1)) \to H^{i+1}(\mathcal{O}_{\tilde{X}}(a,2)).$$

By (9.1) and (9.2), we conclude that $H^i(\mathcal{O}_E(a,1)) \simeq H^i(N^\vee_{L,X}(a)) = 0$ for $a \geq (n-3d)/2$ and we have (iii). (iv) follows from (iii) and Serre duality. □

Theorem 9.11 ([E1]) *Suppose that X is a smooth n-dimensional projective variety in $\mathbb{P}^N$ with $d = \operatorname{def} X > 0$. Let $L = \mathbb{P}^d$ be a generic contact locus. Assume that $K_X = \mathcal{O}_X(b)$ for some $b \in \mathbb{Z}$. Then*

$$\operatorname{def} X \leq (n-2)/2.$$

This inequality is strict unless

$$\dim X = 6, \quad \operatorname{def} X = 2, \quad N^\vee_{L,X} = H^1\left(N^\vee_{L,X}(-1)\right) \otimes \Omega^1_L(1)$$

or

$$\dim X = 10, \quad \operatorname{def} X = 4, \quad N^\vee_{L,X} = H^2\left(N^\vee_{L,X}(-2)\right) \otimes \Omega^2_L(2).$$

Proof. Consider the Beilinson spectral sequence with

$$E_1^{pq} = H^q\left(N^\vee_{L,X}(p)\right) \otimes \Omega_L^{-p}(-p),$$

which by Theorem 7.15 converges to $N^\vee_{L,X}$. If $d \geq (n-1)/2$, then $n-3d-2 \leq d-n$. It follows from Theorem 9.10 that $H^q(N^\vee_{L,X}(p)) = 0$ for $-d \leq p \leq 0$. It follows that $N^\vee_{L,X} = 0$. This is nonsense.

If $\operatorname{def} X = (n-2)/2$ then $\dim X = 4m+2$ and $\operatorname{def} X = 2m$ by Theorem 7.4. In this case $H^q(N^\vee_{L,X}(p)) = 0$ for $-2m \leq p \leq 0$ unless $p = -m$. This implies that $N^\vee_{L,X} = H^m(N^\vee_{L,X}(-m)) \otimes \Omega^m_L(m)$. So

$$2m+2 = \operatorname{rank} N^\vee_{L,X} \geq \operatorname{rank} \Omega^m_L(m) = \binom{2m}{m}.$$

It follows that $m \leq 2$. □

Sketch of the proof of Theorem 9.9. Let $\dim X = n$. We may assume that $n \geq 3$ by (1.19) and (7.5). Now $\operatorname{def} X = N - 1 - n$. Since $\operatorname{def} X \leq n - 2$ by (7.6), we conclude that $n \geq \frac{N+1}{2}$. If $n = \frac{N+1}{2}$, then $\operatorname{def} X = n - 2$ and, therefore, we have case (ii) by (7.7). So we may assume that $n \geq N/2+1$. Then $K_X = \mathcal{O}_X(\frac{-N-1}{2})$ by Example 7.11. We conclude that $\operatorname{def} X = N-1-n \leq \frac{n-2}{2}$ by Theorem 9.11. Hence $n \geq 2N/3$. By our assumption $n \leq 2N/3$. Therefore, $n = 2N/3$. Now $\operatorname{def} X = N - 1 - n = \frac{1}{2}n - 1$. Therefore n is even. Since $\operatorname{def} X \equiv n \mod 2$ by Theorem 7.4, we conclude that $n = 4m + 2$ and $\operatorname{def} X = 2m$. It follows that $m \leq 2$ by Theorem 9.11. This leaves two cases: X is either a 6-dimensional variety in $\mathbb{P}^9$ or a 10-dimensional variety in $\mathbb{P}^{15}$. □

9.1.4 Finiteness Theorem

Theorem 9.12 ([Mun1]) *Let p be a fixed positive integer. Then there exists N_0 such that for any $N > N_0$ the following holds. If $X \subset \mathbb{P}^N$ is a smooth non-degenerate nonlinear n-fold and* $\dim X^* = n+(p-1)$ *then either $n > \frac{2}{3}N$ (in which case X should be a complete intersection if Hartshorne's conjecture is true) or* $\operatorname{def} X = k \geq n/2$ *(in which case by Theorem 7.24 X is a projective bundle $X \simeq \mathbb{P}_Y(F)$, where F is a vector bundle of rank $(n+k+2)/2$ on a smooth $(n-k)/2$-dimensional variety Y and the fibers are embedded linearly).*

Proof. Let $k < n/2$ and $n \leq \frac{2}{3}N$. Let L denote a contact locus of X with a generic tangent hyperplane. Then $n > 2(N-p)/3$ and for $N \geq 4p+3$ we have $n > 1 + N/2$. Then $\operatorname{Pic} X = \mathbb{Z}$ is generated by $\mathcal{O}_X(1)$ by Barth's Theorem [Ba]. If $N > 9p - 6$ then $4n \leq \frac{8}{3}N < 3N - 3p + 2$, and therefore $n - 3k - 2 = n - 3(N - p - n) - 2 < 0$ and by Theorem 9.10 we have

$$H^0(N^\vee_{L,X}(a)) = 0 \text{ for } a \leq 0, \quad H^k(N^\vee_{L,X}(a)) = 0 \text{ for } a \geq -k,$$

$$H^i(N^\vee_{L,X}(a)) = 0 \text{ if } 0 < i < k \text{ and } \left|a + \frac{k}{2}\right| \geq \frac{n}{2} - k.$$

We have $\frac{n}{2} - k = \frac{3n-2N+2p}{2} \leq p$. This means that the first term

$$E_1^{pq} = H^q\left(N^\vee_{L,X}(p)\right) \otimes \Omega_L^{-p}(-p)$$

of the Beilinson spectral sequence has nontrivial entries only in the $2p$ central columns. Since the Beilinson spectral sequence converges to $N^\vee_{L,X}$, $\operatorname{rk}(N^\vee_{L,X}) = n-k$ can be expressed as a linear combination with integer coefficients of ranks of terms of E_1. Therefore we have an equality of the form

$$n - k = \sum_{i=\frac{3k-n}{2}}^{\frac{n-k}{2}} \binom{k}{i} x_i, \quad x_i \in \mathbb{Z}.$$

If N is large enough, this equality is impossible by purely number-theoretic reasons. It is sufficient to show that there exists a prime number p between $\frac{n-k}{2}$ and k: then it will divide the RHS of our equality but not $n-k$. By the result of Nagura [Na] for $m \geq 25$ there always exists a prime number p such that $m < p < 6m/5$. If N is big enough, then this result is applicable. □

The cases $p = 2$, $p = 3$ where studied by Muñoz.

Theorem 9.13 ([Mun1]) *Let X be a non-degenerate n-fold in $\mathbb{P}^N$ with $n \leq \frac{2}{3}N$, positive defect k, and such that* $\dim X = \dim X^* - 1$. *Then X is either a scroll over a smooth curve in $\mathbb{P}^{2n}$ or a hyperplane section of* $\mathrm{Gr}(2,5)$ *or a hyperplane section of $\mathbb{S}_5$.*

Theorem 9.14 ([Mun2]) *Let X be a non-degenerate n-fold in $\mathbb{P}^N$ with $n \leq \frac{2}{3}N$, positive defect k, and such that* $\dim X = \dim X^* - 2$. *Then X is one of the following:*

- *a scroll over a smooth curve in $\mathbb{P}^{2n+1}$,*
- *a scroll over a smooth surface in $\mathbb{P}^{2n-1}$,*
- *a section of* $\mathrm{Gr}(2,5)$ *by two hyperplanes,*
- *a Fano variety X with* $\mathrm{Pic}\, X = \mathbb{Z}$ *such that (N, n, k, K_X) is either* $(15, 10, 2, -7H)$, $(21, 14, 4, -10H)$, $(27, 18, 6, -13H)$, *or* $(33, 22, 8, -16H)$. *It is not known whether these Fano varieties exist.*

9.2 Self-Dual Nilpotent Orbits

If $X \subset \mathbb{P}^n$ is a smooth projective variety then X is almost never self-dual. Moreover, up to the Hartshorne conjecture, Theorem 9.9 provides a complete list of self-dual X. However, there are many non-smooth self-dual varieties. Many equivariant self-dual varieties are provided by the Pyasetskii Theorem 2.4. Perhaps the most interesting examples of self-dual varieties are the Kummer surface in $\mathbb{P}^3$ [GH1] and the Coble quartic in $\mathbb{P}^7$ [Pau].

Other interesting examples were found in [Pop]. Let G be a connected reductive group acting linearly on a vector space V. Suppose that there exists a G-invariant non-degenerate scalar product $(\cdot,\cdot)$ on V, in particular we have an isomorphism $V \simeq V^\vee$. Finally, suppose that G acts on the null-cone

$$\mathcal{R}(V) = \{v \in V \,|\, \overline{G \cdot v} \ni 0\}$$

with finitely many orbits. In particular, any G-orbit $\mathcal{O} \subset \mathcal{R}(V)$ is conical.

Theorem 9.15 ([Pop]) *Let $\mathcal{O} = G \cdot v \subset \mathcal{R}(V)$ be a non-zero orbit and let $X = \overline{\mathbb{P}(\mathcal{O})}$ be the closure of its projectivization. Then $X = X^*$ if and only if $(\mathfrak{g} \cdot v)^\perp \subset \mathcal{R}(V)$.*

Proof. Since $\mathcal{O}$ is conical, $v \in \mathfrak{g} \cdot v$. Therefore $\mathfrak{g} \cdot v$ is the affine cone $\mathrm{Cone}(\hat{T}_{[v]}X)$ over the embedded tangent space $\hat{T}_{[v]}X$. Hence X^* is equal to $\overline{\mathbb{P}(G \cdot (\mathfrak{g} \cdot v)^\perp)}$.

Since $(\cdot,\cdot)$ is G-invariant, we have $v \in (\mathfrak{g} \cdot v)^{\perp}$, and therefore $X \subset X^*$. If $X = X^*$ then, obviously, $(\mathfrak{g} \cdot v)^{\perp} \subset \mathcal{R}(V)$. Suppose that $(\mathfrak{g} \cdot v)^{\perp} \subset \mathcal{R}(V)$. Then $X^* \subset \mathbb{P}(\mathcal{R}(V))$. Therefore $X^* = \overline{\mathbb{P}(\mathcal{O}')}$ for some orbit $\mathcal{O}' \subset \mathcal{R}(V)$. Applying the same arguments we find that $X^* \subset X^{**}$. By the Reflexivity Theorem, we finally obtain $X = X^*$. □

Let G be semisimple and let $V = \mathfrak{g}$ be the adjoint representation. Then the Killing form $(\cdot,\cdot)$ is G-invariant and $\mathcal{R}(\mathfrak{g})$ is the cone of nilpotent elements. Since there are finitely many nilpotent orbits, we can apply Theorem 9.15. For any $x \in \mathfrak{g}$ the orthogonal complement $(\mathrm{ad}(\mathfrak{g})x)^{\perp}$ is identified with the centralizer $\mathfrak{g}_x$ of x in $\mathfrak{g}$. Therefore, a nilpotent orbit $\mathcal{O} = \mathrm{Ad}(G)x$ has a self-dual projectivization if and only if the centralizer $\mathfrak{g}_x$ belongs to $\mathcal{R}(\mathfrak{g})$, i.e. has no semi-simple elements. These orbits are called the distinguished nilpotent orbits. For example, the regular nilpotent orbit (the orbit dense in $\mathcal{R}(\mathfrak{g})$) is distinguished, and hence the null-cone $\mathcal{R}(\mathfrak{g})$ itself has a self-dual projectivization.

The distinguished nilpotent orbits are, in a sense, the building blocks for the set of all nilpotent orbits. Namely, due to the Bala–Carter correspondence [BCa1, BCa2], the set of all nilpotent orbits in a semisimple Lie algebra $\mathfrak{g}$ is in the natural bijection with the set of isoclasses of pairs $(\mathfrak{l}, \mathcal{O})$, where $\mathfrak{l} \subset \mathfrak{g}$ is a Levi subalgebra and $\mathcal{O} \subset \mathfrak{l}$ is a distinguished nilpotent orbit in $[\mathfrak{l}, \mathfrak{l}]$.

Now let us describe self-dual nilpotent orbits (i.e. distinguished nilpotent orbits) in simple Lie algebras more explicitly. We follow the exposition in [Pop]. Normality of closures of nilpotent orbits was studied in [KP, Kos, Kr, Br1, Br2], etc.

Let $G = \mathrm{SL}_n$ and

$$\mathfrak{g} = \mathfrak{sl}_n = \mathrm{Mat}_n^0 = \{A \in \mathrm{Mat}_n \mid \mathrm{Tr}\, A = 0\}.$$

The non-degenerate symmetric form on $\mathfrak{g}$ is defined by $(A, B) = \mathrm{Tr}\, AB$. The nilpotent cone is

$$\mathcal{R}(\mathfrak{sl}_n) = \{A \in \mathfrak{g} \,|\, A^n = 0.$$

The algebra $\mathbb{C}[\mathfrak{sl}_n]^{\mathrm{SL}_n}$, and hence the ideal of $\mathcal{R}(\mathfrak{sl}_n)$, is generated by the coefficients of the characteristic polynomial of the generic traceless $n \times n$ matrix, i.e. by the forms $f_2, f_3, \ldots, f_n \in \mathbb{C}[\mathfrak{sl}_n]^{\mathrm{SL}_n}$ of degrees $2, 3, \ldots, n$ respectively, such that $(t^n + f_2(A)t^{n-2} + \ldots + f_n(A) = \det(t\,\mathrm{Id} - A)$ for any $A \in \mathfrak{sl}_n$. Up to a sign, f_i is the sum of all principal minors of order i of the generic traceless $n \times n$ matrix. Thus the variety $\mathbb{P}(\mathcal{R}(\mathfrak{sl}_n))$ is an $(n^2 - n - 1)$-dimensional self-dual normal complete intersection of degree $n!$ defined (scheme-theoretically) by the equations $f_2 = f_3 = \ldots = f_n = 0$ see also [PV].

The SL_n-orbits in $\mathbb{P}(\mathcal{R}(\mathfrak{sl}_n))$ are described by Jordan's theory. Namely, for a partition $\sigma = (\sigma_1, \ldots, \sigma_t)$ of n, denote by $\mathcal{R}(\mathfrak{sl}_n)_{\sigma}$ the SL_n-orbit consisting of all nilpotent matrices $A \in \mathrm{Mat}_n^0$ such that $\sigma_1, \ldots, \sigma_t$ are the sizes of the blocks of the Jordan normal form of A. In particular, $\mathcal{R}(\mathfrak{sl}_n)_{(n)}$ is the regular orbit. The rule determining, in terms of partitions, when one orbit lies in the closure of another is the following [He]:

$$\mathcal{R}(\mathfrak{sl}_n)_\sigma \subset \overline{\mathcal{R}(\mathfrak{sl}_n)_\nu} \Leftrightarrow \sum_{i=1}^{j} \sigma_j \leq \sum_{i=1}^{j} \nu_j \ \forall j.$$

Only regular nilpotent elements in $\mathfrak{sl}_n$ are distinguished [BCa1, BCa2]. Hence the variety $\mathbb{P}(\mathcal{R}(\mathfrak{sl}_n)_\sigma)$ is not self-dual for every $\sigma \neq (n)$.

Let $G = \mathrm{SO}_n$ and

$$\mathfrak{g} = \mathfrak{so}_n = \{A \in \mathrm{Mat}_n^0 \mid A^\perp = -A\}.$$

The non-degenerate symmetric form on $\mathfrak{g}$ is defined as in the SL_n-case. The variety $\mathcal{R}(\mathfrak{so}_n)$ is given by the equality

$$\mathcal{R}(\mathfrak{so}_n) = (\mathcal{R}(\mathfrak{sl}_n) \cap \mathfrak{so}_n)_{red}.$$

Consider the restrictions $h_i = f_i|_{\mathfrak{so}_n}$ of the forms $f_i \in \mathbb{C}[\mathrm{Mat}_n^0]$, $i = 2, 3, \ldots, n$. Thus, up to a sign, $h_i \in \mathbb{C}[\mathfrak{so}_m]^{\mathrm{SO}_n}$ is the sum of all principal minors of order i of the generic skew-symmetric $(n \times n)$ matrix S. Since every skew-symmetric matrix of an odd size is degenerate, $h_i = 0$ if i is odd. If $n = 2r$, then $h_n = \det S = \mathrm{Pf}^2(S)$, where $\mathrm{Pf} \in \mathbb{C}[\mathfrak{so}_m]^{\mathrm{SO}_n}$ is the Pfaffian of S. The algebra $\mathbb{C}[\mathfrak{so}_m]^{\mathrm{SO}_n}$, and hence the ideal of $\mathcal{R}(\mathfrak{so}_n)$, is generated by $h_2, h_4, \ldots, h_{2r}$ if $n = 2r+1$ and by $h_2, h_4, \ldots, h_{2r-2}, \mathrm{Pf}$ if $n = 2r$. Thus the variety $\mathbb{P}(\mathcal{R}(\mathfrak{so}_n))$ is a self-dual normal complete intersection; see also [PV, HP, Gan]. We have $\dim \mathbb{P}(\mathcal{R}(\mathfrak{so}_{2r+1})) = 2r^2 - 1$, $\deg \mathbb{P}(\mathcal{R}(\mathfrak{so}_{2r+1})) = 2^r r!$, $\dim \mathbb{P}(\mathcal{R}(\mathfrak{so}_{2r})) = 2r^2 - 2r - 1$, $\deg \mathbb{P}(\mathcal{R}(\mathfrak{so}_{2r})) = 2^{2r-1} r!$

The SO_n-orbits in $\mathbb{P}(\mathcal{R}(\mathfrak{so}_n))$ are described as follows; see e.g. [He]. For a partition $\sigma = (\sigma_1, \ldots, \sigma_t)$ of n, the set $\mathcal{R}(\mathfrak{sl}_n) \cap \mathfrak{so}_n$ is nonempty iff every even integer occurs among the σ_i an even number of times. If it is nonempty, denote

$$\mathcal{R}(\mathfrak{so}_n)_\sigma = (\mathcal{R}(\mathfrak{sl}_n)_\sigma \cap \mathfrak{so}_n)_{red}.$$

The variety $\mathcal{R}(\mathfrak{so}_n)_\sigma$ is a single orbit iff at least one of the σ_i is odd. Otherwise, it is a union of two SO_n-orbits (but it is still a single O_n-orbit). The closure relation on the set of SO_n-orbits is determined by the equality

$$\overline{\mathcal{R}(\mathfrak{so}_n)_\sigma} = (\overline{\mathcal{R}(\mathfrak{sl}_n)_\sigma} \cap \mathfrak{so}_n)_{red}.$$

There are nonregular distinguished nilpotent orbits in this case [BCa1, BCa2, CM]. This gives the new self-dual varieties:

Theorem 9.16 ([Pop]) *Let $\sigma = (\sigma_1, \ldots, \sigma_t)$ be a partition of n and let $\tilde{\sigma}$ be the dual partition. Then*

- $\mathbb{P}(\overline{\mathcal{R}(\mathfrak{so}_n)_\sigma})$ *is a self-dual variety iff all integers σ_i are odd and distinct.*
- $\dim \mathbb{P}(\mathcal{R}(\mathfrak{so}_n)_\sigma) = (n^2 - n - \sum_j \tilde{\sigma}_j^2 + \#\{i \mid \sigma_i \text{ odd}\})/2 - 1.$
- *Every self-dual variety $\mathbb{P}(\overline{\mathcal{R}(\mathfrak{so}_n)_\sigma})$ is normal.*

Let $G = \mathrm{Sp}_n$, $n = 2r$ and

$$\mathfrak{g} = \mathfrak{sp}_n = \left\{ \begin{pmatrix} A & B \\ C & -A^\perp \end{pmatrix} \in \mathrm{Mat}_n^0 \mid A, B, C \in \mathrm{Mat}_r,\ B^\perp = B,\ C^\perp = C \right\}.$$

The non-degenerate symmetric form on $\mathfrak{g}$ is defined as in the SL_n-case. The nilpotent cone $\mathcal{R}(\mathfrak{sp}_n)$ is given by the equality

$$\mathcal{R}(\mathfrak{sp}_n) = (\mathcal{R}(\mathfrak{sl}_n) \cap \mathfrak{sp}_n)_{red}.$$

Consider the restrictions $q_i = f_i|_{\mathfrak{sp}_n}$ of the forms $f_i \in \mathbb{C}[\mathrm{Mat}_n^0]$, $i = 2, 3, \ldots, n$. Thus, up to a sign, $q_i \in \mathbb{C}[\mathfrak{sp}_m]^{\mathrm{Sp}_n}$ is the sum of all principal minors of order i of the generic $(n \times n)$ matrix S as above. We have $h_i = 0$ if i is odd. The algebra $\mathbb{C}[\mathfrak{so}_m]^{\mathrm{Sp}_n}$, and hence the ideal of $\mathcal{R}(\mathfrak{sp}_n)$, is generated by $h_2, h_4, \ldots, h_{2r}$. Thus the variety $\mathbb{P}(\mathcal{R}(\mathfrak{sp}_n))$ is a self-dual normal complete intersection; see also [PV, HP, Gan]. We have $\dim \mathbb{P}(\mathcal{R}(\mathfrak{sp}_{2r})) = 2r^2 - 1$ and $\deg \mathbb{P}(\mathcal{R}(\mathfrak{sp}_{2r})) = 2^r r!$

The Sp_n-orbits in $\mathbb{P}(\mathcal{R}(\mathfrak{sp}_n))$ are described as follows; see e.g. [He]. For a partition $\sigma = (\sigma_1, \ldots, \sigma_t)$ of n, the set $\mathcal{R}(\mathfrak{sl}_n) \cap \mathfrak{sp}_n$ is nonempty iff every odd integer occurs among the σ_i an even number of times. If it is nonempty, denote

$$\mathcal{R}(\mathfrak{sp}_n)_\sigma = (\mathcal{R}(\mathfrak{sl}_n)_\sigma \cap \mathfrak{sp}_n)_{red}.$$

The variety $\mathcal{R}(\mathfrak{so}_n)_\sigma$ is always a single Sp_n-orbit.. The closure relation on the set of Sp_n-orbits is determined by the equality

$$\overline{\mathcal{R}(\mathfrak{sp}_n)_\sigma} = (\overline{\mathcal{R}(\mathfrak{sl}_n)_\sigma} \cap \mathfrak{sp}_n)_{red}.$$

There are nonregular distinguished nilpotent orbits in this case, [BCa1, BCa2, CM]. This gives the new self-dual varieties:

Theorem 9.17 ([Pop]) *Let $\sigma = (\sigma_1, \ldots, \sigma_t)$ be a partition of n and let $\tilde{\sigma}$ be the dual partition. Then*

- $\mathbb{P}(\overline{\mathcal{R}(\mathfrak{sp}_n)_\sigma})$ *is a self-dual variety iff all integers σ_i are even and distinct.*
- $\dim \mathbb{P}(\mathcal{R}(\mathfrak{sp}_n)_\sigma) = (n^2 + n - \sum_j \tilde{\sigma}_j^2 - \#\{i \mid \sigma_i \text{ odd}\})/2 - 1.$
- *Every self-dual variety $\mathbb{P}(\overline{\mathcal{R}(\mathfrak{so}_n)_\sigma})$ is normal.*

Case $\mathfrak{g} = G_2$. The nilpotent cone $\mathcal{R}(\mathfrak{g})$ contains precisely 2 distinguished orbits $\mathcal{R}(\mathfrak{g})_i$,

$$i = \dim \mathcal{R}(\mathfrak{g})_i = 12, 10.$$

They are respectively the regular and the subregular orbits. This gives 2 normal self-dual algebraic varieties in $\mathbb{P}^{13}$ of dimensions 11 and 9. The variety $\mathbb{P}(\mathcal{R}(\mathfrak{g})_{12})$ is a complete intersection of degree 12.

Case $\mathfrak{g} = F_4$. The nilpotent cone $\mathcal{R}(\mathfrak{g})$ contains precisely 4 distinguished orbits $\mathcal{R}(\mathfrak{g})_i$,

$$i = \dim \mathcal{R}(\mathfrak{g})_i = 48, 46, 44, 40.$$

The first two of them are respectively the regular and the subregular orbits. This gives 4 normal self-dual algebraic varieties in $\mathbb{P}^{51}$ of dimensions 47, 45, 43, and 39. The variety $\mathbb{P}(\mathcal{R}(\mathfrak{g})_{48})$ is a complete intersection of degree 1152.

Case $\mathfrak{g} = E_6$. The nilpotent cone $\mathcal{R}(\mathfrak{g})$ contains precisely 3 distinguished orbits $\mathcal{R}(\mathfrak{g})_i$,

$$i = \dim \mathcal{R}(\mathfrak{g})_i = 72, 70, 66.$$

The first two of them are respectively the regular and the subregular orbits. This gives 3 normal self-dual algebraic varieties in $\mathbb{P}^{77}$ of dimensions 71, 69, 65. The variety $\mathbb{P}(\mathcal{R}(\mathfrak{g})_{72})$ is a complete intersection of degree 51840.

Case $\mathfrak{g} = E_7$. The nilpotent cone $\mathcal{R}(\mathfrak{g})$ contains precisely 6 distinguished orbits $\mathcal{R}(\mathfrak{g})_i$,

$$i = \dim \mathcal{R}(\mathfrak{g})_i = 126, 124, 122, 120, 116, 112.$$

The first two of them are respectively the regular and the subregular orbits. This gives 6 self-dual algebraic varieties in $\mathbb{P}^{132}$. Normality so far is proved only for $i = 126, 124, 122, 120$. The variety $\mathbb{P}(\mathcal{R}(\mathfrak{g})_{126})$ is a complete intersection of degree 2903040.

Case $\mathfrak{g} = E_8$. The nilpotent cone $\mathcal{R}(\mathfrak{g})$ contains precisely 11 distinguished orbits $\mathcal{R}(\mathfrak{g})_i$,

$$i = \dim \mathcal{R}(\mathfrak{g})_i = 240, 238, 236, 234, 232, 230, 228, 226, 224, 220, 208.$$

The first two of them are respectively the regular and the subregular orbits. This gives 11 self-dual algebraic varieties in $\mathbb{P}^{247}$. Normality so far is proved only for $i = 240, 238, 236, 234, 232$. The variety $\mathbb{P}(\mathcal{R}(\mathfrak{g})_{240})$ is a complete intersection of degree 1625702400.

10

Singularities of Dual Varieties

Here we study how the topology of the variety is reflected in singularities of the dual variety. We start in Sect. 10.1 by proving the class formula and its variant due to Landman that relates the degree of the dual variety and the Euler characteristic of the variety and its hyperplane sections. In the singular case this formula was proved by Ernström, but the Euler characteristic has to be substituted by the degree of the Chern–Mather class. In Sect. 10.2 we prove theorems of Dimca, Nemethi, Aluffi and others that multiplicities of the dual variety are given by Milnor numbers (or classes). To give an example we follow Aluffi and Cukierman and calculate multiplicities of the dual variety to a smooth surface. Finally, we give some results of Weyman and Zelevinsky about singularities of hyperdeterminants.

10.1 Class Formula

For any projective variety X, we denote by $\chi(X)$ its topological Euler characteristic. If X is smooth then $\chi(X)$ is equal to the degree of the top Chern class of the tangent bundle, $\chi(X) = \int_X c_{\dim X}(T_X)$ (Gauss–Bonnet formula).

Suppose that $X \subset \mathbb{P}^N$ is a smooth projective n-dimensional variety with the dual variety $X^* \subset (\mathbb{P}^N)^\vee$. Consider a generic line $L \subset (\mathbb{P}^N)^\vee$. If $\operatorname{def} X = 0$ then $L \cap X^*$ consists of d smooth points, where d is the degree of X^*. If $\operatorname{def} X > 0$ then this intersection is empty and we set $d = 0$.

L can be considered as a pencil (one-dimensional linear system) of divisors on X, and therefore it defines a rational map $F : X \to \mathbb{P}^1$. More precisely, if L is spanned by hyperplanes H_1 and H_2 with equations $f_1 = 0$ and $f_2 = 0$ then F has a form $x \mapsto (f_1(x) : f_2(x))$ for any $x \in X$. Thus F is not defined along the subvariety $X \cap H_1 \cap H_2$ of codimension 2. If we blow it up we get the variety $\tilde{X}$ and the regular morphism $\tilde{F} : \tilde{X} \to \mathbb{P}^1$. Let $D \subset \tilde{X}$ be the exceptional divisor.

Now let us calculate the topological Euler characteristic $\chi(\tilde{X})$ in two ways. First,

$$\chi(\tilde{X}) = \chi(\tilde{X} \setminus D) + \chi(D) = \chi(X \setminus X \cap H_1 \cap H_2) + \chi(D) =$$
$$\chi(X \setminus X \cap H_1 \cap H_2) + 2\chi(X \cap H_1 \cap H_2) = \chi(X) + \chi(X \cap H_1 \cap H_2).$$

Here we use the fact that D is a $\mathbb{P}^1$-bundle over $X \cap H_1 \cap H_2$, and therefore $\chi(D) = 2\chi(X \cap H_1 \cap H_2)$. Notice that the blowing up is an isomorphism on the complement of the exceptional divisor, and hence $\chi(\tilde{X} \setminus D) = \chi(X \setminus X \cap H_1 \cap H_2)$.

On the other hand, we may use $\tilde{F}$ to calculate $\chi(\tilde{X})$. For $x \in \mathbb{P}^1$, we have $\chi(\tilde{F}^{-1}(x)) = \chi(X \cap H_x)$, where H_x is a hyperplane corresponding to x. Let $x_1, \ldots, x_d \in \mathbb{P}^1$ be points that correspond to the intersection of L with X^*. Then, for any $x \in \mathbb{P}^1 \setminus \{x_1, \ldots, x_d\}$, $X \cap H_x$ is a smooth divisor. For any $x \in \{x_1, \ldots, x_d\}$, $X \cap H_x$ has a simple quadratic singularity (see the proof of Theorem 7.1). It is clear then that $\chi(X \cap H_x) = \chi(X \cap H_y) = \chi(X \cap H)$, where $x, y \in \mathbb{P}^1 \setminus \{x_1, \ldots, x_d\}$ and H is a generic hyperplane. For $x \in \{x_1, \ldots, x_d\}$, near the simple quadratic singularity of $X \cap H_x$, the family of divisors $X \cap H_y$, $y \to x$, looks like the family of smooth $(n-1)$-dimensional quadrics Q_ε with equations $T_1^2 + \ldots + T_{n-1}^2 = \varepsilon T_n^2$ near the unique singular point $(0 : \ldots : 0 : 1)$ of the quadric Q_0 with the equation $T_1^2 + \ldots + T_{n-1}^2 = 0$. Therefore

$$\chi(\tilde{X}) = \chi(\mathbb{P}^1)\chi(X \cap H) + d(\chi(Q_0) - \chi(Q_1)) = 2\chi(X \cap H) + d(\chi(Q_0) - \chi(Q_1)).$$

Combining the two formulas for $\chi(\tilde{X})$, we get

$$\chi(X) + \chi(X \cap H_1 \cap H_2) = 2\chi(X \cap H) + d(\chi(Q_0) - \chi(Q_1)).$$

The Euler characteristic of a k-dimensional smooth quadric is equal to $k+2$ for k even and to $k+1$ for k odd. The Euler characteristic of a k-dimensional quadric with a unique singularity is equal to $k+1$ for k even and to $k+2$ for k odd, since it is a cone over a $(k-1)$-dimensional smooth quadric. Finally, we obtain the well-known *class formula* :

Theorem 10.1

$$d = (-1)^n \left[\chi(X) - 2\chi(X \cap H) + \chi(X \cap H \cap H')\right].$$

Example 10.2 ([Kl1, Lam]) The class formula is the main ingredient in the proof of the following formula of Landman (see [Lan, Kl3]):

$$\deg \Delta_X = \big(b_n(X) - b_{n-2}(X)\big) + 2\big(b_{n-1}(X \cap H) - b_{n-1}(X)\big)$$
$$+ \big(b_{n-2}(X \cap H \cap H') - b_{n-2}(X)\big),$$

where b_i is the i-th Betti number. All three summands are nonnegative numbers due to the strong and weak Lefschetz theorems. Hence the characteristic condition for the positive defect is

$$b_n(X) = b_{n-2}(X), \; b_{n-1}(X \cap H) = b_{n-1}(X), \; b_{n-2}(X \cap H \cap H') = b_{n-2}(X).$$

Example 10.3 Suppose that $X \subset \mathbb{P}^N$ is a smooth projective curve of genus g and degree d. Let $H \subset X$ be a hyperplane section. Then X^* is a hypersurface by (1.19), and its degree is given by

$$\deg \Delta_X = -\chi(X) + 2\chi(X \cap H) = \deg K_X + 2 \deg H = 2g - 2 + 2d,$$

where K_X denotes the canonical divisor.

Example 10.4 ([Kl1]) Let $X \subset \mathbb{P}^2$ be a smooth plane curve of degree d. Consider the Veronese embedding $\mathbb{P}^2 \subset \mathbb{P}^N$, where $N + 1 = \binom{r+2}{2}$, and the corresponding embedding $X \subset \mathbb{P}^N$. By the Bézout Theorem, X has degree $d' = rd$ in this embedding. The dual variety $X^* \subset (\mathbb{P}^N)^\vee$ parametrizes plane curves of degree r tangent to X at some point.

$$\deg \Delta_X = 2g - 2 + 2d' = (d-1)(d-2) - 2 + 2rd = d(d + 2r - 3).$$

For example, the variety of conics tangent to a given conic is a hypersurface in $\mathbb{P}^5$ of degree 6.

Example 10.5 Let $X \subset \mathbb{P}^N$ be a smooth projective surface, and let $H \subset X$ be a smooth hyperplane section. Then X^* is a hypersurface by (7.5), and its degree is given by

$$\deg \Delta_X = \chi(X) - 2\chi(H) + H \cdot H = \deg c_2(T_X^\vee) + 2K_X \cdot H + 3 \deg X.$$

Similar formulas exist if $\operatorname{def} X > 0$:

Theorem 10.6 ([Pa2]) *Let $X \subset \mathbb{P}^N$ be a smooth n-dimensional projective variety and let* $\operatorname{def} X = k$. *Then*

$$\deg X^* = (-1)^{n+k} \left[(k+1)\chi(X) - (k+2)\chi(X \cap H) + \chi(X \cap W_{k+2})\right],$$

where H is a generic hyperplane and W_{k+2} is a generic $(k+2)$-codimensional subspace.

The topological Euler characteristic of a smooth variety X can be defined as $\chi(X) = \int_X c(T_X)$. If X is singular then one can substitute $c(T_X)$ by the *Chern–Mather* class $c_M(X)$ defined as follows. Let $\tilde{X}$ be the *Nash blow-up* of X, i.e. the closure in $X \times \mathrm{Gr}(n, \mathbb{P}^N)$ of the graph of the Gauss map. Then $\tilde{X}$ is equipped with the n-dimensional vector bundle S (the pullback of the tautological vector bundle of the Grassmannian) and $c_M(X)$ is defined as $p_* c(S)$, where $p : \tilde{X} \to X$ is the projection. It turns out that the formula in Theorem 10.6 remains valid if we substitute the topological Euler characteristic by $\int_X c_M(X)$. The codimension p part of the Chern–Mather class is equal to [Pi]

$$c_M^p(X) = \sum_{j=0}^{p} (-1)^{p-j} \binom{n+1-p+j}{j} H^j [\Pi^{p-j}],$$

where H is the hyperplane divisor and $\Pi^{n-i} \subset X$ is the codimension $(n-i)$ *polar variety*, which is the closure of the set of smooth points of X at which the embedded tangent space $\hat{T}_{x,X}$ meets a general linear space of codimension $i+2$ not transversally. Ranks δ_i (see Example 6.5) are equal to

$$\delta_i = \int_X H^i[\Pi^{n-i}].$$

In particular, $\delta_{\operatorname{def} X} = \deg X^*$.

Theorem 10.7 ([Er1]) *For $i = 0, \ldots, n$ we have*

$$\delta_i = (-1)^{n-i}\left[\chi(X^i) - 2\chi(X^{i+1}) + \chi(X^{i+2})\right],$$

where X^k is the intersection of X with generic k-codimensional linear subspace, and χ is the topological Euler characteristic if X is smooth and the degree of the Chern–Mather class if X is arbitrary.

10.2 Singularities of X^*

10.2.1 Milnor Numbers

Let f be a holomorphic function defined in some neighbourhood of 0 in $\mathbb{C}^n$. Suppose that 0 is an isolated singular point of the zero set $Z(f)$ of f. Then the vector field

$$(\partial f/\partial z_1, \ldots, \partial f/\partial z_n)$$

has an isolated singularity at 0. The index of this vector field at 0 is called the *Milnor number* of $Z(f)$ (or of f) at 0 (see [Mi]). More generally, if M is a smooth algebraic variety and $Z \subset M$ is a divisor with an isolated singular point $p \in Z$ then the Milnor number $\mu(Z,p)$ is defined as a Milnor number of a local equation of Z at p. It is easy to calculate that if $Z(f)$ has a simple quadratic singularity at 0 then the Milnor number is equal to 1. It follows that the Milnor number of f at 0 is equal to the number of critical points of a small perturbation of f near 0.

Theorem 10.8 ([Di1]) *Let $X \subset \mathbb{P}^N$ be a smooth algebraic variety such that X^* is a hypersurface. Let $H \in X^*$ and suppose that $(X \cap H)_{sing}$ is finite. Then the multiplicity of X^* at H is given by*

$$\operatorname{mult}_H X^* = \sum_{p \in (X\cap H)_{sing}} \mu(X \cap H, p).$$

Proof. Indeed, let U be a small neighbourhood of H in the complex topology, let $H_0 \in U$ be a generic hyperplane, and let $H_0 + \lambda F$ be a pencil of hyperplanes, where F is generic. If H_0 is sufficiently close to H then this pencil intersects $X^* \cap U$ in $\operatorname{mult}_H X^*$ points. It follows that the function (more accurately, the section of $\mathcal{O}_X(1)$) H_0/F has $\operatorname{mult}_H X^*$ critical points near singular points of $X \cap H$. Indeed, critical points of H_0/F and singular points of divisors of the form $H_0 + \lambda F$ are, obviously, identical. □

Next we consider non-isolated singularities. Let M be a smooth projective variety, $\dim M = n$. Let $s \in H^0(M, \mathcal{L})\setminus\{0\}$ be a global section of a line bundle $\mathcal{L}$ over M, and let Z denote the zero set of s. We fix a Hermitian metric on $\mathcal{L}$ and consider the holomorphic component ∂ of the corresponding metric connection. Let Y be a connected component of Z_{sing} and let U be a small neighbourhood of Y with a smooth boundary Γ in the complex topology. We may assume that ∂s does not vanish anywhere on Γ. Then the intersection index of ∂s with the zero section of the cotangent bundle of U tensored by $\mathcal{L}$ is called the Milnor number $\mu(Z, Y)$. If Y is a point p then we get nothing else but the previous definition of the Milnor number. Similarly, we can define $\mu(Z)$ as the intersection index of ∂s and the zero section of $T_M^\vee \otimes \mathcal{L}$ over a small neighbourhood of Z. Then $\mu(Z) = \sum \mu(Z, Y_i)$, where $Y_1, \ldots, Y_r$ are all connected components of Z_{sing}.

Theorem 10.9 ([Pa1])

$$\mu(Z) = (-1)^n(\chi(Z) - \chi(M, \mathcal{L})),$$

where $\chi(M, \mathcal{L})$ is the expected Euler characteristic of the smooth zero set, i.e. of the zero set of a section transversal to the zero section. We have

$$\chi(M, \mathcal{L}) = \int_M c_1(\mathcal{L})c(T_M)/(1 + c_1(\mathcal{L})).$$

Theorem 10.8 has the following generalization.

Theorem 10.10 ([N]) *Let $X \subset \mathbb{P}^N$ be a smooth algebraic variety such that X^* is a hypersurface. Let $H \in X^*$. Then*

$$\operatorname{mult}_H X^* = \mu(X \cap H) + \mu(X \cap H \cap H_g),$$

where H_g is a generic hyperplane.

If $X \cap H$ has isolated singularities, i.e. $(X \cap H)_{sing}$ is 0-dimensional, then $X \cap H \cap H_g$ is smooth and we recover Theorem 10.8.

If $\operatorname{def} X > 0$ then the following analogue of this theorem is still true.

Theorem 10.11 ([Pa2]) *Let $X \subset \mathbb{P}^N$ be a smooth algebraic variety, and let $H \in X^*$. Then*

$$\operatorname{mult}_H X^* = (-1)^{\operatorname{def} X}\mu(X \cap H) + \mu(X \cap H \cap W_g),$$

where W_g is a generic $(\dim X^*)$-dimensional subspace.

10.2.2 Milnor Class

Suppose that M is a smooth n-dimensional algebraic variety, $\mathcal{L}$ is a line bundle on M, and X is the zero-scheme of a section of $\mathcal{L}$. The *singular scheme* Sing X of X is the scheme supported on the singular locus of X and defined locally by the *Jacobian ideal*

$$\left(\frac{\partial F}{x_1}, \ldots, \frac{\partial F}{x_n}\right),$$

where $x_1, \ldots, x_n$ are local parameters for M and F is the section of $\mathcal{L}$ defining X.

Definition 10.12 Let Y be the singular scheme of a section of a line bundle $\mathcal{L}$ on a smooth variety M. The class

$$\mu_{\mathcal{L}}(Y) = c(T^\vee M \otimes \mathcal{L}) \cap s(Y, M)$$

in the Chow group $A_\bullet Y$ of Y is called *the Milnor class* of Y with respect to $\mathcal{L}$.

Here c denotes the total Chern class and s is the Segre class (in the sense of [Ful]). The following theorem summarizes the main properties of the Milnor class. Proofs and further details can be found in [Al].

Theorem 10.13 ([Al])

(i) *$\mu_{\mathcal{L}}(Y)$ only depends on Y and $\mathcal{L}|_Y$: the Milnor class is independent of the choice of a smooth ambient variety M in which Y is realized as the singular scheme of a section of a line bundle restricting to $\mathcal{L}|_Y$.*

(ii) *Suppose that $\mathcal{L}$ is generated by global sections and let X_g be a generic section of $\mathcal{L}$. Then*

$$\mu_{\mathcal{L}}(Y \cap X_g) = c_1(\mathcal{L}) \cap \mu_{\mathcal{L}}(Y).$$

(iii) *Suppose that Y is non-singular. Then*

$$\mu_{\mathcal{L}}(Y) = c(T^\vee Y \otimes \mathcal{L}) \cap [Y].$$

Example 10.14 If Y is supported on a point P, then $\mu_{\mathcal{L}}(Y) = m[P]$, where m is the Milnor number of X at P. Indeed, in this case $\mu_{\mathcal{L}}(Y) = s(Y, M)$, so m is the coefficient of $[P]$ in $s(Y, M)$, which agrees with the definition of the Milnor number (see [Ful]).

Example 10.15 In general, let $|Y|$ be the support variety of Y, and assume that Y is proper. We claim that $\mu(|Y|)$ defined before Theorem 10.9 is equal to the degree of the 0-dimensional component of $\mu_{\mathcal{L}}(Y)$. Namely, the differential of the local equations of X in M determines a section of $(T^\vee M \otimes \mathcal{L})|_X$. This can be extended to a holomorphic section s_X of $T^\vee M \otimes \mathcal{L}$ over the whole of M. Then $\mu(|Y|)$ is the contribution of $|Y|$ to the intersection of s_X with the zero section of $T^\vee M \otimes \mathcal{L}$. In the neighborhood of $|Y|$ we have the fiber diagram

$$\begin{array}{ccc} Y & \longrightarrow & M \\ \big\downarrow & & \big\downarrow {\scriptstyle s_X} \\ M & \xrightarrow{s_0} & T^*M \otimes \mathcal{L} \end{array}$$

that is, Y is the scheme-theoretic intersection of the two sections, so the contribution of $|Y|$ to the intersection number of the sections is equal (see [Fu1]) to the degree of the 0-dimensional component of

$$c(T^\vee M \otimes \mathcal{L}) \cap s(Y, M),$$

which is the claim. In fact, in view of Theorem 10.13, the numerical information carried by the Milnor class of Y (when Y is proper) is equivalent to μ-numbers of $|Y|$ and its generic sections $|Y \cap X_g|$, $|X \cap X_{g_1} \cap X_{g_2}|$, etc.

Now let us apply this machinery in our usual situation. Suppose that $X \subset \mathbb{P}^N$ is a smooth projective variety, $\mathcal{L} = \mathcal{O}_X(1)$, and $X^* \subset (\mathbb{P}^N)^\vee$ is the dual variety parametrizing hyperplanes H such that the singular scheme (or the contact scheme in this case) $\operatorname{Sing} X \cap H$ is not empty. Then we have the Milnor class $\mu_{\mathcal{L}}(\operatorname{Sing} X \cap H) \in A_\bullet(\operatorname{Sing} X \cap H)$. The problem that we address here is how to calculate the multiplicity of X^* at H in terms of the Milnor class.

Theorem 10.16 ([Al, AC]) *Let $X \cap H$ be any singular hyperplane section of X. Then*

(i) *The defect of the dual variety X^* is the smallest integer $r \geq 0$ such that*

$$\int c_1(\mathcal{L})^r c(\mathcal{L}) \cap \mu_{\mathcal{L}}(\operatorname{Sing} X \cap H) \neq 0;$$

and for $r = \operatorname{def} X$ this number is equal to the multiplicity $m_H X^*$.

(ii) *If $\operatorname{def} X = r$ then $m_H X^*$ is equal to*

$$\int (-1)^r \mu_{\mathcal{L}}(\operatorname{Sing} X \cap H) + c_1(\mathcal{L})^{r+1} \cap \mu_{\mathcal{L}}(\operatorname{Sing} X \cap H).$$

Proof. (i) We argue by induction on r. If $r = 0$ then the integral is equal to

$$\begin{aligned} \int c(\mathcal{L}) \cap \mu_{\mathcal{L}}(\operatorname{Sing} X \cap H) &= \int c(\mathcal{L}) c(T^\vee X \otimes \mathcal{L}) \cap s(\operatorname{Sing} X \cap H, X) \\ &= \int c((T^\vee X \oplus \mathcal{O}) \otimes \mathcal{L}) \cap s(\operatorname{Sing} X \cap H, X). \end{aligned}$$

Thinking of H as a point of X^*, we have $s(H, X^*) = (m_H X^*) H$, and thus

$$m_H X^* = \deg s(H, X^*).$$

The dual variety is the projection from the conormal variety I_X, $X^* = \pi(I_X)$. Since $r = 0$, the degree of this projection is equal to 1. The fiber Y of π over H is identified with the contact locus $\mathrm{Sing}\, H \cap X$ (as a scheme). Therefore we have $s(H, X^*) = \pi_* s(Y, I_X)$. Taking degrees, we get

$$m_H X^* = \deg s(Y, I_X).$$

Now using the second projection $I_X \to X$, which is the projective bundle, it is quite easy to see that

$$s(Y, I_X) = c(J(\mathcal{L})) \cap s(\mathrm{Sing}\, X \cap H, X),$$

where $J(\mathcal{L})$ is the jet bundle of $\mathcal{L}$. It remains to notice that $c(J(\mathcal{L})) = c((T^\vee X \oplus \mathcal{O}) \otimes \mathcal{L})$ by the exact sequence (5.1).

Now let $r > 0$. By a similar argument to that above, we have

$$\int c(\mathcal{L}) \cap \mu_{\mathcal{L}}(\mathrm{Sing}\, X \cap H) = 0.$$

By Theorem 5.3, if X^* is not a hypersurface then the dual variety X'^* of a generic hyperplane section $X' = X \cap H_g$ is the cone over X^* with H_g as a vertex. The multiplicity of X^* at H then evidently equals the multiplicity of X'^* at H; but $\mathrm{def}\, X' = \mathrm{def}\, X - 1$, and so the statement follows by induction because

$$\begin{aligned} &\int c_1(\mathcal{L})^r c(\mathcal{L}) \cap \mu_{\mathcal{L}}(\mathrm{Sing}\, X \cap H) \\ = &\int c_1(\mathcal{L})^{r-1} c(\mathcal{L}) \cap (c_1(\mathcal{L}) \cap \mu_{\mathcal{L}}(\mathrm{Sing}\, X \cap H)) \\ = &\int c_1(\mathcal{L})^{r-1} c(\mathcal{L}) \cap \mu_{\mathcal{L}}(\mathrm{Sing}\, X' \cap H), \end{aligned}$$

where we use Theorem 10.13 in the last equality.

(ii) By (i) we see that $\mathrm{def}\, X \geq r$ if and only if the components of dimension i, $0 \leq i < r$, of $c(\mathcal{L}) \cap \mu_{\mathcal{L}}(\mathrm{Sing}\, X \cap H)$ vanish for all hyperplane sections $X \cap H$. That is, if $\mathrm{def}\, X \geq r$, then for all hyperplane sections $X \cap H$

$$c(\mathcal{L}) \cap \mu_{\mathcal{L}}(\mathrm{Sing}\, X \cap H) = A_{\dim \mathrm{Sing}\, X \cap H} + \ldots + A_r$$

with A_j a class in dimension j, $j = r, \ldots, \dim \mathrm{Sing}\, X \cap H$ (depending on H). Therefore

$$\mu_{\mathcal{L}}(\mathrm{Sing}\, X \cap H) = c(\mathcal{L})^{-1} \cap (A_{\dim \mathrm{Sing}\, X \cap H} + \ldots + A_r)$$

(with $A_r \neq 0$), and

$$\int (-1)^r \mu_{\mathcal{L}}(\mathrm{Sing}\, X \cap H) = \sum_{i=r}^{\dim \mathrm{Sing}\, X \cap H} (-1)^{i-r} c_1(\mathcal{L})^i \cap A_i$$

while

$$\int c_1(\mathcal{L})^{r+1}\mu_{\mathcal{L}}(\operatorname{Sing} X\cap H)=\sum_{i=r+1}^{\dim \operatorname{Sing} X\cap H}(-1)^{i-r+1}c_1(\mathcal{L})^i\cap A_i.$$

Hence

$$\int(-1)^r\mu_{\mathcal{L}}(\operatorname{Sing} X\cap H)+c(\mathcal{L})^{r+1}\cap\mu_{\mathcal{L}}(\operatorname{Sing} X\cap H)=\int c_1(\mathcal{L})^r\cap A_r,$$

which coincides with the formula in (i). □

Remark 10.17 An easy Chern class computation shows that ranks (see Section 6.1) could be expressed via Milnor classes as follows:

$$\delta_i=\int_X c_1(\mathcal{L})^i c(\mathcal{L})\cap\mu_{\mathcal{L}}(X).$$

Moreover, Theorem 6.2 can be viewed as a particular case of Theorem 10.16 applied to the zero-section of $H^0(X,\mathcal{L})$.

Theorem 7.4 shows that if $\operatorname{def} X>0$ then the defect and the dimension have the same parity. Since the defect equals the dimension of a contact locus with a generic tangent hyperplane, this can be reformulated by saying that if the generic contact locus is positive-dimensional then it has the same parity as $\dim X$. It is possible to generalize this result for not necessarily generic contact loci using the machinery of Milnor classes. The proof of the following theorem can be found in [Al].

Theorem 10.18 ([Al]) *Let $X\subset\mathbb{P}^N$ be a smooth projective variety. Let $H\in X^*$ be such that the contact locus $Y=\operatorname{Sing} X\cap H$ is pure-dimensional and non-singular in a neighborhood of some complete curve $C\subset Y$ of odd degree. Then*

$$\dim Y\equiv\dim X \mod 2.$$

For example, if Y is a generic contact locus then $Y=\mathbb{P}^{\operatorname{def} X}$. Therefore, if Y is not a point then it contains a line C and we get Theorem 7.4.

The following result is an immediate corollary of Theorem 10.16:

Theorem 10.19 ([Al]) *Let $X\subset\mathbb{P}^N$ be a smooth projective variety. Suppose that the contact scheme* $\operatorname{Sing} X\cap H$ *of a hyperplane H with X is a projective space $\mathbb{P}^r$. Then* $\operatorname{def} X$ *is equal to r and X^* is smooth at H.*

Recall that by a contact scheme we mean the contact locus endowed with the scheme structure given by the Jacobian ideal. The surprising feature of this result is that we are not assuming a hyperplane to be generic. The dimension of the contact scheme of a hyperplane is not necessarily equal to the dimension of the contact locus of a generic hyperplane. The claim, however, is that it is so if the contact scheme is a projective subspace as a scheme. Moreover, it can be shown (also using Theorem 10.16) that if the contact scheme of a variety with a given hyperplane is smooth then the defect of the variety is equal to the defect of the contact scheme. This 'explains' the previous theorem, because the defect of a projective subspace is equal to its dimension.

If $\operatorname{def} X = 0$, Theorem 10.19 can be reformulated a little more explicitly. Let $X \subset \mathbb{P}^n$ be a smooth variety such that X^* is a hypersurface. Let $I_X \subset \mathbb{P}^n \times (\mathbb{P}^n)^\vee$ be a conormal variety. We denote by π the projection $I_X \to X^*$. Let $U \subset X^*$ be the set of points where π is unramified. In other words, $H \in U$ iff $\operatorname{Sing} X \cap H$ is a single point p as a scheme (the Hessian of a function defining $X \cap H$ is non-zero at p). Clearly, U is non-empty and open in X^*.

Theorem 10.20 ([Ka]) *Suppose that $X \subset \mathbb{P}^N$ is smooth and X^* is a hypersurface. Then U consists of smooth points of X^* and is the biggest open set in X^* for which the projection $\pi : \pi^{-1}(U) \to U$ is an isomorphism.*

In fact, one may expect that $U = X_{sm}$. For example, this is so if $X^* \setminus U$ has codimension 1.

Proposition 10.21 *Suppose that $X \subset \mathbb{P}^N$ is smooth and X^* is a hypersurface. If $Z = X^* \setminus U$ is a divisor then $Z = (X^*)_{sing}$.*

Proof. By Theorem 10.20 ${X^*}_{sing} \subset Z$. We need to prove an opposite inclusion. Let $Y \subset X^*$ be the variety of all points $x \in X^*$ such that $\dim \pi^{-1}(x) > 0$. Then $Y \cap U = \emptyset$, and therefore $\operatorname{codim}_{X^*} Y \geq 2$. If $Z_0 \subset Z$ is an irreducible component then for generic point $H \in Z_0$ the contact locus $\operatorname{Sing} H \cap X$ consists of finitely many points. Therefore, $H \in (X^*)_{sing}$, e.g. by Theorem 10.8. □

10.2.3 Dual Variety of a Surface

Let X be a smooth projective algebraic surface with a very ample line bundle $\mathcal{L}$. By 7.5, $X^* \subset |\mathcal{L}|$ is a hypersurface. We want to find $m_D X^*$ for each $D \in |\mathcal{L}|$. Suppose that

$$D = \sum_{i=1}^{r} n_i D_i,$$

where any curve D_i is reduced and irreducible. Let $D_{red} = \sum D_i$.

Theorem 10.22 ([AC]) *With the above notation, we have*

$$m_D X^* = (D - D_{red}) \cdot (K_X + 2D + D_{red}) + \sum_{p \in D_{red}} \mu(D_{red}, p),$$

where K_X is the canonical divisor of X and μ is the ordinary Milnor number.

Proof. Take a general $H \in |\mathcal{L}|$ and denote by $L \subset |\mathcal{L}|$ the pencil containing D and H. Then

$$m_D X^* = \deg X^* - s, \tag{10.1}$$

where s is the number of singular members of L different from D (each of these singular members has one node as a singular locus). In order to determine s we shall blow up X to construct a family parametrized by L and use Lefschetz's formula [GH1].

For each i $(1 \le i \le r)$ denote by p_{ij} $(1 \le j \le H \cdot D_i)$ the points of intersection of H and D_i. Let $\hat{X}$ denote the surface obtained from X by blowing up each point p_{ij} n_i times (in the direction of H). The induced pencil $\hat{L}$ on $\hat{X}$ is base-point-free and gives a map $f : \hat{X} \to \mathbb{P}^1$. If E_{ij}^k, $k = 1, \dots, n_i$, are the exceptional divisors at p_{ij} then the fiber of f at the point 0 (corresponding to D) is

$$f^*(0) = D + \sum_{i,j} \sum_{1 \le k \le n_i - 1} (n_i - k) E_{ij}^k.$$

In other words, the special fiber is isomorphic to D with strings of $\mathbb{P}^1$'s (each $\mathbb{P}^1$ with a certain multiplicity) attached at the points p_{ij}; each string has $n_i - 1$ components.

We now denote

$$D' = (f^*(0))_{red} = D_{red} + \sum_{i,j} T_{ij}$$

the reduced fiber of f at 0, where $T_{ij} = \sum_{1 \le k \le n_i - 1} E_{ij}^k$ is the reduced string attached at p_{ij}. Then the standard argument involving Lefschetz's formula (see [GH1] or the beginning of this chapter) gives

$$\chi(X) = 2\chi(H) - H \cdot H + (\chi(D') - \chi(H)) + \sum_{1 \le \lambda \le s} (\chi(H_\lambda) - \chi(H)),$$

where χ denotes the topological Euler characteristic and H_λ are the singular fibers of f for $\lambda \ne 0$. Since $\chi(H_\lambda) - \chi(H) = 1$ (see (10.4) below) and $\deg X^* = \chi(X) - 2\chi(H) + H \cdot H$ (Example 10.5), we can rewrite (10.1) as

$$m_D X^* = \chi(D') - \chi(H). \tag{10.2}$$

In order to compute $\chi(D')$, denoting $T = \cap_{i,j} T_{ij}$, we have

$$\chi(D') = \chi(D_{red} \cap T) = \chi(D_{red}) + \chi(T) - \chi(D_{red} \cap T) = \tag{10.3}$$

$$\chi(D_{red}) + \sum_{i,j} \chi(T_{ij}) - \sum_{i,j} \chi(\{p_{ij}\}) =$$

$$\chi(D_{red}) + \sum_{i,j} n_i - \sum_{i,j} 1 = \chi(D_{red}) + D \cdot (D - D_{red}).$$

Now we compute $\chi(D_{red})$. Let $Z = \sum_{1\leq i\leq r} Z_i$ be a reduced connected curve with normalization

$$\rho : \tilde{Z} = \bigsqcup_{1\leq i\leq r} \tilde{Z}_i \to Z.$$

If $p \in Z$ is a singular point, let $B(p) = \rho^{-1}(p)$ denote the set of branches at p. Topologically, Z is obtained from the smooth real surface $\tilde{Z}$ by identifying each of the sets $B(p)$ with a point p. Therefore we have

$$\chi(\tilde{Z}) - \chi(Z) = \sum_{p\in Z} (b(Z,p) - 1), \tag{10.4}$$

where $b(Z,p)$ is the number of branches of Z at p. Also, from the exact sequence of sheaves

$$0 \to \mathcal{O}_Z \to \rho_*\mathcal{O}_{\tilde{Z}} \to \rho_*\mathcal{O}_{\tilde{Z}}/\mathcal{O}_Z \to 0$$

we obtain

$$1 - r + \sum_{p\in Z} \delta(Z,p) - h^1(\mathcal{O}_Z) + h^1(\mathcal{O}_{\tilde{Z}}) = 0,$$

where we set $\delta(Z,p) = \mathrm{length}_p(\rho_*\mathcal{O}_{\tilde{Z}}/\mathcal{O}_Z)$. Combining this with (10.4) we obtain

$$\chi(Z) = 2 - 2p_a(Z) + \sum_{p\in Z} \mu(Z,p), \tag{10.5}$$

where

$$\mu(Z,p) = 2\delta(Z,p) - b(Z,p) + 1$$

is the Milnor number of (Z,p). Now writing together (10.2), (10.3), and (10.5) we get

$$m_D X^* = \chi(D_{red}) - \chi(H) + D \times (D - D_{red}) =$$
$$(2 - 2p_a(D_{red})) - (2 - 2p_a(H) + D\cdot(D - D_{red}) + \sum_{p\in D_{red}} \mu(D_{red},p) =$$
$$-(K_X + D_{red})\cdot D_{red} + (K_X + H)\cdot H + D\cdot(D - D_{red}) + \sum_{p\in D_{red}} \mu(D_{red},p).$$

After rearranging we finally obtain the claim of the theorem. □

Example 10.23 Suppose that $X = \mathbb{P}^2$ and $\mathcal{L} = \mathcal{O}_X(d)$. X^* parametrizes plane singular curves of degree d. If C is such curve then applying Theorem 10.22 we get

$$m_C X^* = (3(d-1) - d')\,d' + \mu,$$

where d' is the degree of $C - C_{red}$ and μ is the sum of the Milnor numbers of the singularities of C_{red}. For example, the multiplicity of the discriminant of plane conics at a double line is 2.

10.2.4 Singularities of Hyperdeterminants

The description of the singular locus of the dual variety X^* is known only in some special cases. For hyperdeterminants, this problem was studied in [WZ]. To formulate this result, we need the following notations. Suppose first that $X \subset \mathbb{P}^N$ is a smooth variety. We define two subvarieties X^*_{node} and X^*_{cusp} in X^*. First, X^*_{node} is the closure of the set of hyperplanes H such that H is tangent to X at two distinct points. In other words, $X^*_{node} = \overline{\mathrm{pr}_1(I^2_X)}$, where

$$I^2_X = \{(H, x, y) \in X^* \times X \times X \mid x \neq y,\ \hat{T}_x X, \hat{T}_y X \in H\}.$$

Second, X^*_{cusp} consists of all hyperplanes H such that there exists a point $x \in (X \cap H)_{sing}$ that is not a simple quadratic singularity. That is, the Hessian of the function f defining $X \cap H$ is equal to zero at x.

For hyperdeterminants, X^*_{node} can be decomposed further. Namely, for any subset $J \subset \{1, \ldots, r\}$ we set $X^*_{node}(J) = \overline{\mathrm{pr}_1(I^2_X(J))}$, where

$$I^2_X(J) = \{(H, x, y) \in I^2_X \mid x^{(j)} = y^{(j)} \text{ if and only if } j \in J\}.$$

Theorem 10.24 ([WZ]) *Suppose that $k_1 \leq k_2 + \ldots + k_r$ and $k_1 \geq \ldots \geq k_r$.*

(i) *If the format is boundary then X^*_{sing} is an irreducible hypersurface in X^*. Furthermore, $X^*_{sing} = X^*_{node}(\emptyset)$ if the format is different from $4 \times 2 \times 2$. In this exceptional case $X^*_{sing} = X^*_{node}(1)$.*

(ii) *If the format is interior and does not belong to the following list of exceptions, then X^*_{sing} has two irreducible components X^*_{cusp} and $X^*_{node}(\emptyset)$, both having codimension 1 in X^*. The list of exceptional cases consists of the following 3- and 4-dimensional formats:*
   - *For $2 \times 2 \times 2$ matrices, $X^*_{sing} = X^*_{cusp}$ and has three irreducible components, all of codimension 2 in X^*.*
   - *For $3 \times 2 \times 2$ matrices, $X^*_{sing} = X^*_{cusp}$ and is irreducible.*
   - *For $3 \times 3 \times 3$ matrices, X^*_{sing} has five irreducible components X^*_{cusp}, $X^*_{node}(\emptyset)$, $X^*_{node}(1)$, $X^*_{node}(2)$, and $X^*_{node}(3)$.*
   - *For $m \times m \times 3$ matrices with $m > 3$, X^*_{sing} has three irreducible components X^*_{cusp}, $X^*_{node}(\emptyset)$, and $X^*_{node}(3)$.*
   - *For $2 \times 2 \times 2 \times 2$ matrices, X^*_{sing} has eight irreducible components X^*_{cusp}, $X^*_{node}(\emptyset)$, and $X^*_{node}(i, j)$ for $1 \leq i < j \leq 4$.*
   - *For $m \times m \times 2 \times 2$ matrices with $m > 2$, X^*_{sing} has three irreducible components X^*_{cusp}, $X^*_{node}(\emptyset)$, and $X^*_{node}(3, 4)$.*

   In all cases all irreducible components of X^*_{sing} have codimension 1 unless written otherwise.

References

[AH] J. Alexander, A. Hirschowitz, *Polynomial interpolation in several variables*, J. of Alg. Geom., **4** (1995), 201–222

[Al] P. Aluffi, *Singular schemes of hypersurfaces*, Duke Math. J., **80** (1995), 325–351

[AC] P. Aluffi, F. Cukierman, *Multiplicities of discriminants*, Manuscripta Math., **78** (1993), 245–258

[An] A.Z. Anan'in, *Quasiderivations of the algebra of diagonal matrices*, Tret'ya Sibirskaya shkola po algebre i analizu, Irkutsk University, 1990, 8–15 (in Russian)

[AVE] E.M. Andreev, E.B. Vinberg, A.G. Elashvili, *Orbits of greatest dimension in semisimple linear Lie groups*, Functional Anal. Appl., **1** (1967), 257–261

[ACGH] E. Arbarello, M. Cornalba, P.A. Griffiths, J. Harris, *Geometry of Algebraic Curves, Vol. 1*, Springer–Verlag, New York, 1985

[Ar] V.I. Arnold, *Mathematical Methods of Classical Mechanics*, Springer–Verlag, New York–Heidelberg–Berlin, 1978

[At] M.D. Atkinson, *Primitive spaces of matrices of bounded rank, II*, Jour. Australian Math. Soc., **34** (1983), 306–315

[AW] M.D. Atkinson, R. Westwik, *Spaces of linear transformations of equal rank*, Linear and Multilinear Algebra, **13** (1983), 231–239

[BCa1] P. Bala, R.W. Carter, *Classes of unipotent elements in simple algebras. I*, Math. Proc. Camb. Phil. Soc., **79** (1976), 401–425

[BCa2] P. Bala, R.W. Carter, *Classes of unipotent elements in simple algebras. II*, Math. Proc. Camb. Phil. Soc., **80** (1976), 1–18

[BC] E. Ballico, L. Chiantini, *On smooth subcanonical varieties of codimension 2 in $\mathbb{P}^n$, $n \geq 4$*, Ann. Mat. Pure Appl., **135:4** (1983), 99–117

[Bar] F. Bardelli, *The Sylvester pentahedron and some applications to moduli spaces*, preprint

[Ba] W. Barth, *Transplanting cohomology class in complex projective space*, Amer. J. Math., **92** (1970), 951–961

[BH] W. Barth, K. Hulek, *Monads and moduli of vector bundles*, Man. Math., **25** (1978), 323–347

[Be] L.B. Beasley, *Null spaces of matrices of bounded rank*, Current Trends in Matrix Theory (R. Grone and F. Uhlig, Eds.), North–Holland, Amsterdam, 1987, 45–50

[Bea] A. Beauville, *Fano contact manifolds and nilpotent orbits*, Comm. Math. Helv., **73:4** (1998), 566–583

[Bek] N.D. Beklemishev, *Classification of quaternary cubic forms not in general position*, In: Voprosy teorii grupp i gomologicheskoy algebry, Yaroslavl, 1981 (in Russian)

[BFS1] M.C. Beltrametti, M.L. Fania, A.J. Sommese, *On the discriminant variety of a projective manifold*, Forum Math, **4** (1992), 529–547

[BFS2] M.C. Beltrametti, M.L. Fania, A.J. Sommese, *On the adjunction theoretic classification of projective varieties*, Math. Ann., **290** (1991), 31–62

[BS] M.C. Beltrametti, A.J. Sommese, *The adjunction theory of complex projective varieties*, Expositions in Math., **16** (1995), W. de Gruyter, Berlin

[BSW] M.C. Beltrametti, A.J. Sommese, J.A. Wiśniewski, *Results on varieties with many lines and their applications to adjunction theory*, Proceedings of the Bayreuth Conference "Complex Algebraic Varieties", April 1990. Lecture Notes in Math. 1507, Springer, 1992, 16–38

[Berg] G.M. Bergman, *The diamond lemma for ring theory*, Adv. in Math., **29/2** (1978), 178–218

[BGG] I.N. Bernstein, I.M. Gelfand, S.I. Gelfand, *Schubert cells and the cohomology of* G/P, Uspekhi Matematicheskih Nauk, **28/3** (1973), pp. 3–26

[Ber] E. Bertini, *Introduzione alla geometria proiettriva degli iperspazi*, Enrico Spaerri, Piza, 1907

[Bot] R. Bott, *Homogeneous vector bundles*, Ann. Math., **66/2** (1957), pp. 203–248

[Bo] N. Bourbaki, *Groupes et Algébres de Lie*, Hermann, Paris 1968

[Br] M. Brion, *Représentations exceptionnelles des groupes semi-simples*, Ann. Sci. Ec. Norm. Sup., **18** (1985), 345–387

[Br1] A. Broer, *Line bundles on the cotangent bundle of the flag variety*, Invent. Math., **113** (1993), 1–20

[Br2] A. Broer, *Normality of some nilpotent varieties and cohomology of line bundles on the cotangent bundle of flag variety*, in "Lie Theory and Geometry. In Honor of Bertram Kostant". Progress in Math. Birkhäuser, Boston, **123** (1994), 1–19

[Ca] A. Cayley, *On the theory of elimination*, Collected papers Vol. 1, Cambridge Univ. Press, 1889, pp. 370–374

[Ca1] A. Cayley, *On the theory of linear transformations*, Collected papers Vol. 1, Cambridge Univ. Press, 1889, pp. 80–94

[Ca2] A. Cayley, *On linear transformations*, Collected papers Vol. 1, Cambridge Univ. Press, 1889, pp. 95–112

[CR] T. Cecil, P. Ryan, *Focal sets and real hypersurfaces in complex projective space*, Trans. Amer. Math. Soc., **269** (1982)

[C] A. Clebsch, *Über Curven vierter Ordnung*, J. Reine Angew. Math., **59** (1861), pp. 125–145

[CKM] H. Clemens, J. Kollar, S. Mori, *Higher dimensional complex geometry*, Asterisque, **166** (1988)

[CM] D. Collingwood, W. McGovern, *Nilpotent Orbits in Semisimple Lie Algebras*, Van Nostrand Reinhold Company, 1993

[CW] C. De Concini, J. Weiman, *A formula with nonnegative terms for the degree of the dual variety of a homogeneous space*, Proc. AMS, **125:1** (1997), 1–8

[De] L. Degoli, *Due Nouvi Teoremi Sui Sistemi Lineari Di Quadriche A Jacobiana Identicamente Nulla*, Collect. Math., **35** (1984), 125–139

[Del] P. Deligne, *Le déterminant de la cohomologie*, In: Current trends in arithmetic algebraic geometry, Contemporary Math., No. 67,Amer. Math. Soc., 1987, pp. 93–177

[DK] P. Deligne, N.Katz, *Groupes de monodromie en geometrie algébrique*, Lecture Notes in Math. 340, Springer, 1973

[Di1] A. Dimca, *Milnor numbers and multiplicities of dual varieties*, Rev. Roumaine Math. Pures (1986), 535–538

[Di2] A. Dimca, *Topics on real and complex singularities*, Vieweg Advanced Lectures (1987)

[Dix] A. Dixon, *The canonical forms of the ternary sextic and the quaternary quartic*, Proc. London Math. Soc., **4** (1906), 160–168

[D] R. Donagi, *On the geometry of Grassmannians*, Duke Math. J., **44** (1977), 795–837

[E1] L. Ein, *Varieties with small dual varieties, I*, Invent. Math., **86** (1986), 63–74

[E2] L. Ein, *Varieties with small dual varieties, II*, Duke. Math. J., **52** (1985), 895–907

[E3] L. Ein, *Stable vector bundles on projective space in char $p > 0$*, Math. Ann., **254** (1980), 53–72

[E4] L. Ein, *The ramification divisors for branched coverings of* $\mathbb{P}^n_k$, Math. Ann., **261** (1982), 483–485

[ES] L. Ein, N. Shepherd–Barron, *Some special Cremona transformations*, Amer. J. Math., **111** (1989), 783–800

[EH] D. Eisenbud, J. Harris, *Vector spaces of matrices of low rank*, Adv. Math., **70** (1988), 135–155

[El1] A.G. Elashvili, *Canonical form and stationary subalgebras of points of general position for simple linear Lie groups*, Functional Anal. Appl., **6** (1972), 44–53

[El2] A.G. Elashvili, *Stationary subalgebras of points of general position for irreducible linear Lie groups*, Functional Anal. Appl., **6** (1972), 139–148

[EHS] G. Elencwajg, A. Hirschowitz, M. Schneider, *Les fibres uniformes de rang au plus n sur* $\mathbb{P}_n(\mathbb{C})$, Proceedings of the Nice Conference 1979 on Vector bundles and Differential equations

[Er1] L. Ernström, *A Plücker formula for singular projective varieties*, Comm. Algebra, **25(9)** (1997), 2897–2901

[Er2] L. Ernström, *Topological Radon transforms and the local Euler obstruction*, Duke Math. J., **76(1)** (1994), 1–21

[FaL1] M.L. Fania, E.L. Livorni, *Degree nine manifolds of dimension* ≥ 3, Math. Nachr., **169** (1994), 117–134

[FaL2] M.L. Fania, E.L. Livorni, *Degree ten manifolds of dimension* ≥ 3, preprint (1992)

[Fa] B. Fantechi, *On the superadditivity of secant defects*, Bull. Soc. Math. France, **118** (1990), 85–100

[Fi] E. Fisher, *Über die Cayleyshe Eliminationsmethode*, Math. Zeit., **26** (1927), 497–550

[FF] L.R. Ford, D.R. Fulkerson, *Flows in Networks*, Princeton Univ. Press, Princeton, NJ, 1962

[Fra] W. Franz, *Die torsion einer Überdeckung*, J. Reine u. angew. Math., **173** (1935), 245–254

[Fub] G. Fubini, *Il problema della deformazione proiettiva delle ipersuperficie*, Rend. Acad. Naz. dei Lincei, **27** (1918), 147–155

[F] T. Fujita, *Projective threefolds with small secant varieties*, Sci. Papers College of General Education. Univ. Tokio, **32:1** (1982), 33–46

[FR] T. Fujita, J. Roberts, *Varieties with small secant varieties: the extremal case*, Amer. J. Math., **103** (1981), 953–976

[Fu1] W. Fulton, *Intersection Theory*, Ergebnisse der Mathematik 2, Springer, 1984

[Fu2] W. Fulton, *On the topology of algebraic varieties*, Proc. Symp. Pure Math., **46** (1987), 15–46

[FH] W. Fulton, J. Hansen, *A connectedness theorem for projective varieties, with applications*, Ann. Math., **110:1** (1979), 159–166

[FL] W. Fulton, R. Lazarsfeld, *Connectivity and its applications in algebraic geometry*, Proc. Midwest Algebraic Geometry Conf. Univ. Illinois, 1980. Lect. Notes Math. Springer–Verlag, Berlin–Heidelberg–New York, **862** (1981), 26–92

[FL1] W. Fulton, R. Lazarsfeld, *On the connectedness of degeneracy loci and special divisors*, Acta Math., **146** (1981), 271–283

[Ga] P. Gabriel, *Unzerlegbare Darstellungen, I*, Manuscripta Math., **6** (1972), 71–103

[Gan] F.R. Gantmacher, *The Theory of Matrices Vol.2*, Chelsea Publishing Company, New York, 1959

[GK] I.M. Gelfand, M.M. Kapranov, *On the dimension and degree of projectively dual varieties: a q-analogue of the Katz–Kleiman formula*, in: Gelfand Mathematical Seminars 1990–1992, Birkhaüser 1993, pp. 27–34

[GKZ1] I.M. Gelfand, M.M. Kapranov, A.V. Zelevinski, *A-discriminants and the Cayley–Koszul complexes*, Dokl. Akad. Nauk SSSR, **307** (1989), 1307–1311 (Russian)

[GKZ2] I.M. Gelfand, M.M. Kapranov, A.V. Zelevinski, *Discriminants, Resultants, and Multidimensional Determinants*, Birkhauser, Boston, 1994

[GKZ3] I.M. Gelfand, M.M. Kapranov, A.V. Zelevinski, *Hyperdeterminants*, Adv. Math., **84** (1990), pp. 237–254

[Go] N. Goldstein, *Ampleness and connectedness in complex G/P*, Trans. AMS, **274** (1982), 361–374

[GR] H. Grauert, O. Rimenschneider, *Verschwindungssätze für Annalische Kogomologiegruppen auf Komplexen Räumen*, Inv. Math., **11** (1970), 263–292

[GH] P. Griffiths, J. Harris, *Algebraic geometry and local differential geometry*, Ann. Sci. Ec. Norm. Super., **12** (1979), 355–432

[GH1] P. Griffiths, J. Harris, *Principles of Algebraic Geometry*, Wiley–Interscience, London–New York, 1979

[HT] J. Harris, L.W. Tu, *The connectedness of symmetric degeneracy loci: odd ranks, Topics in Algebra: Banach Center Publications, Vol. 26, Part 2,* PWN–Polish Scientific Publishers, Warsaw, 1990, pp. 249–256

[HT1] J. Harris, L.W. Tu, *On symmetric and skew-symmetric determinantal varieties*, Topology, **23** (1984), 71–84

[Ha1] R. Hartshorne, *Algebraic Geometry*, Graduate Texts in Mathematics, **52**, Springer–Verlag, New York–Heidelberg–Berlin, 1977

[Ha2] R. Hartshorne, *Varieties of low codimension in projective space*, Bull. Amer. Math. Soc., **80** (1974), 1017–1032

[Ha3] R. Hartshorne, *Ample vector bundles*, IHES Publ. Math., **29** (1966), 63–94

[HK] A. Hefez, S. Kleiman, *Notes on duality for projective varieties*

[He] W. Hesselink, *Singularities in the nilpotent scheme of a classical group*, Trans. Amer. Math. Soc., **222** (1976), 1–32

[H] D. Hilbert, *Letter adresseé à M. Hermite*, Gesam. Abh., **vol. II** (1976), pp. 148–153

[HRo] L. Hille, G. Röhrle, *A classification of parabolic subgroups of classical groups with a finite number of orbits on the unipotent radical*, Transformation Groups, **4:1** (1999), 35–52

[Hir] F. Hirzebruch, *Topological Methods in Algebraic Geometry*, Grund. Math. Wiss. **131**, 1966, Springer–Verlag

[HP] W.V.D. Hodge, D. Pedoe, *Methods of Algebraic Geometry*, Cambridge University Press, Cambridge, 1994

[Ho1] A. Holme, *On the dual of the smooth variety*, Lect. Notes in Math., **732** (1979), Springer–Verlag, 144–156

[Ho2] A. Holme, *The geometric and numerical properties of duality in projective algebraic geometry*, Manuscripta Math., **61** (1988), 145–162

[HR] A. Holme, J. Roberts, *Pinch points and multiple locus for generic projections of singular varieties*, Adv. Math., **33** (1979), 212–256

[Hu] J. Humphreys, *Linear Algebraic Groups* Springer-Verlag, 1975

[IL] B. Ilic, J.M. Landsberg, *On symmetric degeneracy loci, spaces of symmetric matrices of constant rank and dual varieties*, Math. Ann., **314** (1999), 159–174

[I] P. Ionescu, *An enumeration of all smooth projective varieties of degree* 5 *and* 6, Increst Preprint Series Math., **74** (1981)

[I1] P. Ionescu, *Embedded projective varieties of small invariants*, in: Proceedings of the 1982 Week of Algebraic Geometry, Bucharest, Lecture Notes in Math., **1056** (1984), 142–187

[I2] P. Ionescu, *Embedded projective varieties of small invariants, III*, in: Algebraic Geometry, Proceedings of Conference on Hyperplane Sections, L'Aquila, Italy, 1988, Lecture Notes in Math., **1417** (1990), 138–154

[J] N. Jacobson, *Structure and representations of Jordan algebras*, Publ. Amer. Math. Soc., **39** (1968)

[Kac] V.G. Kac, *Some remarks on nilpotent orbits*, J. Algebra, **64** (1980), no. 1, 190–213

[KOY] H. Kaji, M. Ohno, O. Yasukura, *Adjoint varieties and their secant varieties*, Indag. Math., **10** (1999), 45–57

[Kan] I.L. Kantor, *Models of exceptional Lie algebras*, Soviet Math. Dokl., **14** (1973), 254–258

[Kas] V.V. Kashin, *Orbits of adjoint and coadjoint actions of Borel subgroups of semisimple algebraic groups* (Russian), Problems in Group Theory and Homological Algebra, Yaroslavl', (1997), 141–159

[Ka] N. Katz, *Pinceaux de Lefschetz; Théorème d'existence*, in: SGA 7, Lecture Notes in Math., **340**, 212–253

[KMM] Y. Kawamata, K. Matsuda, K. Matsuki, *Introduction to the minimal model program*, Algebraic Geometry, Sendai 1985. Adv. Stud. Pure Math., **10** (1987), 283–360

[Kaw] Y. Kawamata, *The cone of curves of algebraic varieties*, Ann. Math, **119** (1984), 603–633

[Ke] G. Kempf, *On the collapsing of homogeneous bundles*, Invent. Math. **37** (1976), 229–239

[Kl1] S. Kleiman, *Enumerative theory of singularities*, Real and Complex Singularities (Proc. Ninth Nordic Summer School/NAVF Sympos. Math., Oslo, 1976), 297–396

[Kl2] S. Kleiman, *About the conormal scheme*, "Complete intersections", Proceedings of Acireale (1983), Lecture Notes in Math. 1092, Springer, 1984, 161–197

[Kl3] S. Kleiman, *Tangency and duality*, Proceedings of the 1984 Vancouver Conference in Algebraic Geometry, July 2–12, 1984, ed. by J. Carrell, A.V. Geramita, and R. Russell. Canadian Math. Soc. Conference Proceedings **6**, (1986), 163–225

[Kl4] S. Kleiman, *Concerning the dual variety*, Proc. 18th Scand. Congr. Math. Aarhus, 1981, Birkhäuser Verlag, Basel–Boston–Stuttgart, 386–396

[Kl5] S. Kleiman, *Plane forms and multiple point formulas*, Algebraic threefolds (Varenna, 1981), pp. 287–310, Lecture Notes in Math., **947** (1982)

[Kl6] S. Kleiman, *Toward a numerical theory of ampleness*, Ann. Math, **84** (1966), 293–344

[KZ] K. Knight, A. Zelevinsky, *Representations of quivers of type A and the multisegment duality*, Adv. Math, **117** (1996), 273–293

[KM] F. Knop, G. Menzel, *Duale Varietäten von Fahnenvarietäten*, Comm. Math. Helv., **62** (1987), 38–61

[KnMu] F. Knudsen, D. Mumford, *Projectivity of moduli spaces of stable curves I. Preliminaries on* Det *and* Div, Math. Scand., **39** (1976), 19–55

[KO] S. Kobayashi, T. Ochiai *Characterization of complex projective spaces and hyperquadrics*, J. Math. Kyoto Univ., **13:1** (1973), 31–47

[Ko] Y. Kollar, *The structure of algebraic threefolds. An introduction to Mori's program*, Bull. AMS, **17:2** (1987), 211–273

[Kos] B. Kostant, *Lie group representations on polynomial rings*, Amer. J. Math., **86** (1963), 327–402

[KP] H. Kraft, C. Procesi, *On the conjugacy classes in classical groups*, Comment. Math. Helvetici, **57** (1982), 539–602

[Kr] H. Kraft, *Closures of conjugacy classes in G_2*, J. Algebra, **126** (1989), 454–465

[KS] A. Kumpera, D. Spencer, *Lie Equations, Vol. 1: General Theory*, Ann. of Math. Stud., **73** (1972), Princeton University Press

[Lam] I. Lamotke, *The topology of complex projective varieties after S. Lefschetz*, Topology, **20** (1981), 15–51

[Land] A. Landman, *Examples of varieties with small dual varieties. Picard–Lefschetz theory and dual varieties*, Two lectures at Aarhus Univ., (1976)

[Lan] J.M. Landsberg, *On degenerate secant and tangential varieties and local differential geometry*, Duke Math. J., **85** (1996), 605–634

[Lan1] J.M. Landsberg, *Differential-geometric characterization of complete intersections*, J. of Diff. Geom., **44** (1996), 32–73

[Lan2] J.M. Landsberg, *On second fundamental form of projective varieties*, Invent. Math., **117** (1994), 303–315

[Lan3] J.M. Landsberg, *On degenerate secant and tangential varieties and local differential geometry*, Duke Math. J., **85** (1996), 1–30

[Lan4] J.M. Landsberg, *Algebraic geometry and projective differential geometry*, preprint, (1998)

[LM] J.M. Landsberg, L. Manivel, *On the projective geometry of homogeneous varieties*, preprint, (2000)

[LP] A. Lanteri, M. Palleschi, *Projective manifolds containing many rational curves*, Indiana Math. J., **36** (1987), 857–865

[LPS] A. Lanteri, M. Palleschi, A.J. Sommese, *Discriminant loci of varieties with smooth normalization*, Comm. Algebra, **28** (2000), 4179–4200

[LS] A. Lanteri, D. Struppa, *Projective 7-folds with positive defect*, Compositio Math., **61** (1987), 329–337

[L] M.E. Larsen, *On the topology of complex projective manifolds*, Invent. Math., **19** (1972), 251–260

[Las] A. Lascoux, *Degree of the dual Grassmann variety*, Comm. Algebra, **9/11** (1981), 1215–1225

[La] R. Lazarsfeld, *An example of* 16*-dimensional projective variety with a* 25*-dimensional secant variety*, Math. Letters, **7** (1981), 1–4

[La1] R. Lazarsfeld, *Some applications of the theory of positive vector bundles*, LNM 1092, Complete Intersections (Acireale 1983), Springer–Verlag, Berlin, 1984, pp. 29–61

[LV] R. Lazarsfeld, A. Van de Ven, *Topics in the Geometry of Projective Space, Recent Work of F.L. Zak*, DMV Seminar, Birkhauser, 1984

[Li] W. Lichtenstein, *A system of quadrics describing the orbit of the highest weight vector*, Proc. A.M.S., **84** (1982), 605–608

[Ma] Yu.I. Manin, *Cubic Forms. Algebra, Geometry, Arithmetic*, M., Nauka, 1972. English transl.: North–Holland, Amsterdam–London 1974

[MM] H. Matsumura, P. Monsky, *On the automorphisms of hypersurfaces*, J. Math. Kyoto Univ., **3:3** (1964), 347–361

[Me] R. Meshulam, *On two-parameter families of symmetric matrices*, Lin. Alg. Appl., **216** (1995), 93–96

[Mi] J. Milnor, *Singular points of complex hypersurfaces*, Ann. of Math. Studies **61**, Princeton Univ. Press (1968)

[MW] C. Moeglin, J.L. Waldspurger, *Sur l'involution de Zelevinski*, J. Reine Angew. Math, **372** (1986), 136–177

[M] B.G. Moishezon, *Algebraic homology classes of algebraic varieties*, Izvestiya AN SSSR, **31** (1967) (in Russian)

[Mo] E.H. Moore, *General Analysis - Part I*, Mem. Amer. Phil. Soc., **1** (1935)

[Mor] S. Mori, *Threefolds whose canonical bundle is not numerically effective*, Ann. Math., **116** (1982), 133–176

[Mor1] S. Mori, *Projective manifolds with ample tangent bundle*, Ann. Math., **110** (1979), 593–606

[Mu] S. Mukai, *New classification of Fano threefolds and Fano manifolds of coindex* 3, preprint (1988)

[MRS] I. Muller, H. Rubenthaler, G. Schiffmann, *Structure des espaces préhomogènes associés à certaines algébres de Lie graduées*, Math. Ann., **274** (1986), 95–123

[Mum] D. Mumford, *Some footnote of the work of C.P. Ramanujam*, in: C.P. Ramanujam, a Tribute, pp. 247–262, Springer, 1978

[Mum1] D. Mumford, *Complex Projective Varieties*, Springer–Verlag, Berlin–Heidelberg–New York, 1979

[MFK] D. Mumford, J. Fogarty, F. Kirwan, *Geometric Invariant Theory*, Ergebnisse der Mathematik und Grenzgebiete, vol. 34, Springer–Verlag, New York–Heidelberg–Berlin, 1994. (3rd edition)

[Mun1] R. Muñoz, *Varieties with low dimensional dual variety*, Manuscripta Math., **94** (1997), 427–435

[Mun2] R. Muñoz, *Varieties with almost maximal defect*, Rend. -Classe di Scienze Mat. e Nat.-, **133** fasc. 1 (1999) dell'Instituto Lombardo Accad. di Scienze e Lettere, 103–114

[Mun3] R. Muñoz, *Varieties with degenerate dual variety*, Forum Math., **13** (2001), 757–779

[Na] J. Nagura, *On the interval containing at least one prime number*, Proc. Japan Acad., **28** (1952), 177–181

[N] A. Némethi, *Lefschetz theory for complex hypersurfaces*, Rev. Roumaine Math. Pures (1988), 233–250

[O1] C. Okonek, *3-Mannigfaltigkeiten im* $\mathbb{P}^5$ *und ihre zugehörigen stabilen Garben*, Manuscripta Math. **38** (1982), 175–199

[O2] C. Okonek, *Moduli reflexiver Garben und Flachem von kleinem Grad im* $\mathbb{P}^4$, Math. Z. **184** (1983), 549–572

[O3] C. Okonek, *Über 2-codimensionale Untermannigfaltigkeiten vom Grad 7 in* $\mathbb{P}^4$ *und* $\mathbb{P}^5$, Math. Z. **187** (1984), 209–219

[O4] C. Okonek, *Flachem von Grad* 8 *im* $\mathbb{P}^4$, Math. Z. **191** (1986), 207–223

[OSS] C. Okonek, H. Spindler, M. Schneider, *Vector bundles on complex projective spaces*, Prog. Math. **3**, Basel–Boston, Birkhäuser, 1980

[OV] A.L. Onischik, E.B. Vinberg, *Lie groups and Lie algebras III*, Encyclopedia of Mathematical Sciences **41**, Springer–Verlag 1994

[OV1] A.L. Onischik, E.B. Vinberg, *Lie Groups and Algebraic Groups*, Springer–Verlag 1990

[Pa] F. Palatini, *Sulla rappresentazione delle forme ternarie mediante la somma di potenze di forme lineari*, Rom. Acc. L. Rend., **12** (1903), 378–384

[Pan] D.I. Panyushev, *Parabolic subgroups with abelian unipotent radical as a testing site for invariant theory*, Canad. J. Math., **51:3** (1999), 616–635

[Pa1] A. Parusiǹski, *A generalization of the Milnor number*, Math. Ann., **281** (1988), 247–254

[Pa2] A. Parusiǹski, *Multiplicity of the dual variety*, Bull. London Math. Soc., **23** (1991), 428–436

[Pau] C. Pauly, *Self-duality of Coble's quartic hypersurface and applications*, preprint

[Pe] R.A. Penrose, *A generalized inverse for matrices*, Proc. Camb. Phil. Soc., **51** (1955)

[P] R. Piene, *Plücker formulas*, Doctoral thesis, M.I.T. Cambridge, MA, 1976

[Pi] R. Piene, *Cycles Polaires et classes de Chern pour les variétés projectives singulières*, Introduction à la théorie des singularités II, Travaux en Cours, vol. 37, Hermann, Paris, 1988, 7–34

[Po] S. Poljak, *Maximum rank of powers of a matrix of a given pattern*, Proc. Amer. Math. Soc., **106:4** (1989), 1137–1144

[Pop] V. Popov, *Self-dual algebraic varieties and nilpotent orbits*, Proceedings of the International Colloquium "Algebra, Arithmetic, and Geometry", Tata Institute of fundamental Research, Bombay, 2001, 496–520

[PR] V. Popov, G. Röhrle, *On the number of orbits of a parabolic subgroup on its unipotent radical*, Algebraic Groups and Lie Groups, Lehrer G.I., Ed., Austr. Math. Soc. Lecture Series, **9** (1997)

[PV] V.L. Popov, E.B. Vinberg, *Invariant Theory*, in: Algebraic Geometry IV, Encyclopedia of Mathematical Sciences, Springer–Verlag, **55** (1994), 123–278

[Py] V.S. Pyasetskii, *Linear Lie groups acting with finitely many orbits*, Funct. Anal. Appl. **9** (1975), 351–353

[Q] D. Quillen, *Determinants of Cauchy–Riemann operators on Riemann surfaces*, Funct. Anal. Appl., **19** (1985), 31–34

[RS] N. Ray, I. Singer, *R-torsion and the Laplacian for Riemannian manifolds*, Adv. Math., **7** (1971), 145–210

[Ra1] Z. Ran, *The (dimension+2)-secant Lemma*, Invent. Math., **106** (1991), 65–71

[Ra2] Z. Ran, *On projective varieties of codimension two*, Invent. Math., **73** (1983), 333–336

[Ra3] Z. Ran, *The structure of Gauss-like maps*, Compos. Math., **52** (1984), 171–177

[RaS] K. Ranestad, F.-O. Schreyer, *Varieties of sums of powers*, e-print math.AG/9801110 (1998)

[Re1] T. Reye, *Erweiterung der Polarentheory algebraische Flächen*, J. Reine Angew. Math., **78** (1874), 97–114

[Re2] T. Reye, *Geometrischer Beweis des Sylversterschen Satzes: "Jede quaternäre cubische Form ist darstellnar als Summe von fünf Cuben linearer Formen"*, J. Reine Angew. Math., **78** (1874), 114–122

[Re3] T. Reye, *Darstellung quaternärer biquadratischer Formen als Summen von zehn Biquadraten*, J. Reine Angew. Math., **78** (1874), 123–129

[Ri] R.W. Richardson, *Finiteness theorems for orbits of algebraic groups*, Indag. Math., **88** (1985), 337–344

[Ric] H.W. Richmond, *On canonical forms*, J. Pure Appl. Math., **33** (1904), 331–340

[Ri1] R.W. Richardson, *Conjugacy classes of n-tuples in Lie algebras and algebraic groups*, Duke Math. J., **57** (1988), 1–35

[RRS] R.W. Richardson, G. Röhrle, R. Steinberg, *Parabolic subgroups with abelian unipotent radical*, Invent. Math., **110** (1992), 649–671

[Rob] J. Roberts, *Generic projections of algebraic varieties*, Amer. J. Math., **93** (1971), 191–214

[Rog] E. Rogora, *Varieties with many lines*, Man. Math., **82** (1994), 207–226

[R] T. Room, *A Synthesis of Clifford matrices and its generalization*, Amer. J. Math., **74** (1952), 967–984

[Ro] B. Rosenfeld, *Geometry of Lie groups*, Mathematics and its Applications **393**, Kluwer Academic Publishes 1997

[Ru] W. Rudin, *Real and Complex Analysis*, Mc Graw Hill, 1987

[Sa] E. Sato, *Uniform vector bundles on a projective space*, J. Math. Soc. Japan, **28** (1976), 123–132

[Schl] L. Schläfli, *Gesammelte Abhandlungen*, Band. 2, pp. 9–112L, Birkhäuser Verlag, Basel, 1953

[Se] B. Segre, *Bertini forms and hessian matrices*, J. London Math. Soc., **26** (1951), 164–176

[Sn] D. Snow, *The nef value and defect of homogeneous line bundles*, Trans. AMS, **340:1** (1993), 227–241

[Sn1] D. Snow, *On the ampleness of homogeneous vector bundles*, Trans. AMS, **294:2** (1986), 227–241

[S] A.J. Sommese, *Hyperplane section of projection surface I — the adjunction mapping*, Duke Math. J., **46** (1979), 377–401

[S1] A. Sommese, *Submanifolds of abelian varieties*, Math. Ann., **233** (1978), 229–256

[Sw] H.P.F. Swinnerton-Dyer, *An enumeration of all varieties of degree* 4, Amer. J. Math., **05** (1973), 403–418

[Sy] J. Sylvester, *On the dimension of spaces of linear transformations satisfying rank conditions*, Linear Algebra and its Applications, **76** (1986), 1–10

[Syl1] J.J. Sylvester, *An essay on canonical forms supplemented by a sketch of a memoir on elimination, transformation, and canonical forms*, Collected Works, Vol. 1, Cambridge Univ. Press, Linear Algebra and its Applications (1904), 184–197, 203–216

[Syl2] J.J. Sylvester, *Sur une extension d'un théorème de Clebsch relatif aux courbes de quartième degré*, Compte Rendus de l'Acad de Science, **102** (1886), 1532–1534

[Ta] R. Takagi, *On homogeneous real hypersurfaces in a complex projective space*, Osaka J. Math., **10** (1973), 495–506

[Tan] H. Tango, *Remark on varieties with small secant varieties*, Bul. Kyoto Univ. Ed. Ser. B, **60** (1982), 1–10

[Te] Terracini, *Alcune questioni sugli spazi tangenti e osculatori ad una varieta, I, II, III*, Societa dei Naturalisti e Matematici (1913), 214–247

[T1] E. Tevelev, *Subalgebras and discriminants of anticommutative algebras*, Izvestiya RAN, Ser. Mat., **63:3** (1999), 169–184

[T2] E. Tevelev, *Generic algebras*, Transformation Groups, **1** (1996), 127–151

[T3] E. Tevelev, *Isotropic subspaces of multilinear forms*, Matematicheskie Zametki, 2001

[T4] E. Tevelev, *Moore–Penrose inverse, parabolic subgroups, and Jordan pairs*, preprint, 2000

[T5] E. Tevelev, *Generic algebras*, PhD Thesis

[T6] E. Tevelev, *Discriminants and quasi-derivations of commutative algebras*, Uspechi Matematicheskih Nauk, **54:2** (1999), 189–190

[Tu] L.W. Tu, *The connectedness of symmetric and skew symmetric degeneracy loci: even ranks*, Trans. AMS, **313** (1989), 381–392

[V] A. Van de Ven, *On the 2-connectedness of the very ample divisor on a surface*, Duke Math. J., **46** (1979), 403–407

[Vi] E.B. Vinverg, *The Weyl group of a graded Lie algebra*, Math. USSR Izv., **10** (1976), 463–495

[Vi1] E.B. Vinberg, *Generalized derivations of algebras*, AMS Transl., **163** (1995), 185–188

[VP] E.B. Vinberg, V.L. Popov, *On a class of quasi-homogeneous affine varieties*, Math. USSR Izv., **6** (1972), 743–758

[Wa] A.H. Wallace, *Tangency and duality over arbitrary fields*, Proc. London Math. Soc., **6:3** (1956), 321–342

[Wei] A. Weil, *Foundations of algebraic geometry*, A.M.S. Coloquium Publ., **29** (1962), Providence, R.I.

[XXX] XXX, *Correspondence*, Amer. J. Math., **79** (1957), 951–952

[W] R. Westwick, *Spaces of matrices of fixed rank*, Linear and Multilinear Algebra, **20** (1987), 171–174

[W1] R. Westwick, *Spaces of matrices of fixed rank, II*, Linear Algebra and its Applications, **235** (1996), 163–169

[We] J. Weyman, *Calculating discriminants by higher direct images*, Trans. Amer. Math. Soc., **343** (1994), 367–389

[WZ] J. Weyman, A. Zelevinsky, *Singularities of hyperdeterminants*, Ann. Inst. Fourier, **46:3** (1996), 591–644

[WZ1] J. Weyman, A. Zelevinsky, *Multiplicative properties of projectively dual varieties*, Manuscripta Math., **82:2** (1994), 139–148

[WZ2] J. Weyman, A. Zelevinsky, *Determinantal formulas for multigraded resultants*, J. Alg. Geom., **3** (1994), 569–597

[Wi] P. Wilson, *Towards birational classification of algebraic varieties*, Bull. London Math. Soc., **19** (1987), 1–48

[Wis] J.A. Wiśniewski, *Length of extremal rays and generalized adjunction*, Math. Z., **200** (1989), 409–427

[Wis1] J.A. Wiśniewski, *On a conjecture of Mukai*, Manuscripta Math., **68** (1990), 135–141

[Z1] F.L. Zak, *Projection of algebraic varieties*, Mat. Sbornik, **116** (1981), 593–602 (in Russian). English Translation: Math. USSR Sbornik, **44** (1983), 535–544

[Z2] F.L. Zak, *Severi varieties*, Mat. Sbornik, **126** (1985), 115–132 (in Russian).

[Z3] F.L. Zak, *The structure of Gauss maps*, Funct. Anal. i Prilozh., **21** (1987), 39–50 (in Russian).

[Z4] F.L. Zak, *Tangents and secants of algebraic varieties*, Translations of Math. Monographs, **127**, AMS 1993

[Z5] F.L. Zak, *Linear systems of hyperplane sections of varieties of low codimension*, Funct. Anal. i Prilozh., **19** (1985), 165–173

[Z6] F.L. Zak, *Some properties of dual varieties and their applications in projective geometry*, in Algebraic Geometry, LNM 1479, 273–280, Springer–Verlag, New York, 1991

[Z7] F.L. Zak, *Determinants of projective varieties and their degrees*, preprint

[Za] O. Zariski, *Theory and applications of holomorphic functions on algebraic varieties over arbitrary ground fields*, Mem. A.M.S. No. 5 (1951), Providence, R.I.

Index

Printing: Mercedes-Druck, Berlin
Binding: Stein + Lehmann, Berlin